Lecture Notes in Computer Science   9989

Commenced Publication in 1973
Founding and Former Series Editors:
Gerhard Goos, Juris Hartmanis, and Jan van Leeuwen

Editorial Board

More information about this series at http://www.springer.com/series/7411

Harald Sack · Giuseppe Rizzo
Nadine Steinmetz · Dunja Mladenić
Sören Auer · Christoph Lange (Eds.)

The Semantic Web

ESWC 2016 Satellite Events
Heraklion, Crete, Greece, May 29 – June 2, 2016
Revised Selected Papers

 Springer

Editors

Harald Sack
Hasso-Plattner-Institut für
  Softwaresystemtechnik
Universität Potsdam
Potsdam
Germany

Giuseppe Rizzo
Innovation Development
Istituto Superiore Mario Boella
Turin
Italy

Nadine Steinmetz
Technical University of Ilmenau
Ilmenau
Germany

Dunja Mladenić
Artificial Intelligence Laboratory
J. Stefan Institute
Ljubljana
Slovenia

Sören Auer
Institut für Informatik III
Universität Bonn
Bonn
Germany

Christoph Lange
Institut für Informatik III
Universität Bonn
Bonn
Germany

ISSN 0302-9743   ISSN 1611-3349   (electronic)
Lecture Notes in Computer Science
ISBN 978-3-319-47601-8   ISBN 978-3-319-47602-5   (eBook)
DOI 10.1007/978-3-319-47602-5

Library of Congress Control Number: 2016954127

LNCS Sublibrary: SL5 – Computer Communication Networks and Telecommunications

Printed on acid-free paper

This Springer imprint is published by Springer Nature
The registered company is Springer International Publishing AG
The registered company address is: Gewerbestrasse 11, 6330 Cham, Switzerland

Preface

The 13th edition of ESWC took place in Heraklion, Crete, Greece, from May 29 to June 2, 2016. Its program included three keynotes by: Jim Hendler (Rensselaer Polytechnic Institute), Ernesto Damiani (Università degli Studi di Milano), and Eleni Pratsini (IBM Research Ireland).

The main scientific program of the conference comprised 47 papers, which have been published by Springer as LNCS volume 9678: 39 research and eight in-use, selected out of 204 submissions, which corresponds to an acceptance rate of 21 % for the 184 research papers submitted, and of 40 % for the 20 in-use papers submitted. This program was completed by a demonstration and poster session, in which researchers had the chance to present their latest results and advances in the form of live demos. In addition, the conference program included 12 workshops, nine tutorials, a PhD symposium, a hackshop, a challenge track with five challenges (revised selected papers published by Springer as CCIS volume 641), an EU project networking session, and a "Minute of Madness." The PhD symposium program included ten contributions, selected out of 21 submissions.

This volume includes the accepted contributions to the demonstration and poster track: 19 poster and 19 demonstration papers, selected out of 62 submissions (28 demos and 34 posters), which corresponds to an overall acceptance rate of 61 %. Each submission was reviewed by at least two, and on average three, Program Committee members. During the poster session the students from the PhD symposium were invited to display a poster about their work. This resulted in ten additional posters being presented during the session.

Additionally, this book includes a selection of the best papers from the workshops co-located with the conference, which are distinguished meeting points for discussing ongoing work and the latest ideas in the context of the Semantic Web. The ESWC 2016 Workshops Program Committee carefully selected 12 workshops focusing on specific research issues related to the Semantic Web, organized by internationally renowned experts in their respective fields:

- The Workshop on Emotions, Modality, Sentiment Analysis and the Semantic Web (EMSA-SW 2016), published by CEUR-WS.org as a part of volume 1613
- The 5th Workshop on Knowledge Discovery and Data Mining Meets Linked Open Data (Know@LOD 2016), which joined forces with the International Workshop on Completing and Debugging the Semantic Web (CoDeS 2016) and has been published by CEUR-WS.org as volume 1586
- The Third Workshop on Linked Data Quality (LDQ 2016), published by CEUR-WS.org as a part of volume 1585
- The 4th Workshop on Linked Media (LiME 2016), published by CEUR-WS.org as a part of volume 1615
- The Second Workshop on Managing the Evolution and Preservation of the Data Web (MEPDaW 2016), published by CEUR-WS.org as a part of volume 1585

- The Third International Workshop on Dataset PROFIling and fEderated Search for Linked Data (PROFILES 2016), published by CEUR-WS.org as volume 1597
- The Workshop on Extraction and Processing of Rich Semantics from Medical Texts (RichMedSem), published by CEUR-WS.org as a part of volume 1613
- The Workshop on Services and Applications Over Linked APIs and Data (SALAD 2016), published by CEUR-WS.org as volume 1629
- The Workshop on Semantic Web Technologies in Mobile and Pervasive Environments (SEMPER 2016), published by CEUR-WS.org as volume 1588
- The International Workshop on Summarizing and Presenting Entities and Ontologies (SumPre 2016), published by CEUR-WS.org as volume 1605
- The Second International Workshop on Semantic Web for Scientific Heritage (SW4SH 2016), published by CEUR-WS.org as volume 1595
- The First Workshop on Humanities in the Semantic web (WHiSe 2016), published by CEUR-WS.org as volume 1608

From the overall set of 77 papers that were accepted for these workshops, a selection of the best papers has been included in this volume. Each workshop's Organizing Committee evaluated the papers accepted in their workshop to propose those to be included in this volume. The authors of the selected papers improved their original submissions, taking into account the comments and feedback obtained during the workshop and the conference. As a result, 16 papers were selected for this volume.

As general chair, poster and demo chairs, and workshop chairs, we would like to thank everybody who has been involved in the organization of ESWC 2016.

Special thanks go to the Poster and Demo Program Committee and to all the workshop organizers and their respective Program Committees who contributed to making the ESWC 2016 workshops a real success.

We would also like to thank the Organizing Committee and especially the local organizers and the program chairs for supporting the day-to-day operation and execution of the workshops.

A special thanks also to our proceedings chair, Christoph Lange, who did a remarkable job in preparing this volume with the kind support of Springer.

Last but not least, thanks to all our sponsors listed herein, for their trust in ESWC.

August 2016

Harald Sack
Giuseppe Rizzo
Nadine Steinmetz
Dunja Mladenić
Sören Auer

Organization

Organizing Committee

General Chair

Harald Sack — Hasso Plattner Institute (HPI), Germany

Program Chairs

Mathieu d'Aquin — Knowledge Media Institute KMI, UK
Eva Blomqvist — Linköping University, Sweden

Poster and Demo Chairs

Giuseppe Rizzo — Istituto Superiore Mario Boella, Italy
Nadine Steinmetz — Technische Universität Ilmenau, Germany

Workshops Chairs

Dunja Mladenic — Jožef Stefan Institute, Slovenia
Sören Auer — University of Bonn and Fraunhofer IAIS, Germany

Tutorials Chairs

Tommaso di Noia — Politecnico di Bari, Italy
H. Sofia Pinto — INESC-ID, Instituto Superior Técnico, Universidade de Lisboa, Portugal

PhD Symposium Chairs

Simone Paolo Ponzetto — Universität Mannheim, Germany
Chiara Ghidini — Fondazione Bruno Kessler, Italy

Challenge Chairs

Stefan Dietze — L3S, Germany
Anna Tordai — Elsevier, The Netherlands

Semantic Technologies Coordinators

Andrea Giovanni Nuzzolese — University of Bologna/STLab ISTC-CNR, Italy
Anna Lisa Gentile — University of Mannheim, Germany

EU Project Networking Session Chairs

Erik Mannens — Data Science Lab – iMinds – Ghent University, Belgium

Mauro Dragoni	Fondazione Bruno Kessler, Italy
Lyndon Nixon	Modul Universität Vienna, Austria
Oscar Corcho	Universidad Politécnica de Madrid, Spain

Publicity Chair

| Heiko Paulheim | University of Mannheim, Germany |

Sponsorship Chair

| Steffen Lohmann | Fraunhofer IAIS, Germany |
| Freddie Lecue | IBM, Ireland |

Web Presence

| Venislav Georgiev | STI International, Austria |

Proceedings Chair

| Christoph Lange | University of Bonn and Fraunhofer IAIS, Germany |

Treasurer

| Ioan Toma | STI International, Austria |

Local Organization and Conference Administration

| Katharina Haas | YouVivo GmbH, Germany |

Program Committee

Program Chairs

| Mathieu d'Aquin | Knowledge Media Institute KMI, UK |
| Eva Blomqvist | Linköping University, Sweden |

Track Chairs

Vocabularies, Schemas, Ontologies

| Krzysztof Janowicz | University of California, Santa Barbara, USA |
| Rinke Hoekstra | VU University Amsterdam, The Netherlands |

Reasoning

| Uli Sattler | University of Manchester, UK |
| Thomas Schneider | Universität Bremen, Germany |

Linked Data

| Monika Solanki | University of Oxford, UK |
| Aidan Hogan | Universidad de Chile, Santiago de Chile, Chile |

Social Web and Web Science

Claudia Müller-Birn	Freie Universität Berlin, Germany
Steffen Staab	Universität Koblenz, Germany

Semantic Data Management, Big Data, Scalability

Philippe Cudré-Mauroux	University of Fribourg, Switzerland
Katja Hose	Aalborg University, Denmark

Natural Language Processing and Information Retrieval

Nathalie Aussenac Gilles	IRIT, Université Toulouse, France
Pablo N. Mendes	IBM, USA

Machine Learning

Claudia d'Amato	University of Bari, Italy
Jens Lehmann	Universität Leipzig, Germany

Mobile Web, Sensors, and Semantic Streams

Raúl García Castro	Universidad Politécnica de Madrid, Spain
Jean-Paul Calbimonte	École Polytechnique Fédérale de Lausanne, Switzerland

Services, APIs, Processes, and Cloud Computing

Maria Maleshkova	AIFB Karlsruhe Institute of Technology, Germany
Karthik Gomadam	Accenture Technology Labs, USA

In-Use and Industrial Track

Mike Lauruhn	Elsevier Labs, The Netherlands
Jacco van Ossenbruggen	Centrum Wiskunde & Informatica (CWI), Amsterdam, The Netherlands

Trust and Privacy

Sabrina Kirrane	Wirtschaftsuniversität Wien, Austria
Pompeu Casanovas	Universidad Autónoma de Barcelona, Spain

Smart Cities, Urban, and Geospatial Data

Carsten Kessler	Hunter College, CUNY, New York, USA
Vanessa Lopez	IBM, Ireland

Steering Committee

Chair

John Domingue	The Open University, UK and STI International, Austria

Members

Claudia d'Amato	Universià degli Studi di Bari, Italy
Philipp Cimiano	Bielefeld University, Germany
Oscar Corcho	Universidad Politécnica de Madrid, Spain
Fabien Gandon	Inria, W3C, Ecole Polytechnique de l'Université de Nice Sophia, Antipolis, France
Axel Polleres	Vienna University of Economics and Business, Austria
Valentina Presutti	CNR, Italy
Marta Sabou	Vienna University of Technology, Austria
Elena Simperl	University of Southampton, UK

Workshops Organization

Workshops Chairs

Dunja Mladenic	Jožef Stefan Institute, Slovenia
Sören Auer	University of Bonn and Fraunhofer IAIS, Germany

EMSA-SW Workshop Organizers

Mauro Dragoni	Fondazione Bruno Kessler, Trento, Italy
Diego Reforgiato Recupero	University of Cagliari, Italy

Know@LOD Workshop Organizers

Heiko Paulheim	University of Mannheim, Germany
Jens Lehmann	University of Bonn, Germany
Vojtech Svatek	University of Economics, Prague, Czech Republic
Craig Knoblock	University of Southern California, USA

CoDeS Workshop Organizers

Matthew Horridge	Stanford University, USA
Patrick Lambrix	Linköping University, Sweden
Bijan Parsia	University of Manchester, UK
Heiko Paulheim	Mannheim University, Germany

LDQ Workshop Organizers

Anisa Rula	University of Milano-Bicocca, Italy
Amrapali Zaveri	Stanford University, USA

| Magnus Knuth | Hasso Plattner Institute, University of Potsdam, Germany |
| Dimitris Kontokostas | AKSW, University of Leipzig, Germany |

LiME Workshop Organizers

Raphaël Troncy	EURECOM, France
Thomas Kurz	Salzburg Research, Austria
Lyndon Nixon	New Media Technology Group, MODUL University, Austria
Kai Schlegel	University of Passau, Germany

MEPDaW Workshop Organizers

Jeremy Debattista	Enterprise Information Systems, University of Bonn and Fraunhofer IAIS, Germany
Jürgen Umbrich	Vienna University of Economics and Business, Austria
Javier D. Fernández	Vienna University of Economics and Business, Austria

PROFILES Workshop Organizers

Elena Demidova	WAIS, University of Southampton, UK
Stefan Dietze	L3S Research Center, Germany
Julian Szymanski	Gdansk University of Technology, Poland
John Breslin	NUI Galway, Ireland

RichMedSem Workshop Organizers

Kerstin Denecke	Bern University of Applied Sciences, Switzerland
Yihan Deng	University of Leipzig, Germany
Thierry Declerck	Saarland University and German Research Center for Artificial Intelligence, Germany

SALAD Workshop Organizers

Maria Maleshkova	Karlsruhe Institute of Technology, Germany
Ruben Verborgh	Ghent University, Belgium
Felix Leif Keppmann	Karlsruhe Institute of Technology, Germany

SEMPER Workshop Organizers

Georgios Meditskos	Information Technologies Institute, Centre for Research and Technology – Hellas, Greece
Thanos G. Stavropoulos	Information Technologies Institute, Centre for Research and Technology – Hellas, Greece
Antonis Bikakis	University College London, UK

SumPre Workshop Organizers

Gong Cheng   Nanjing University, China
Kalpa Gunaratna   (Kno.e.sis) Wright State University, USA
Andreas Thalhammer   Karlsruhe Institute of Technology (KIT), Germany

SW4SH Workshop Organizers

Isabelle Draelants   CNRS, IRHT, France
Catherine Faron Zucker   Université Nice Sophia Antipolis, France
Alexandre Monnin   Inria, France
Arnaud Zucker   Université Nice Sophia Antipolis, France

WHiSe Workshop Organizers

Alessandro Adamou   The Open University, UK
Enrico Daga   The Open University, UK
Leif Isaksen   University of Lancaster, UK

SemDev Hackshop Organizers

Ruben Verborgh   Multimedia Lab, Ghent University – iMinds, Belgium
Miel Vander Sande   Multimedia Lab, Ghent University – iMinds, Belgium

Sponsoring Institutions

Gold Sponsors

http://www.eutravelproject.eu/

Silver Sponsors

http://byte-project.eu/

http://entropy-project.eu/

http://project-hobbit.eu/

http://www.springer.com/lncs

Contents

Demo Papers

Workshop on Emotions, Modality, Sentiment Analysis and the Semantic Web

5th Workshop on Knowledge Discovery and Data Mining Meets Linked Open Data (Know@LOD)

LDQ: 3rd Workshop on Linked Data Quality

Fourth Workshop on Linked Media (LiME-2016)

Managing the Evolution and Preservation of the Data Web

1st Workshop on Humanities in the Semantic Web (WHiSe 2016)

Poster Papers

Moving Real-Time Linked Data Query Evaluation to the Client

Ruben Taelman[(✉)], Ruben Verborgh, Pieter Colpaert, and Erik Mannens

imec – Ghent University – IDLab,
Sint-Pietersnieuwstraat 41, 9000 Ghent, Belgium
{ruben.taelman,ruben.verborgh,pieter.colpaert,
erik.mannens}@ugent.be

Abstract. Traditional RDF stream processing engines work completely server-side, which contributes to a high server cost. For allowing a large number of concurrent clients to do continuous querying, we extend the low-cost Triple Pattern Fragments (TPF) interface with support for time-sensitive queries. In this poster, we give the overview of a client-side RDF stream processing engine on top of TPF. Our experiments show that our solution significantly lowers the server load while increasing the load on the clients. Preliminary results indicate that our solution moves the complexity of continuously evaluating real-time queries from the server to the client, which makes real-time querying much more scalable for a large amount of concurrent clients when compared to the alternatives.

Keywords: Linked data · Linked data fragments · SPARQL · Continuous querying · Real-time querying

1 Introduction

Several use cases need updating query results over time, and may thus require the re-execution of entire queries over and over again (i.e., polling). The problem is that polling can be very inefficient when not knowing when the data will change. An additional problem is that many public (even static) SPARQL query endpoints suffer from low availability [2]. This is partially caused by the unrestricted complexity of SPARQL queries [5] combined with the public character of SPARQL endpoints. RDF stream processing engines like C-SPARQL [1] and CQELS [4] offer combined access to dynamic data streams and static background data through continuously executing queries. Because of this continuous querying, the cost for these servers is *even higher* than with static querying.

In this work we present a client-side RDF stream processing engine based on Triple Pattern Fragments (TPF) [6]. TPF is a low-cost server interface for retrieving triple patterns, this makes it possibly for a client to evaluate any query by breaking it up into several triple patterns and joining them locally. We focus on non-high-frequency dynamic data, for example, information on train delays, which updates in the order of minutes. Because some dynamic data might

© Springer International Publishing AG 2016
H. Sack et al. (Eds.): ESWC 2016 Satellite Events, LNCS 9989, pp. 3–7, 2016.
DOI: 10.1007/978-3-319-47602-5_1

have a frequency that is too high for clients to efficiently poll. The resulting framework requires the server to *annotate* its data with a predicted expiration time. Using this expiration time, the client can efficiently determine when to retrieve fresh data. The generic approach in this paper is applied to the use case of public transit route planning. It can be used in various other domains with continuously updating data, such as smart city dashboards, business intelligence, or sensor networks.

2   Query Streamer

Our solution consists of a partial redistribution of query evaluation workload from the server to the client. This requires the client to be able to access the server data so that the query evaluation can be done client-side. There needs to be a distinction between regular static data and continuously updating dynamic data in the server's dataset. By *annotating* dynamic data with a time interval or expiration time using a *temporal vocabulary* [3], the client can detect for how long a certain fact remains valid. The data could however still remain the same after its expiration. When dynamic data expires in time, the client knows that it has to evaluate the query again to fetch the latest version of the data.

We have added an extra layer, which is called the *Query Streamer*, on top of the TPF client. This query streamer is able to transform a regular SPARQL query to a separate *static* and *dynamic* query. This rewriting is done by exchanging metadata with the server. The Query Streamer continuously evaluates this dynamic query based on the time annotation it can find on the dynamic data. The Query Streamer exploits these time annotations by only initiating new queries when the dynamic facts have expired. Every time results from the dynamic query are finalized, they are combined with the static query to form a *materialized static query*. This materialized static query will then either be evaluated or its results will be retrieved from a local cache. The combined results of the dynamic query and materialized static query are then continuously returned to the user who initiated the original query.

3   Preliminary Evaluation

In order to analyze the effects of our solution, we set up an experiment to measure the impact of our proposed redistribution of workload between the client and server by simultaneously executing a set of queries against a TPF server using our proposed solution. We repeat this experiment for two state-of-the-art server-side solutions: C-SPARQL and CQELS, in which the clients simply register the query to the respective server engine and get a stream of results.

To test the client and server performance, our experiment consists of one server and ten physical clients. Each of these clients can execute from one to twenty unique concurrent queries derived from the query in Listing 1.1. This results in a series of 10 to 200 concurrent query executions.

```
SELECT ?delay ?platform ?headSign ?routeLabel ?departureTime
WHERE {
    _:id t:delay            ?delay.
    _:id t:platform         ?platform.
    _:id t:departureTime    ?departureTime.
    _:id t:headSign         ?headSign.
    _:id t:routeLabel       ?routeLabel.
    FILTER (?departureTime > "2015-12-08T10:20:00"^^xsd:dateTime).
    FILTER (?departureTime < "2015-12-08T11:20:00"^^xsd:dateTime).
}
```

Listing 1.1. The basic SPARQL query for retrieving all upcoming train departure information in a certain station. The two first triple patterns are dynamic, the last three are static.

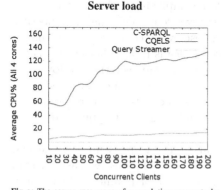

Fig. a: The server CPU usage of our solution proves to be influenced less by the number of clients.

Fig. b: In the case of 200 concurrent clients, client CPU usage initially is high after which it converges to about 5%. The usage for C-SPARQL and CQELS is almost non-existing.

Fig. 1. Average server and client CPU usage for one query stream for C-SPARQL, CQELS and the proposed solution. Our solution effectively moves complexity from the server to the client.

Our solution was implemented[1] in JavaScript using Node.js to allow for easy communication with the existing TPF client. The tests[2] were executed on machines having two Hexacore Intel E5645 (2.4 GHz) CPUs with 24 GB RAM and were running Ubuntu 12.04 LTS.

The server performance results from our main experiment can be found in Fig. 1a. This plot shows an increasing CPU usage for C-SPARQL and CQELS for higher numbers of concurrent query executions. On the other hand, our solution never reaches more than 1 % of server CPU usage.

The results for the average CPU usage across the duration of the query evaluation of all clients that send queries to the server in our main experiment can

---

[1] The source code for this implementation is available at https://github.com/rubensworks/TPFStreamingQueryExecutor.

[2] The code used to run these experiments with the relevant queries can be found at https://github.com/rubensworks/TPFStreamingQueryExecutor-experiments/.

be found in Fig. 1b. The clients that send C-SPARQL and CQELS queries to the server have a client CPU usage of nearly zero percent for the whole duration of the query evaluation. The clients using the client-side query streamer solution that is presented in this work have an initial CPU peak reaching about 80 %, which drops to about 5 % after 4 s. This initial peak is caused by the preprocessing done by our query streamer.

4   Conclusions

In this paper, we presented a solution for doing client-side query evaluation over dynamic data, with the goal of lowering the server load. Our preliminary evaluation shows that for queries of limited complexity with limited dataset sizes, our solution significantly reduces the server load. This makes it possible for the server to handle much more client requests when compared to alternative approaches. This lower server load consequently leads to a higher client load in our experiments. Future research should show how this solution performs for larger datasets and different query types. The movement from query registration at the server as is done by C-SPARQL and CQELS to client-side query evaluation is important for reducing the server load and for being able to publish dynamic data at a low cost.

This low-cost publication of dynamic Linked Data opens up a whole new range of possibilities. Dynamic data that currently requires expensive server infrastructure for its publication to a large number of potential clients or is somehow being rate-limited to avoid server overloading, can now be exposed through a low-cost interface where clients are required to do part of the work. This can be used, for example, for public access to real-time information on public transport scheduling and non-high frequency sensors.

Acknowledgments. The described research activities were funded by iMinds and Ghent University, the Institute for the Promotion of Innovation by Science and Technology in Flanders (IWT), the Fund for Scientific Research Flanders (FWO Flanders), and the European Union. Ruben Verborgh is a Postdoctoral Fellow of the Research Foundation Flanders.

References

1. Barbieri, D.F., Braga, D., Ceri, S., Valle, E.D., Grossniklaus, M.: Querying RDF streams with C-SPARQL. SIGMOD Rec. **39**(1), 20–26 (2010)
2. Buil-Aranda, C., Hogan, A., Umbrich, J., Vandenbussche, P.-Y.: SPARQL web-querying infrastructure: ready for action? In: Alani, H., Kagal, L., Fokoue, A., Groth, P., Biemann, C., Parreira, J.X., Aroyo, L., Noy, N., Welty, C., Janowicz, K. (eds.) ISWC 2013. LNCS, vol. 8219, pp. 277–293. Springer, Heidelberg (2013). doi:10.1007/978-3-642-41338-4_18
3. Gutierrez, C., Hurtado, C., Vaisman, A.: Introducing time into RDF. IEEE Trans. Knowl. Data Eng. **19**(2), 207–218 (2007)

4. Le-Phuoc, D., Dao-Tran, M., Xavier Parreira, J., Hauswirth, M.: A native and adaptive approach for unified processing of linked streams and Linked Data. In: Aroyo, L., Welty, C., Alani, H., Taylor, J., Bernstein, A., Kagal, L., Noy, N., Blomqvist, E. (eds.) ISWC 2011, Part I. LNCS, vol. 7031, pp. 370–388. Springer, Heidelberg (2011). doi:10.1007/978-3-642-25073-6_24
5. Pérez, J., Arenas, M., Gutierrez, C.: Semantics and complexity of SPARQL. In: Cruz, I., Decker, S., Allemang, D., Preist, C., Schwabe, D., Mika, P., Uschold, M., Aroyo, L.M. (eds.) ISWC 2006. LNCS, vol. 4273, pp. 30–43. Springer, Heidelberg (2006). doi:10.1007/11926078_3
6. Verborgh, R., et al.: Querying datasets on the Web with high availability. In: Mika, P., et al. (eds.) ISWC 2014, Part I. LNCS, vol. 8796, pp. 180–196. Springer, Heidelberg (2014). doi:10.1007/978-3-319-11964-9_12

Building a General Knowledge Base of Physical Objects for Robots

Valerio Basile[1(✉)], Elena Cabrio[2], and Fabien Gandon[1]

[1] Inria Sophia Antipolis, Valbonne, France
{valerio.basile,fabien.gandon}@inria.fr
[2] University of Nice Sophia Antipolis, Nice, France
elena.cabrio@unice.fr

Abstract. In this paper we present an ongoing work on building a repository of knowledge about objects typically found in homes, their usual locations and usage. We extract an RDF knowledge base by automatically reading text on the Web and applying simple inference rules. The obtained common sense object relations are ready to be used in a domestic robotic setting, e.g. "a `frying pan` is usually located in the `kitchen`".

1 Introduction

When working with and for humans, robots and autonomous systems must know about the objects involved in human activities. While great progress has been made in object instance and class recognition, a robot is always limited to knowing about the objects it has been trained to recognize. To overcome this issue, robots should be enabled to exploit the vast amount of knowledge on the Web to learn about previously unseen objects and to use this knowledge when acting in the real world. Methods to extract basic ontological knowledge about objects by analyzing unstructured and structured information sources on the Web are needed, to retrieve information about objects properties and functionalities.

In this paper, we address precisely this issue, that can be further divided into the following two subquestions: *(i)* How much general knowledge is already present on the Web?, and *(ii)* How to represent this knowledge in a way that robots can use for their real-world tasks?

In this paper, we propose a method based on "machine reading" [2] to extract formally encoded knowledge from unstructured text. Given the robot domestic setting scenario, we focus on extracting the following information: *(i)* the type of the object, *(ii)* where it is typically located, and *(iii)* common semantic frames involving the object. Existing resources as ConceptNet or Cyc[1] attempt to assemble an ontology and knowledge base of everyday common sense knowledge, but these models are not sufficient to meet our objectives since they do not contain extensive object knowledge or the right level of abstraction. ConceptNet, for instance, provides for the `AtLocation, knife` relation of the concept `knife` a list of 34 candidates, including rooms and containers (`in_kitchen`, `backpack`)

---

[1] http://conceptnet5.media.mit.edu/, http://www.opencyc.org/.

© Springer International Publishing AG 2016
H. Sack et al. (Eds.): ESWC 2016 Satellite Events, LNCS 9989, pp. 8–11, 2016.
DOI: 10.1007/978-3-319-47602-5_2

but also other kinds of locations, not necessarily prototypical (`pocket`, `oklahoma`, `your_back`). OpenCyc, on the other hand, contains a great deal of taxonomical knowledge about the entity `knife`, but nothing about its possible locations.

Our approach combines linguistic and semantic analysis of natural language with entity linking and formal reasoning to create a meaning bank of common sense knowledge. We leverage curated resources (e.g. DBpedia[2], BabelNet [8]) and learn from the unstructured Web when a knowledge gap occurs.

2   Information Extraction from Natural Language

To extract useful knowledge from unstructured data on the Web, two main types of language analysis are needed: semantic parsing (the process of extracting some formal representation of the meaning of the target text) and entity linking. Concerning the first one, we are mostly interested in *thematic roles*, i.e. the role played by a specific entity in a given event. We use the C&C tools pipeline plus Boxer [3,5], in a similar way to the FRED information extraction API [9]. For the latter analysis, we use Babelfy [7], an online API that, given an arbitrary text, returns a list of the word senses, represented as BabelNet synsets, and the URIs of DBpedia entities linked to the words that mention them.

Once the system has extracted the entities, properly linked to the LOD cloud, and decided the roles they play in the situations they are involved in, the only missing piece is a formal description of the situations themselves. For this, we resort to frame semantics [6], a theory of meaning that describes events and situations, along with the possible roles for each involved entity. FrameNet [1] provides a mappings to other linguistic and semantic resources, therefore we are able to link the events in the meaning representation returned by the semantic parsing stage to FrameNet frames. By integrating all these sources of information, from a text like "Annie eats an apple", we extract RDF triples such as: `dpb:Apple vn:Patient fn:Ingestion`.[3] Since these resources are already in the LOD (FrameNet relations are modeled through LEMON[4]), this method produces new triples that can be directly published back to the Web.

To automatically augment the knowledge base, we apply inference rules to the triples to extract new information on top of the language analysis. We exploit the relation of *co-mention* (i.e., two entities being mentioned together in a sentence or similar context) together with the DBpedia type hierarchy, to define the rule: if X is subject of a DBpedia category that is subsumed by `dbp:Tool` and Y is a category that is subsumed by `dbp:Room` and the two entities are co-mentioned, then a likely location for X is Y ($isTool(X) \wedge isRoom(Y) \wedge comention(X,Y) \Rightarrow location(X,Y)$). For example: `dbp:Knife` is `purl:subject`[5] of `dbp:Category: Blade_weapons`, which is `skos:narrower`[6]

---

[2] http://dbpedia.org.

[3] Prefixes `dbp`, `vn`, `fn` respectively link back to DBpedia, VerbNet and FrameNet.

[4] http://lemon-model.net/lexica/uby/fn/.

[5] http://dublincore.org/documents/dcmi-terms/.

[6] http://www.w3.org/TR/swbp-skos-core-spec.

than `dbp:Category:Tools`. At the same time, `dbp:Kitchen` is `purl:subject` of `dbp:Category:Rooms`. Combining this information with the fact that `dbp:Knife` is co-mentioned with `dbp:Kitchen`, we infer that the likely location of a knife is the kitchen.

Evaluation. To test the quality of the created knowledge base, we applied the system to text from the Web and evaluated its output. We created two corpora of written English: *(i)* the first comprises five short documents (952 words in total) extracted from the documentation of the RoCKIn@Home challenge[7], that describe typical tasks for a domestic robot; *(ii)* then, to learn common knowledge about objects in general, we also experimented with open domain text. Language learners material is a fit option, since it is usually made of short, simple sentences about concrete, day-to-day situations. The ESL YES website[8] contains a collection of 1,600 free short stories and dialogues for English learners. We extracted and converted to plain text 725 short stories (83,532 tokens).

Running the pipeline on the two corpora, we obtained two knowledge bases composed of co-mentions and semantic role triples: 3,184 triples from RoCKIn-@home (57 of which semantic roles) and 49,165 from ESL YES (2,953 semantic roles). 91.2 % of semantic role triples from RoCKIn@home and 76.4 % from ESL YES are not aligned with any frame in FrameNet. Next we apply the rule defined in Sect. 2 to extract location relations between tools and rooms. We implemented the rule in CoReSe [4] obtaining 5 location relations from RoCKIn@home (5 objects in 2 rooms) and 101 from ESL YES (49 objects in 14 rooms).

We manually inspected the set of location relation extracted: of the five <objects, room> pairs extracted from RoCKIn@home, two are definitely accurate, two are disputable, depending on the situation, and one is too generic, thus not very informative. Following the same methodology, out of 101 location relations extracted from the ESL YES corpus, we judged 42 of them correct (e.g., `dbp:Frying_pan`, `dbp:Kitchen`), 31 questionable (e.g., `dbp:Suitcase`, `dbp:Bathroom`), and 28 not informative, i.e. they contain the entities `dbp:Tool` or `dbp:Room` (these can be easily filtered out).

Discussion. A great amount of text is needed to build a large scale knowledge base. Several steps in the processing pipeline introduce mismatches, especially with respect to the coverage of the lexical resources involved, ultimately resulting in the extraction of relatively few triples, considering the size of the input text. On average, we extract roughly a little less than one semantic role triple with a proper frame per sentence. This issue can be circumvented by adding more text from the Web. However, the problem remains to find and retrieve text that contains the right kind of information, i.e., common knowledge about objects.

Another issue highlighted by this work is the difficulty of categorizing DBpedia entities, especially entries that are not named entities, such as the household items of our knowledge base. Inference rules like the one we defined are dependent on the correct classification of entities (in tools and rooms, in our case), therefore a high-coverage taxonomy of objects is needed.

---

[7] http://rockinrobotchallenge.eu/RoCKIn_D1.1_short.pdf.

[8] http://www.eslyes.com/.

Finally, a unified tokenization does not solve all the alignment problems. Our system maps discourse referents to entity URIs or FrameNet frames, but these in turns rely on the respective mappings between text and semantics made by Boxer and Babelfy, which are not always aligned.

3   Conclusions and Future Work

This paper presents a novel method to extract information from natural language text based on the combination of semantic parsing and entity linking. The knowledge extracted is represented as RDF triples, to exploit other Web resources to augment the result of the information extraction (e.g. we infer location relations between objects and rooms).

The availability of large quantities of text is crucial to build a large common sense knowledge base following our approach. However, not all source text is the same, as the outcome is influenced by genre, style, and most important, the topics covered in the text. As future work we aim at building a corpus similar to the language learners one, but orders of magnitude larger. Moreover, besides the issues we found with ConceptNet, its sheer size and variety of semantic links makes for a valuable resource that we plan to incorporate in our work.

References

1. Baker, C.F., Fillmore, C.J., Lowe, J.B.: The Berkeley FrameNet project. In: Proceedings of the 36th Annual Meeting of ACL, ACL 1998, pp. 86–90 (1998)
2. Barker, K., Agashe, B., Chaw, S.-Y., Fan, J., Friedland, N., Glass, M., Hobbs, J., Hovy, E., Israel, D., Kim, D.S., Mulkar-Mehta, R., Patwardhan, S., Porter, B., Tecuci, D., Yeh, P., Learning by reading: a prototype system, performance baseline and lessons learned. In: Proceedings of the 22nd National Conference on AI, AAAI 2007, vol. 1, pp. 280–286. AAAI Press (2007)
3. Bos, J.: Wide-coverage semantic analysis with boxer. In: Bos, J., Delmonte, R. (eds.) STEP 2008 Conference Proceedings on Semantics in Text Processing. Research in Computational Semantics, vol. 1, pp. 277–286. College Publications (2008)
4. Corby, O., Dieng-kuntz, R., Faron-zucker, C.: Querying the semantic web with the corese search engine (2004)
5. Curran, J., Clark, S., Bos, J.: Linguistically motivated large-scale NLP with C&C and boxer. In: Proceedings of the 45th Annual Meeting of ACL Companion (Demo and Poster Sessions), Prague, Czech Republic, pp. 33–36. Association for Computational Linguistics, June 2007
6. Fillmore, C.: Frame semantics. In: Linguistics in the Morning Calm
7. Moro, A., Raganato, A., Navigli, R.: Entity linking meets word sense disambiguation: a unified approach. Trans. Assoc. Comput. Linguist. (TACL) **2**, 231–244 (2014)
8. Navigli, R., Ponzetto, S.P.: BabelNet: the automatic construction, evaluation and application of a wide-coverage multilingual semantic network. Artif. Intell. **193**, 217–250 (2012)
9. Presutti, V., Draicchio, F., Gangemi, A.: Knowledge extraction based on discourse representation theory and linguistic frames. In: ten Teije, A., Völker, J., Handschuh, S., Stuckenschmidt, H., d'Acquin, M., Nikolov, A., Aussenac-Gilles, N., Hernandez, N. (eds.) EKAW 2012. LNCS, vol. 7603, pp. 114–129. Springer, Heidelberg (2012)

Some Thoughts on OWL-Empowered SPARQL Query Optimization

Vassilis Papakonstantinou[1]([✉]), Giorgos Flouris[1], Irini Fundulaki[1], and Andrey Gubichev[2]

[1] Institute of Computer Science-FORTH, Heraklion, Greece
papv@ics.forth.gr
[2] TU Munich, Munich, Germany

Abstract. The discovery of optimal or close to optimal query plans for SPARQL queries is a difficult and challenging problem for query optimisers of RDF engines. Despite the growing volume of work on optimising SPARQL query answering, using heuristics or data statistics (such as cardinality estimations) there is little effort on the use of OWL constructs for query optimisation. OWL axioms can be the basis for the development of *schema-aware* optimisation techniques that will allow significant improvements in the performance of RDF query engines when used in tandem with data statistics or other heuristics. The aim of this paper is to show the potential of this idea, by discussing a diverse set of cases that depict how schema information can assist SPARQL query optimisers.

1 Introduction

The Linked Data paradigm, which is now the prominent enabler for sharing huge volumes of data using Semantic Web technologies, has created novel challenges for non-relational data management technologies such as RDF and graph database systems. Semantics of Linked Data are expressed in terms of the RDF Schema Language (RDFS) and the OWL Web Ontology Language. RDFS and OWL vocabularies are used from nearly all data sources in the LOD cloud. Moreover, according to a recent study[1], 36.49 % of LOD use various OWL fragments, so it becomes critical to optimize RDF engines by considering OWL features.

Commercial RDF engines implement RDFS and OWL rules by performing *forward* or *backward* reasoning. Regardless of the way of reasoning, they basically store RDF data in a large *triple table* and consequently the evaluation of SPARQL queries boils down to performing a query with a large number of costly *self-joins*. To evaluate such difficult SPARQL queries a number of prototypes have been proposed. Many of these approaches propose a mapping of regular schema-conforming part of the RDF dataset into a set of relational tables [1,2,4], and rely on the optimization techniques of the underlying DBMSs for query evaluation. Other approaches [6,7,9,13] propose main-memory resident extensive indexes for RDF triples. In either case, information residing in OWL

---

[1] http://linkeddatacatalog.dws.informatik.uni-mannheim.de/state/.

© Springer International Publishing AG 2016
H. Sack et al. (Eds.): ESWC 2016 Satellite Events, LNCS 9989, pp. 12–16, 2016.
DOI: 10.1007/978-3-319-47602-5_3

schemas is rarely taken into account as in [3,5,8,11], so it is our belief that an *an OWL schema-aware SPARQL query optimizer* could complement those approaches since many datasets (especially in the LOD Cloud) come with such good quality schemas.

In this paper we discuss how schema information expressed in terms of OWL ontologies can be used to perform interesting, possibly complex, optimizations in order to improve SPARQL query execution plans, and, consequently, the performance of the RDF engines. Such optimizations can be employed in a complementary fashion to traditional ones to further improve query planners' performance. Our intention in this work *is not* to provide full solutions, but to present the *potential* of the idea (fully described in [10]) by discussing some possible types of optimizations (Sect. 2) that can be performed. Many more may exist.

2   Schema Based Optimization Techniques

2.1   Constraint Violation

An RDF engine could be able at compile time to take advantage of class and property constraints as expressed in an OWL schema; these include *equivalence* (owl:equivalentClass, owl:equivalentProperty) and *disjointness* (owl:-disjointWith, owl:propertyDisjointWith) of classes and properties as well as constraints relevant to the property's *domain* and *range* (rdfs:domain and rdfs:range resp.). For instance, a query looking for an instance of two disjoint classes (owl:disjointWith construct) is certain to return no answers, so it should be answerable in constant time, without having the query engine evaluate it. This kind of information is important for RDF engines that follow either a *forward* or *backward* reasoning approach for computing the inferred knowledge.

2.2   Selectivity Estimation

Cardinality Constraints: OWL allows defining cardinality restrictions through the *min* (owl:minCardinality), *max* (owl:maxCardinality) and *exact* (owl:cardinality) cardinality constraints for object and datatype properties, which state how many instances of said property a resource can have. These schema-level constraints can be used to guide the optimizer into selecting a possibly efficient join ordering without resorting to statistics [3,5]. To do so, triple patterns that refer to more selective properties (e.g. *functional* properties, owl:-FunctionalProperty could be pushed down in the plan to reduce intermediate results).

Complex Class Expressions: Selectivity of triple patterns in a SPARQL query can be estimated through OWL constructs that define classes through *set operations*, such as *intersection* (owl:intersectionOf) and *union* (owl:unionOf). For example, consider a query that requests instances ?x of a class <C>, the latter defined as an intersection of classes <C1> and <C2>, in conjunction with triple patterns that relate instances ?y and ?z of the intersected classes, with triple

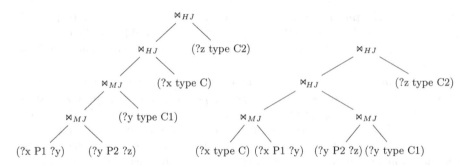

Fig. 1. Optimal plan, considers $C = C1$ `owl:intersectionOf` $C2$ (right) and suboptimal ignores the rule (left)

patterns with predicates <P1> and <P2>. The class <C>, being more selective, should be considered first in a bushy plan with two sub-trees (around ?x and ?y, respectively) being joined with a hash join (right side of Fig. 1). Without the knowledge of schema constraints, the query optimizer would put the three triple patterns with `rdf:type` predicate at the end, since those usually match a lot of triples (left side of Fig. 1) [12]. An analogous line of thought can be followed for the `owl:unionOf` construct.

Class and Property Hierarchies: Hierarchies of classes and properties (through `rdfs:subClassOf` and `rdfs:subPropertyOf`) can also improve selectivity estimation. In this case, the triple patterns that request for instances of classes found lower in a class hierarchy should be considered first in a query plan (depending on the form of the query), when deciding join ordering.

2.3 Advanced Optimizations

In this section we present a set of cases where schema information can help the query engine determine the optimal plan in a more sophisticated way.

Inference: In backward reasoning systems, the inferred knowledge obtained through OWL reasoning rules is computed at query time. Is some cases, the same information may be obtained in various ways. For example, assume that we have a long hierarchy where $<B_i>$ is a subclass (`rdfs:subClassOf`) of $<B_{i+1}>$, $i = 1, \ldots, n$. Consider also that the domain (`rdfs:domain`) of property <P> is class <A> and *all its values* (`owl:allValuesFrom`) come from root class $<B_{n+1}>$. In a query that asks for instances ?v of class $<B_{n+1}>$ that are also values of property <P>, there are two ways to obtain instances ?v: one through the `owl:allValuesFrom` (*cls-avf* axiom[2]), and another through the transitivity of `rdfs:subClassOf` (*cax-sco* axiom[4]). For large n, class $<B_{n+1}>$ is positioned high in the hierarchy, so the engine should use the `owl:allValuesFrom` construct to

---

[2] https://www.w3.org/TR/owl2-profiles/#Reasoning_in_OWL_2_RL_and_RDF_Graphs_using_Rules.

obtain the values for ?v. The alternative may be better if the two classes are sufficiently "close" in the hierarchy, especially given the fact that subsumption-related inference is the most optimized type (due to its widespread use).

Star Query Transformation: Schema information can also be used by the query optimizer to rewrite SPARQL queries to equivalent ones that have a form for which already known optimization techniques are easily applicable. For example, when a triple pattern, involving a *symmetric property* (`owl:SymmetricProperty`), "breaks" a star-shaped query pattern (subject values of remaining triple patterns appear as an object value), a schema-aware optimizer, should rewrite this query into its equivalent one, where all triple pattern's subject values are the same, according to the semantics of `owl:SymmetricProperty`.

3   Conclusions

We advocated on the use of OWL schema information for improving SPARQL query planning, and described some optimizations that can be employed in this direction. Our proposal is meant to be complementary to well-known optimizations (e.g., based on statistics) for query planning, and is most appropriate for datasets and benchmarks that use a rich schema structure (e.g., UOBM). In the future, we plan to work further on understanding the different possible optimizations and potential trade-offs, so that they can be implemented on top of an RDF store in order to quantify the achieved speed-up.

Acknowledgements. This work was partially funded by the EU projects LDBC (FP7 GA No. 317548) and HOBBIT (H2020 GA No. 688227).

References

1. Abadi, D.J., Marcus, A., Madden, S.R., Hollenbach, K.: SW-Store: a vertically partitioned DBMS for Semantic Web data management. VLDBJ **18**(2), 385–406 (2009)
2. Bornea, M.A., Dolby, J., Kementsietsidis, A., Srinivas, K., Dantressangle, P., Udrea, O., Bhattacharjee, B.: Building an efficient RDF store over a relational database. In: SIGMOD, pp. 121–132. ACM (2013)
3. Bursztyn, D., GoasdouT, F., Manolescu, I.: Optimizing reformulation-based query answering in RDF. In: EDBT (2015)
4. Chong, E.I., Das, S., Eadon, G., Srinivasan, J.: An efficient SQL-based RDF querying scheme. In: VLDB (2005)
5. Bursztyn, D., Frantois GoasdouT, I.M.: Efficient query answering in DL-Lite through FOL reformulation. In: DL (2015)
6. Erling, O., Mikhailov, I.: RDF support in the virtuoso DBMS. In: Pellegrini, T., Auer, S., Tochtermann, K., Schaffert, S. (eds.) Networked Knowledge-Networked Media. SCI, pp. 7–24. Springer, Heidelberg (2009)
7. Harth, A., Umbrich, J., Hogan, A., Decker, S.: YARS2: a federated repository for querying graph structured data from the web. In: Aberer, K., et al. (eds.) ASWC 2007 and ISWC 2007. LNCS, vol. 4825, pp. 211–224. Springer, Heidelberg (2007)

8. Kollia, I., Glimm, B.: Optimizing SPARQL query answering over OWL ontologies. JAIR **48**, 253–303 (2013)
9. Neumann, T., Weikum, G.: The RDF-3X engine for scalable management of RDF data. VLDBJ **19**(1), 91–113 (2010)
10. Papakonstantinou, V., Fundulaki, I., Flouris, G., Alexiev, V.: Benchmark design for reasoning. Technical report D4.4.2, LDBC Council (2014)
11. Rodriguez-Muro, M., Rezk, M.: Efficient SPARQL-to-SQL with R2RML mappings. Web Semant. Sci. Serv. Agents World Wide Web **33**, 141–169 (2015). Elsevier
12. Tsialiamanis, P., Sidirourgos, L., Fundulaki, I., Christophides, V., Boncz, P.: Heuristics-based query optimisation for SPARQL. In: EDBT (2012)
13. Weiss, C., Karras, P., Bernstein, A.: Hexastore: sextuple indexing for Semantic Web data management. PVLDB **1**(1), 1008–1019 (2008)

Edinburgh Associative Thesaurus as RDF and DBpedia Mapping

Jörn Hees[1,2(✉)], Rouven Bauer[1,2], Joachim Folz[1,2], Damian Borth[1,2], and Andreas Dengel[1,2]

[1] Computer Science Department, University of Kaiserslautern, Kaiserslautern, Germany
[2] Knowledge Management Department, DFKI GmbH, Kaiserslautern, Germany
{joern.hees,rouven.bauer,joachim.folz,damian.borth,andreas.dengel}@dfki.de

Abstract. Associations, which are one of the key ingredients of human intelligence and thinking, are not easily accessible to the Semantic Web community. High quality RDF datasets of this kind are missing. In this paper we generate such a dataset by transforming 788 K free-text associations of the Edinburgh Associative Thesaurus (EAT) into RDF. Furthermore, we provide a verified mapping of strong textual associations from EAT to DBpedia Entities with the help of a semi-automatic mapping approach. Both generated datasets are made publicly available and can be used as a benchmark for cross-type link prediction and pattern learning.

1 Introduction

Associations as one of the building blocks of human intelligence, thinking, context forming and everyday communication [4] are not well represented in currently published Linked Data datasets. This impedes AI research: due to the missing ground truth of semantic entities which are associated by humans, we can neither analyse human associations in existing datasets, nor train machines to learn graph patterns for them.

2 Related Work

Previously, we developed semantic games with a purpose to collect a semantic association ground truth (Linked Data Games [7], KnowledgeTestGame [5]) or to rank existing triples by association strengths (BetterRelations [6]). Along the lines of fact ranking ground truth datasets, other works such as WhoKnows [9] and more recently FRanCo [2] have been published. While fact ranking in general only focuses on existing facts, FRanCo in its first step also collected free-text fact input about the entity in question, resulting in ∼ 7.8 K raw free-text facts and a

---

The original version of this chapter was revised: Two percentages in the listing on page 19 contained a mistake which has been corrected. The erratum to this chapter is available at https://doi.org/10.1007/978-3-319-47602-5_55

© Springer International Publishing AG 2016
H. Sack et al. (Eds.): ESWC 2016 Satellite Events, LNCS 9989, pp. 17–20, 2016.
DOI: 10.1007/978-3-319-47602-5_4

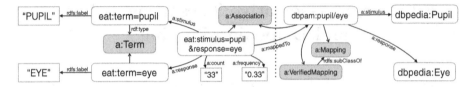

Fig. 1. Example of the EAT associations "pupil - eye" as RDF (left) and its mapping to the semantic association dbpam:pupil/eye between the DBpedia entities dbpedia:Pupil and dbpedia:Eye (right). Prefixes: a: <https://w3id.org/associations/vocab#>, eat: <http://www.eat.rl.ac.uk/#>, dbpedia <http://dbpedia.org/resource/>, dbpam <https://w3id.org/associations/mapping_eat_dbpedia#>

NER mapping back to semantic entities[1]. While these works can help collecting new associations, the datasets generated in this paper are orders of magnitude larger, published in RDF, and each of their mappings has been manually verified in order to provide high precision ground truth for machine learning.

3   Edinburgh Associative Thesaurus as RDF

EAT [8] was created in the 1970s and is a dataset of single free-text associations collected directly from humans. It consists of a well connected network of $\sim$ 788 K *raw associations* which form $\sim$ 326 K *unique associations* (unique stimulus-response-pairs) between 8200 unique stimuli and $\sim$ 22700 unique responses.

About 5000 unique associations occur more than 20 times (167 K raw associations). In the remainder of this paper we will refer to them as *strong associations*. An example for such a strong association is the one between stimulus "dog" and response "cat" which occurred 57 out of 100 times.

As the EAT dataset[2] is not available as RDF, we create an association vocabulary[3] and use it to transform EAT into RDF (see example in Fig. 1). We formally model EAT as a multi-set of raw associations. Each raw association $a \in EAT$ is a free-text stimulus-response-pair: $a = (s, r), s \in S, r \in R$. The union of all stimuli S and responses R forms the set of terms $T = S \cup R$. Further, we can define the count $c_{s,r}$ as the number of occurrences of the raw association and the relative frequency $f_{s,r}$ as the relative count of response r with respect to a fixed stimulus s over all responses to that stimulus. The resulting transformation of EAT into RDF consists of $1\,674\,376$ triples[4].

4   Mapping EAT to DBpedia

This section describes the process of mapping associations from EAT to equivalent semantic associations between pairs of DBpedia [1] entities. If we find such two entities, we call the relation between them a *semantic association*.

---

[1] http://s16a.org/node/13.

[2] http://www.eat.rl.ac.uk/.

[3] https://w3id.org/associations/vocab#.

[4] https://w3id.org/associations/eat.nt.gz.

For example, let's focus on the association "pupil - eye", with URI eat:stimulus=pupil&response=eye in Fig. 1. We can identify two DBpedia entities, namely dbpedia:Pupil and dbpedia:Eye with the intended meaning of the association and create a new semantic association dbpam:pupil/eye with the corresponding links.

For the mapping we focused on the ∼ 5000 unique strong associations occurring more than 20 times (167 K raw associations), as they are more robust with respect to subjectivity, location and time dependency.

Based on experiences gained from a manual mapping of a random sample[5], we were able to develop an automatic mapping approach with the following scoring component (non-exclusive likelihoods and examples in brackets) which uses the Wikipedia API[6]:

- **Composite phrases** (22 %, e.g., "port - wine"): As a composite phrase is a name for a single semantic entity it is a bad candidate for a semantic association (between two different semantic entities). Hence, if searching for Wikipedia articles (or redirect pages) containing stimulus and response in their title is successful, the mapping's score receives a strong punishment.
- **Adjectives & verbs vs. nouns** (22 %, e.g., "unbound - free"): Due to Wikipedia's nature of being an encyclopaedia, adjectives and verbs are under-represented in contrast to nouns. To identify such cases, the stimulus and response are searched in Wordnet [3], potentially resulting in multiple synset candidates for each. Mappings containing only synset candidates with the given type "noun" are preferred. The more synset candidates with types unequal to "noun" are found, the stronger the punishment for the mapping's score.
- **Reflexive mappings/synonyms** (18 %, e.g., "children - kids"): If the mapping of both the stimulus and the response result in the same semantic entity, the score is strongly punished.
- **Plural words** (20 %, e.g., "thumbs - fingers"): A simple stemming approach is used to compare the stimulus/response to the identified Wikipedia article titles after following redirects. If the match is close to perfect and only differs in singular/plural, the score only receives a slight punishment.
- **Disambiguation pages** (16 %, e.g., "pod - pea"): If the mappings of stimulus or response result in a Wikipedia disambiguation page, the mapping's score receives a strong punishment.

After applying the automatic mapping to the ∼ 5000 strong associations, the top scoring 1066 semantic association candidates (corresponding to ∼ 34.2 K raw associations) were selected for human verification.

In order to quickly verify the 1066 mapping candidates, a small web application was used, which shows the textual association from EAT on top (stimulus - response) and the abstracts of both mapped Wikipedia articles below and asks the user if both stimulus and response are correctly mapped.

---

[5] The manual mapping showed that about 12–28 % of the 5000 strong associations are mappable to DBpedia entities (depending on the amount of human labour and intelligence involved).

[6] http://www.mediawiki.org/wiki/API:Main_page.

The web application was used by 10 reviewers and allowed the verification (3 independent "Yes" ratings) of 790 of 1066 mappings (corresponding to ~ 25.5 K raw associations).

For each of the 790 verified mapped semantic associations a mapping URI is created analogously to Fig. 1. The resulting mapping dataset consisting of 4740 triples can be downloaded[7] or simply dereferenced.

5   Conclusion and Outlook

In this paper we presented a transformation of 788 K free-text associations from the Edinburgh Associative Thesaurus into a RDF dataset. Further, we presented a first mapping of its strong associations to semantic associations between DBpedia entities, resulting in 790 manually verified mappings corresponding to ~ 25.5 K raw associations.

In the future we plan to conduct pattern learning based on the mapped semantic associations. As all generated datasets are publicly available, we also look forward to them being used as benchmark or ground truth datasets, for example for link prediction tasks.

This work was financed by the University of Kaiserslautern PhD scholarship program and the BMBF project MOM (Grant 01IW15002).

References

1. Bizer, C., Lehmann, J., Kobilarov, G., Auer, S., Becker, C., Cyganiak, R., Hellmann, S.: DBpedia - a crystallization point for the web of data. Web Semant. Sci. Serv. Agents World Wide Web **7**(3), 154–165 (2009)
2. Bobić, T., Waitelonis, J., Sack, H.: FRanCo - a ground truth corpus for fact ranking evaluation. In: SumPre 2015 at ESWC 2015, pp. 1–12 (2015)
3. Fellbaum, C. (ed.): WordNet: An Electronic Lexical Database. MIT Press, Cambridge (1998)
4. Gerrig, R.J., Zimbardo, P.G.: Psychology and Life, 19th edn. Allyn & Bacon, Pearson, Boston (2010)
5. Hees, J., Khamis, M., Biedert, R., Abdennadher, S., Dengel, A.: Collecting links between entities ranked by human association strengths. In: Cimiano, P., Corcho, O., Presutti, V., Hollink, L., Rudolph, S. (eds.) ESWC 2013. LNCS, vol. 7882, pp. 517–531. Springer, Heidelberg (2013). doi:10.1007/978-3-642-38288-8_35
6. Hees, J., Roth-Berghofer, T., Biedert, R., Adrian, B., Dengel, A.: BetterRelations: using a game to rate linked data triples. In: Bach, J., Edelkamp, S. (eds.) KI 2011. LNCS (LNAI), vol. 7006, pp. 134–138. Springer, Heidelberg (2011). doi:10.1007/978-3-642-24455-1_12
7. Hees, J., Roth-Berghofer, T., Dengel, A.: Linked data games: simulating human association with linked data. In: LWA 2010, Kassel, Germany, pp. 255–260 (2010)
8. Kiss, G.R., Armstrong, C., Milroy, R., Piper, J.: An associative thesaurus of English and its computer analysis. In: The Computer and Literary Studies, pp. 153–165. Edinburgh University Press, Edinburgh (1973)
9. Kny, E., Kölle, S., Töpper, G., Wittmers, E.: WhoKnows? (2010)

---

[7] https://w3id.org/associations/mapping_eat_dbpedia.nt.gz.

StaRe: Statistical Reasoning Tool for 5G Network Management

Kasper Apajalahti[1]([✉]), Eero Hyvönen[1], Juha Niiranen[2], and Vilho Räisänen[3]

[1] Semantic Computing Research Group (SeCo), Aalto University, Espoo, Finland
kasper.apajalahti@aalto.fi
[2] Department of Mathematics and Statistics, University of Helsinki, Helsinki, Finland
[3] Nokia Networks Research, Espoo, Finland

Abstract. In operations of increasingly complex telecommunication networks, characterization of a system state and choosing optimal operation in it are challenges. One possible approach is to utilize statistical and uncertain information in the network management. This paper gives an overview of our work in which a Markov Logic Network model (MLN) is used for mobile network analysis with an RDF-based faceted search interface to monitor and control the behavior of the MLN reasoner. Our experiments, based on a prototype implementation, gives promising results of utilizing an ontology and MLN model in network status characterization, optimization and visualization.

1 Introduction

The growing complexity of telecommunication networks requires more automation from the network management layer. Currently researched and standardized technology in the telecommunication field is Self-Organizing Networks (SON) [1] which solves automatically some management tasks in a limited context using a fixed rule base. However, advanced uncertainty management beyond simple static rule bases is required to combine high service quality with optimization of operational expenses [4]. For this goal, we present a prototype tool StaRe that provides the user with a possibility to understand the characterization of the autonomic network management system and its uncertainties. The novel idea is to apply MLN [7] for mobile network analysis and management under uncertainty. We have examined how an ontology-based MLN model can be utilized by a human operator using a SPARQL endpoint and a faceted browser GUI. It is crucial that the operator monitors and controls the system behavior even when the autonomic system solves the majority of management tasks [8].

2 Prototype Architecture and Data Sequence

StaRe is a runtime environment tool that integrates dynamically an MLN model, an ontology based on it, and a GUI for mobile network data analysis and management. A Long-Term Evolution (LTE) simulator is used for simulating an urban mobile network environment.

© Springer International Publishing AG 2016
H. Sack et al. (Eds.): ESWC 2016 Satellite Events, LNCS 9989, pp. 21–25, 2016.
DOI: 10.1007/978-3-319-47602-5_5

Figure 1 depicts our architecture and its data sequence for managing a mobile network. The data sequence starts from the right where the simulator data is retrieved to the MLN model in every 15 min of simulation time. This data contains key performance indicators (KPI) for measurement cases, such as channel quality indicator (CQI) and radio link failures (RLF). In return, the MLN model reasons configuration management parameters for the mobile network (i.e., the LTE simulator) that contain needed changes in the transmission power (TXP) and angle (remote electrical tilt, RET) of a cell antenna. These parameters are critical for the quality of service of the network and operation optimization.

The simulator data is used as the evidence of the MLN model to infer posterior probabilities for action proposals. Network cells in the simulator are then configured based on the action proposal distributions. In order to make this process manageable to the operator, the ontology processor retrieves and parses the evidence, rules, action proposals, and configurations, and populates the ontology with network- and MLN-related instances. The ontology is then uploaded into a SPARQL server based on Fuseki[1]. The server dynamically generates facets from the ontology (with SPARQL update scripts) for the GUI and acts as a data storage both for the GUI and MLN model. The GUI interacts with the SPARQL endpoint to retrieve semantic data from the ontology and to update the semantic MLN rule base. Similarly, the MLN model queries the SPARQL endpoint in order to retrieve updated rule base.

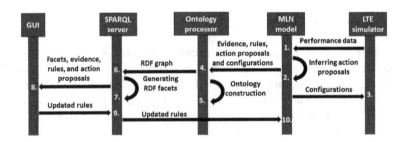

Fig. 1. Data sequence diagram for managing a mobile network (simulator)

3   Faceted Browser Interface

The GUI is an HTML5 application which is built by using faceted browsing and interactive visualization methods in order to (1) determine needed network management actions in a situation and to (2) manage the MLN rules. Figure 2 shows how facet selections can be used to search for recommended actions. Here cells with a high RLF value and low CQI value are selected on the facets on the left, and the tool suggests as an action proposal to increase the TXP of the cells on the right with a varying uncertainty.

---

[1] https://jena.apache.org/documentation/serving_data/.

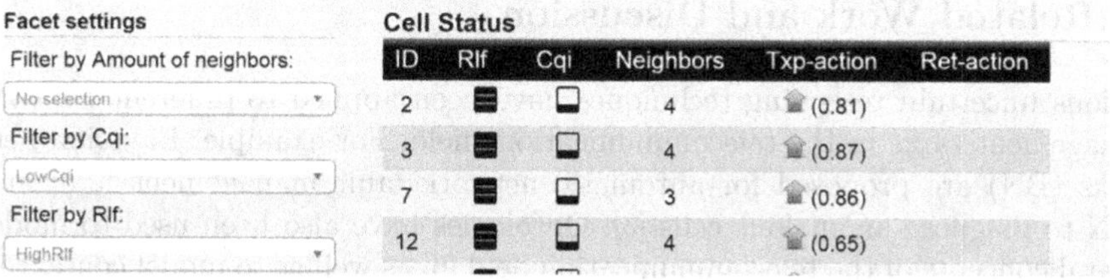

Facet settings	Cell Status					
Filter by Amount of neighbors:	ID	Rlf	Cqi	Neighbors	Txp-action	Ret-action
No selection ▾	2	▤	▢	4	🏆 (0.81)	
Filter by Cqi:	3	▤	▢	4	🏆 (0.87)	
LowCqi ▾	7	▤	▢	3	🏆 (0.86)	
Filter by Rlf:	12	▤	▢	4	🏆 (0.65)	
HighRlf ▾						

Fig. 2. Faceted browsing for the filtering reasoning outcome

Figure 3 shows a view for managing the uncertain MLN rules by dividing each rule into a rule weight and to semantically defined rule classes: context (current network status), objective (desired change in the network status) and action (configurations for the network). The operator uses this view to investigate contents of the rules and to manipulate the rule base in order to change the behaviour of the MLN reasoner.

The facets are generated as a combination of rule classes (context, objectives and actions) and their objects (CQI, RLF, TXP, and RET). Here the operator has filtered out rules containing low CQI in the context part and increase CQI in the objective part. The result indicates that every single rule can be removed or its weight can be updated. The possibility to remove a set of rules that satisfies current facet selections can be seen above the result table.

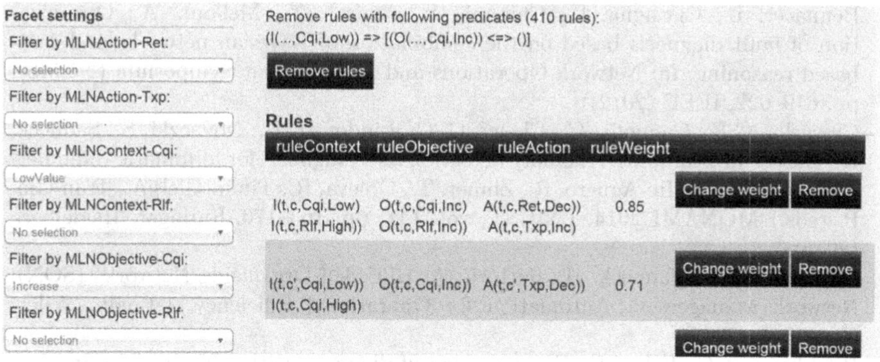

Facet settings	Remove rules with following predicates (410 rules):
Filter by MLNAction-Ret:	(I(...,Cqi,Low)) => [(O(...,Cqi,Inc)) <=> ()]
No selection ▾	**Remove rules**
Filter by MLNAction-Txp:	
No selection ▾	**Rules**
Filter by MLNContext-Cqi:	ruleContext ruleObjective ruleAction ruleWeight
LowValue ▾	Change weight | Remove
Filter by MLNContext-Rlf:	I(t,c,Cqi,Low) O(t,c,Cqi,Inc) A(t,c,Ret,Dec)) 0.85
No selection ▾	I(t,c,Rlf,High)) O(t,c,Rlf,Inc)) A(t,c,Txp,Inc)
Filter by MLNObjective-Cqi:	Change weight | Remove
Increase ▾	I(t,c',Cqi,Low)) O(t,c,Cqi,Inc)) A(t,c',Txp,Dec)) 0.71
Filter by MLNObjective-Rlf:	I(t,c,Cqi,High)
No selection ▾	Change weight | Remove

Fig. 3. Faceted search for uncertain rules

In StaRe, the operator can also create new rules with a rule creation form which enables a productive way to create MLN rules without writing the actual MLN syntax.

4   Related Work and Discussion

Various uncertain reasoning techniques have been applied to different network management tasks in the telecommunications field. For example, Bayesian networks (BN) are proposed for automatic network fault management [2,5] and MLN to diagnose anomalous cells [3]. Ontologies have also been used to model general concepts of the telecommunication field [6] as well as to model context in mobile network management [9,11]. The Linked Open Data (LOD)[2] paradigm has also been addressed in [10], which models cells and terminals, and combines them with other data sources, for example with event data. However, there exists no research of using ontologies and statistical reasoning together to analyze and configure the mobile network, as in this paper.

Altogether, StaRe has proven to be a useful tool in complex network management tasks as it contains a reasoner for processing uncertain information and a semantic faceted browser interface for information exploration. As this paper shows, the StaRe ontology can be used as a semantic data storage between the MLN reasoner and GUI.

References

1. 3GPP: Telecommunication management; Self-Organizing Networks (SON); Concepts and requirements, [3Gpp. TS 32.500 Release 13, modified]. Technical report, 3rd Generation Partnership Project (3GPP), Feburary 2016. http://www.3gpp.org/DynaReport/32500.htm
2. Bennacer, L., Ciavaglia, L., Chibani, A., Amirat, Y., Mellouk, A.: Optimization of fault diagnosis based on the combination of Bayesian networks and case-based reasoning. In: Network Operations and Management Symposium (NOMS), pp. 619–622. IEEE (2012)
3. Ciocarlie, G.F., Connolly, C., Cheng, C.-C., Lindqvist, U., Nováczki, S., Sanneck, H., Naseer-ul-Islam, M.: Anomaly detection and diagnosis for automatic radio network verification. In: Agüero, R., Zinner, T., Goleva, R., Timm-Giel, A., Tran-Gia, P. (eds.) MONAMI 2014. LNICST, vol. 141, pp. 163–176. Springer, Heidelberg (2015)
4. Hämäläinen, S., Sanneck, H., Sartori, C.: LTE Self-Organising Networks (SON): Network Management Automation for Operational Efficiency, 1st edn. Wiley, New York (2012)
5. Hounkonnou, C., Fabre, E.: Empowering self-diagnosis with self-modeling. In: Proceedings of the 8th International Conference on Network and Service Management, pp. 364–370. International Federation for Information Processing (2012)
6. Qiao, X., Li, X., Chen, J.: Telecommunications service domain ontology: semantic interoperation foundation of intelligent integrated services. In: Telecommunications Networks-Current Status and Future Trends pp. 183–210 (2012)
7. Richardson, M., Domingos, P.: Markov logic networks. Mach. Learn. **62**(1–2), 107–136 (2006)
8. Russell, D.M., Maglio, P.P., Dordick, R., Neti, C.: Dealing with ghosts: managing the user experience of autonomic computing. IBM Syst. J. **42**(1), 177–188 (2003)

---

[2] http://linkeddata.org/.

9. Stamatelatos, M., Grida Ben Yahia, I., Peloso, P., Fuentes, B., Tsagkaris, K., Kaloxylos, A.: Information model for managing autonomic functions in future networks. In: Pesch, D., Timm-Giel, A., Calvo, R.A., Wenning, B.-L., Pentikousis, K. (eds.) Monami. LNICST, vol. 125, pp. 259–272. Springer, Heidelberg (2013)
10. Uzun, A., Küpper, A.: OpenMobileNetwork: extending the web of data by a dataset for mobile networks and devices. In: Proceedings of the 8th International Conference on Semantic Systems, pp. 17–24. ACM (2012)
11. Yahia, B., Grida, I., Bertin, E., Crespi, N.: Ontology-based management systems for the next generation services: state-of-the-art. In: Third International Conference on Networking and Services, ICNS 2007, p. 40. IEEE (2007)

Dem@Home: Ambient Monitoring and Clinical Support for People Living with Dementia

Thanos G. Stavropoulos[✉], Georgios Meditskos, Thodoris Tsompanidis, Stelios Andreadis, and Ioannis Kompatsiaris

Information Technologies Institute, Center for Research and Technologies - Hellas, Thessaloniki, Greece
{athstavr,gmeditsk,thtsompa,andreadisst,ikom}@iti.gr

Abstract. Dem@Home is an Ambient Assisted Living framework to support intelligent dementia care, by integrating a variety of ambient and wearable sensors together with sophisticated, interdisciplinary methods, such as image and semantic analysis. Semantic Web technologies, such as OWL, are used to represent sensor observations and application domain specifics as well as to implement hybrid activity recognition and problem detection solutions. Complete with tailored user interfaces, Dem@Home supports accurate monitoring of multiple aspects, such as physical activity, sleep, complex daily activities and problems, leading to adaptive interventions for the optimal care of dementia, validated in four home pilots.

Keywords: Ambient assisted living · Sensors · Semantic web · Ontologies · Reasoning · Context-awareness · Dementia

1   Introduction

The increase of the average lifespan across the world has been accompanied by an unprecedented upsurge in the occurrence of dementia, with high socio-economic costs, reaching 818 billion US dollars worldwide, in 2015. Nevertheless, its prevalence is increasing as the number of people aged 65 and older with Alzheimer's disease may nearly triple by 2050, from 46.8 million to 131 million people around the world, the majority of which, living in an institution [1].

Assistive technology is called upon to contribute to cognitive and physical state improvement and to cost reduction, by prolonging independent living. However, not many existing solutions follow a holistic approach, in order to achieve greater impact. Pervasive technology solutions have already been employed in several ambient environments, either homes or clinics, but most of them focus on a single domain to monitor, using only a single or a few devices. Such applications include wandering behavior prevention with geolocation devices, monitoring physical activity, sleep, medication and performance in daily chores [2, 3].

In order to assess cognitive state, activity modelling and recognition appears to be a critical task, common amongst existing assistive technology. OWL has been widely used for modelling human activity semantics, reducing complex activity definitions to the

H. Sack et al. (Eds.): ESWC 2016 Satellite Events, LNCS 9989, pp. 26–29, 2016.
DOI: 10.1007/978-3-319-47602-5_6

intersection of their constituent parts. In most cases, activity recognition involves the segmentation of data into snapshots of atomic events, fed to the ontology reasoner for classification. Time windows [4] and slices [5] provide background knowledge about the order or duration [6] of activities are common approaches for segmentation. In this paradigm, ontologies are used to model domain information, whereas rules, which are widely embraced to compensate for OWL's expressive limitations, aggregate activities, describing the conditions that drive the derivation of complex activities e.g. temporal relations.

The proposed system, Dem@Home, complements these developments by providing a holistic approach for context-aware monitoring and personalized care of dementia at homes, prolonging independent living. To begin with, the system integrates a wide range of sensor modalities and high-level analytics to support accurate monitoring of all aspects of daily life including physical activity, sleep and activities of daily living (ADLs), based on a service-oriented middleware [7]. After integrating them in a uniform knowledge representation format, Dem@Home employs semantic interpretation techniques to infer complex activity recognition from atomic events and highlight clinical problems. Specifically, it follows a hybrid reasoning scheme, using DL reasoning for activity detection and SPARQL to extract clinical problems. Utterly, Dem@Home presents information to applications tailored to clinicians, and patients, endorsing technology-aided clinical interventions to improve care. Dem@Home has been deployed and evaluated in four home pilots showing positive results.

2   The Dem@Home Framework

Dem@Home proposes a multidisciplinary approach that brings into effect the synergy of the latest advances in sensor technologies addressing a multitude of complementary modalities, large-scale fusion and mining, knowledge representation and intelligent decision-making support. In detail, the framework integrates several heterogeneous sensing modalities, such as physical activity and sleep sensor measurements, combined input from lifestyle sensors and higher-level image analytics, providing their unanimous semantic representation and interpretation.

The current selection of sensors, shown on Fig. 1, is comprised of proprietary, low-cost devices, originally intended for lifestyle monitoring, repurposed to a medical context. Each device is integrated by using dedicated modules that wrap their respective API, retrieve data and process them accordingly to generate atomic events from sensor observations e.g. through aggregation. In the case of image data, computer vision techniques are employed to extract information about humans performing activities, such as opening the fridge, holding a cup or drinking [8]. All atomic events and observations are mapped to a uniform semantic representation for interoperability and stored to Dem@Home's Knowledge Base.

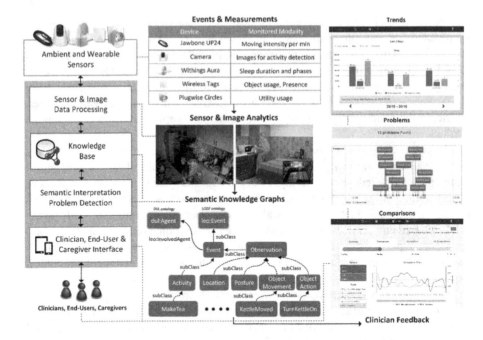

Fig. 1. Dem@Home architecture, sensors and clinical applications

To obtain a more comprehensive image of an individual's condition and its progression, driving clinical interventions, Dem@Home employs semantic interpretation to perform intelligent fusion and aggregation of atomic, sensor events to complex ones and identify problematic situations. It does so, with a hybrid combination of OWL 2 reasoning and SPARQL queries.

Dem@Home provides a simple pattern for modelling the context of complex activities, in other words, the semantics for activity recognition. Each activity context is described through class equivalence axioms that link them with lower-level observations of domain models (Fig. 1). The instantiation of this pattern is used by the underlying reasoner to classify context instances, generated during the execution of the protocol, as complex activities. The instantiation involves linking IADLs with context containment relations through class equivalence axioms. For example, given that the activity MakeTea involves the observations TurnKettleOn, CupMoved, KettleMoved, TeaBagMoved, TurnKettleOff, HoldingCup, its semantics are defined as:

$$MakeTea \equiv Context \sqcap \exists contains.TurnKettleOn \sqcap \exists contains.CupMoved$$
$$\sqcap \exists contains.KettleMoved \sqcap \exists contains.TeaBagMoved$$
$$\sqcap \exists contains.TurnKettleOff \sqcap \exists contains.HoldingCup$$

According to clinical experts involved in the development of Dem@Home, highlighting problematic situations next to the entire set of monitored activities and metrics would further facilitate and accelerate clinical assessment. Dem@Home uses a set of pre-defined rules (expressed in SPARQL) with numerical thresholds that clinicians can adjust and personalize to each of the individuals in their care. Furthermore, each analysis is invoked for a period of time allowing different thresholds for different intervals e.g. before and after a clinical intervention. Problematic situations supported so far regard night sleep (short duration, many interruptions, too long to fall asleep), physical activity (low daily activity totals), missed activities (e.g. skipping daily lunch) and reoccurring problems (problems for consecutive days).

Dem@Home was evaluated in four home installations, in the residences of individuals living alone, clinically diagnosed with mild cognitive impairment or mild dementia, and maintained for four months. Complex activity recognition evaluated over the first month, achieved an average of 84 % recall and 88 % precision for daily tasks. Losses are attributed to the openness and uncertainty in the environment, e.g. when interleaving activities. However, with Dem@Home supporting clinical interventions, significant improvement was found in post-pilot clinical assessment in multiple domains, such as increase in physical condition, sleep duration decreased insomnia, utterly bringing about improvement in mood and cognitive state.

Acknowledgement. This work has been funded by the EU FP7 project Dem@Care (contract No. 288199).

References

1. Prince, M., Wimo, A., Guerchet, M., Ali, G., Wu, Y.T., Prina, M.: World Alzheimer Report 2015: The global impact of dementia. An analysis of prevalence, incidence, cost and trends. Alzheimer's Disease International, London (2015)
2. Kerssens, C., Kumar, R., Adams, A.E., Knott, C.C., Matalenas, L., Sanford, J.A., Rogers, W.A.: Personalized technology to support older adults with and without cognitive impairment living at home. Am. J. Alzheimers Dis. Other Demen. **30**, 85–97 (2015). doi: 10.1177/1533317514568338
3. Dawadi, P.N., Cook, D.J., Schmitter-Edgecombe, M., Parsey, C.: Automated assessment of cognitive health using smart home technologies. Technol. Health Care Off. J. Eur. Soc Eng. Med. **21**, 323 (2013)
4. Okeyo, G., Chen, L., Wang, H., Sterritt, R.: Dynamic sensor data segmentation for real-time knowledge-driven activity recognition. Pervasive Mob. Comput. **10**, 155–172 (2014)
5. Riboni, D., Pareschi, L., Radaelli, L., Bettini, C.: Is ontology-based activity recognition really effective? In: Pervasive Computing and Communications Workshops, pp. 427–431. IEEE (2011)
6. Patkos, T., Chrysakis, I., Bikakis, A., Plexousakis, D., Antoniou, G.: A reasoning framework for ambient intelligence. In: Konstantopoulos, S., Perantonis, S., Karkaletsis, V., Spyropoulos, C.D., Vouros, G. (eds.) SETN 2010. LNCS, vol. 6040, pp. 213–222. Springer, Heidelberg (2010)
7. Stavropoulos, T.G., Meditskos, G., Kontopoulos, E., Kompatsiaris, I.: The DemaWare Service-Oriented AAL Platform for People with Dementia★. In: Artificial Intelligence and Assistive Medicine (AI-AM/NetMed 2014), November 2014
8. Avgerinakis, K., Briassouli, A., Kompatsiaris, I.: Recognition of activities of daily living for smart home environments. In: 2013 9th International Conference on Intelligent Environments (IE), pp. 173–180. IEEE (2013)

Pattern-Based Keyword Search on RDF Data

Hanane Ouksili[1,2]([✉]), Zoubida Kedad[1], Stéphane Lopes[1], and Sylvaine Nugier[2]

[1] DAVID Lab., Univ. Versailles St Quentin, Versailles, France
{hanane.ouksili,zoubida.kedad,stephane.lopes}@uvsq.fr
[2] EDF R&D, Departement STEP, Chatou, France
sylvaine.nugier@edf.fr

Abstract. An increasing number of RDF datasets are available on the Web. Querying RDF data requires the knowledge of a query language such as SPARQL; it also requires some information describing the content of these datasets. The goal of our work is to facilitate the interrogation of RDF datasets, and we present an approach for enabling users to search in RDF data using keywords. We propose the notion of pattern to include some external knowledge during the search process which increases the quality of the results.

Keywords: RDF graph · Keyword search · Patterns · Domain knowledge

1 Introduction

A huge volume of RDF datasets is available on the Web, enabling the design of novel intelligent applications. In order to query these datasets, users must have information about their schema to target the relevant resources and properties. They must also be familiar with a formal query language such as SPARQL. An alternative approach to query RDF data is keyword search which is a straightforward and intuitive way of querying datasets.

We can distinguish between two categories of approaches for keyword search in RDF data. The first one uses information retrieval techniques on documents which are previously defined. For example, a document can be defined as a triple [1]. Documents are indexed, then keywords are mapped into the graph to identify relevant documents. Finally, the result is returned, consisting in a ranked list of documents. These works do not try to combine parts of the graph which correspond to different keywords. In the second category, the structure of the graph is used to construct the results [2,3]. Keyword-to-graph mapping functions are used to identify a set of elements which contains the query keywords. The ranked list of subgraphs is then returned as a result. In these works, only the structure of the graph is considered during the construction of the result, unlike our approach, which integrates external knowledge formalized as patterns.

In this paper, we introduce a novel keyword search approach which returns graphs as a result to a keyword query (see Fig. 1). The key contribution of

This work is supported by EDF and french ANR project CAIR.

H. Sack et al. (Eds.): ESWC 2016 Satellite Events, LNCS 9989, pp. 30–34, 2016.
DOI: 10.1007/978-3-319-47602-5_7

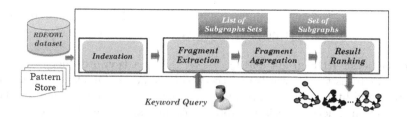

Fig. 1. Keyword search process

this paper is the use of external knowledge, expressed as patterns, during the extraction of the relevant graph fragments, the construction of the result and the ranking step. Experiments show that our approach delivers high-quality search results.

2   Expressing Knowledge with Patterns

According to the knowledge available for a specific domain, or to the user's point of view, some equivalence relations can be defined between properties and paths in the dataset. In our approach, they are expressed through patterns.

• A *pattern* represents an equivalence between one property expression and one path expression. It is defined as a pair $[exp, exP]$ where $exp = (X, p, Y)$ is a property expression, $exP = (X, P, Y)$ is a path expression, p is a property, P is a SPARQL 1.1 path expression[1], and X and Y are either resources or variables.

For example, consider the following pattern: $[(X, swrc:isAbout, Y), (X, swrc:isAbout/(owl:sameAs|^owl:sameAs)^+, Y)]$. This pattern means that the property *swrc:isAbout* is equivalent to a path composed of a property *swrc:isAbout* followed by a sequence of *owl:sameAs* properties. Its evaluation on the RDF graph of Fig. 2 extracts the value *D.B.* for the property *swrc:isAbout* of the resource *Art1*, *Database* is therefore considered as value for this property.

Fig. 2. Subgraph from SWRC   **Fig. 3.** Final results

3   Extracting Relevant Fragments

We use an inverted index to extract the relevant fragments. To this end, we consider a representative keyword from the local name of URI for resources and properties as document content; for example, a document is created for the

---

[1] http://www.w3.org/TR/sparql11-property-paths/.

property *swrc:hasAuthor*, containing the word *Author* (it can be generated using Information Extraction techniques). For literals, we consider their content as a document.

Let G be an RDF graph and $Q = \{k_1, ..., k_n\}$ a keyword query. The system first matches the keywords and the graph elements (i.e. classes, instances, properties and literals) using the index, a mapping function and some standard transformation functions such as abbreviation and synonym. This step returns a set of lists $El=\{El_1, ..., El_n\}$ where each El_i is a list of elements el_{ij} that matches the keyword k_i. We refer to these as *keyword elements*. Patterns are then used to extract, for each keyword element el_i, subgraphs that are relevant to the query. We denote these subgraphs and keyword element as relevant fragments. If no pattern is applicable, the relevant fragment is reduced to el_i.

We use two kinds of patterns: generic and domain specific. Generic patterns are valid for any dataset as they are related to RDF, RDFS or OWL vocabularies. For example, the pattern $[(X, \textit{pat:sameResult}, Y), (X, (\textit{owl:sameAs} |\textasciicircum\textit{owl:sameAs})^+, Y)]$ is defined to consider X as a relevant fragment if it is related to one keyword element Y with a path composed of a sequence of *owl:sameAs* properties. This is true for any dataset. The evaluation of this pattern in the graph of Fig. 2 allows to consider that the resource *database* is relevant when the resource *D.B.* is extracted as a keyword element.

Similarly, domain specific patterns can be defined. For example, the property *swrc:cooperateWith* between two researchers in the ontology SWRC[2] states that they have already collaborated. However, this property is not always defined. If we know that they have authored the same paper, we can infer that they have collaborated, even if the property *swrc:cooperateWith* is missing. We formalize this by the following pattern:

$$[(X, \textit{swrc:cooperateWith}, Y), (X, \textit{swrc:publication}/\textasciicircum\textit{swrc:publication}, Y)].$$

This pattern is used during the *fragment extraction* module by considering the subgraph corresponding to the path $(\textit{swrc:publication}/\textasciicircum \textit{swrc:publication})$ as a relevant fragment if the property *swrc:cooperateWith* is a keyword element. Consider that *"Studer Cooperator"* is a query issued to find the collaborators of the researcher named *Studer* in the graph of Fig. 2. Let us consider the following two extracted keyword elements: the node *Rudi Studer* and the property *swrc:cooperateWith*. Without using patterns, the only result will be the subgraph G_1 of Fig. 3 representing *Daniel Deutch*, a collaborator of *Rudi Studer*, both linked by the *swrc:cooperateWith* property. However, *Victor Vianu* is also a collaborator of *Rudi Studer* as they published a paper together (*Art1*). Using the pattern defined above, our approach will also return the subgraph G_2 of Fig. 3 as result.

Note that if the keyword element is a property as in our example, our approach will not extract all occurrences of the property in the graph but only the ones for which either the object or the subject is a relevant fragment. Other-

---

[2] http://ontoware.org/swrc/.

wise, the result might include irrelevant answers. In our example, the system will return only cooperators of *Studer* and not cooperators of others researchers.

4   Result Construction and Ranking

The result for a keyword query Q is a list of subgraphs. The goal of the aggregation is to merge relevant fragments to form the set of results. Each result is a connected minimal subgraph which contains for each keyword k_i one corresponding relevant fragment. We use a Cartesian product between the different sets of relevant fragments to construct the combinations. Finally, for each combination, we perform a bidirectional expansion search strategy to join different relevant fragments and construct the result. We introduce the defined patterns to calculate the path between resources. If a pattern indicates that one path is equivalent to one property, the distance is reduced to 1.

Since several elements may contain the same keyword, several subgraphs are returned to the user. Hence, we define a ranking function to evaluate the relevance degree of each subgraph. Our approach combines two criteria: (i) *Compactness Relevance* where compact answers are preferred. When the size of the subgraph is smaller, the compactness is higher. Patterns are taken into account to calculate the size. (ii) *Matching Relevance* is calculated using standard IR approaches. For our approach, we have used the TF-IDF function.

5   Evaluations

We have implemented a prototype which provides an interactive interface to keyword search and allows users to express some external knowledge, which is formalized by our system using patterns. We have used AIFB dataset[3] and we have compared our approach to two baseline algorithms which contain the same steps but do not use patterns; the first one considers both node and edge content without any condition on their extraction and the second one considers node content only.

We have used 10 queries that are randomly constructed (with 2 to 5 keywords) and asked four users to assess the top-k results of each query using the three algorithms on a 3-levels scale: 3 (perfectly relevant), 1 (relevant) and 0 (irrelevant). We use the two metrics NDCG@k and precision@k. To calculate the precision@k, we assume that the scores 3 or 1 correspond to a relevant result and 0 to an irrelevant one. Table 1 reports average Precision@k and NDCG@k values over the 10 queries with k = 5, 10 and 20. Our approach significantly outperforms the two other algorithms in terms of NDCG and precision values at all levels. Furthermore, since NDCG includes the ranking position of the results, the ranking algorithm of our approach is better because it includes patterns during the ranking step.

---

[3] http://www.aifb.kit.edu/web/Hauptseite/en.

Table 1. Evaluation results

Approach	Our Approach			Approach (i)			Approach (ii)		
k	5	10	20	5	10	20	5	10	20
Precision@k	0.99	0.98	0.97	0.73	0.73	0.69	0.66	0.66	0.66
NDCG@k	0.92	0.90	0.90	0.64	0.63	0.58	0.47	0.46	0.46

References

1. Elbassuoni, S., Blanco, R.: Keyword search over RDF graphs. In: CIKM, pp. 237–242 (2011)
2. Wang, H., Zhang, K., Liu, Q., Tran, T., Yu, Y.: Q2Semantic: a lightweight keyword interface to semantic search. In: Bechhofer, S., Hauswirth, M., Hoffmann, J., Koubarakis, M. (eds.) ESWC 2008. LNCS, vol. 5021, pp. 584–598. Springer, Heidelberg (2008)
3. Yang, S., Wu, Y., Sun, H., Yan, X.: Schemaless and structureless graph querying. VLDB Endow. **7**, 565–576 (2014)

Connecting the Dots: Explaining Relationships Between Unconnected Entities in a Knowledge Graph

Nitish Aggarwal[1(✉)], Sumit Bhatia[2], and Vinith Misra[2]

[1] Insight Centre for Data Analytics, National University of Ireland, Galway, Ireland
nitish.aggarwal@insight-center.org
[2] IBM Research, Almaden, USA
{sumit.bhatia,vmisra}@us.ibm.com

Abstract. We discuss the problem of explaining relationships between two unconnected entities in a knowledge graph. We frame it as a path ranking problem and propose a path ranking mechanism that utilizes features such as specificity, connectivity, and path cohesiveness. We also report results of a preliminary user evaluation and discuss a few example results.

1   Introduction

The advent of semantic knowledge bases like DBpedia, Freebase, etc. has led to the development of smart search systems that produce rich and enhanced results by providing additional related information about the entities/concepts being queried by the users. Further, increasing efforts are being made to build *knowledge discovery systems* that help users to navigate/explore the semantic graph and discover hitherto unknown, yet extremely useful information. For example, Nagarajan et al. [6] describe a discovery system that uses a semantic network built out of medical literature and helps researchers in discovering previously unknown protein-protein interactions. Likewise, web search engines like Google, Bing, etc. also incorporate data from their knowledge graphs to provide a list of entities that are related to the user search query [3] and users often navigate through these recommended entities to discover new non-trivial information about their search topics.

Despite providing a greatly simplified knowledge discovery process by recommending related entities of interest, such systems often fail to provide explanations for such recommendations to users, especially for less popular entities and entities that are not directly connected to the input entity. For example, for an entity query "Abu Bakr al-Baghdadi" (leader of the terrorist organization Islamic State of Iraq and the Levant (ISIL)), Google recommends entities such as "Musab al-Zarqawi", "Qasem Soleimani", etc., but fails to provide any explanation about how these entities are related to the input entity. Previous research efforts [5,7] have tried to explain the relatedness between entities by deriving

H. Sack et al. (Eds.): ESWC 2016 Satellite Events, LNCS 9989, pp. 35–39, 2016.
DOI: 10.1007/978-3-319-47602-5_8

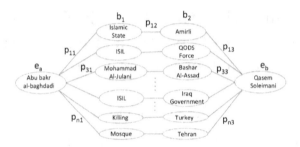

Fig. 1. Paths between "Abu bakr al-baghdadi" and "Qasem Soleimani"

important paths between entities in the knowledge graph. However, these methods generally focus either on popular entities in the graph or rely on query log data from the search engines that may not be available always, especially in enterprise domains.

In this work, we address the problem of *ranking all the paths between any two entities in a knowledge graph*. A solution to this problem can help in explaining relationships between seemingly unconnected entities as well as in finding interesting, non-obvious paths between two given entities. Given two entities in the graph, there could be hundreds and thousands of possible paths connecting the two entities. We posit that each such path represents a fact (or a hidden connection), and thus, provides a potential explanation of the relationship between the two entities. Not all the paths connecting the two entities are equally important and therefore, a mechanism is required to *rank* these paths based on their usefulness. For instance, Fig. 1 shows several different paths existing in our knowledge graph between entities "Abu bakr al-baghdadi" and "Qasem Soleimani" (a major general in the Iranian Army of the Guardians of the Islamic Revolution). We observe that all the paths may not be considered equally informative as some of the paths involve very generic entities like "Mosque", "Tehran", etc. and some paths involve very specific entities like "Islamic State" and "Amirli", representing the fact that Islamic State and Iranian Army led by Qasem Soleimani were involved in a battle in the city of Amirli. Thus, we see how these paths can provide deep insights into different type of relations between given entities.

2   Proposed Path Ranking Algorithm

Let e_a and e_b be the two input entities and let $P = \{P_1, P_2, \ldots, P_n\}$ be the set of all possible paths connecting e_a and e_b. Let $P_i = \{p_{i1}, p_{i2}, \ldots, p_{im}\}$ be a path of length m between e_a and e_b, consisting of $m-1$ bridging nodes $b_1, b_2 \ldots b_{m-1}$, such that p_{i1} is the edge between e_a and b_1, p_{i2} is the path between b_1 and b_2, and so on. Our task is to produce a ranked list of paths in P ordered by their relevance scores. Our proposed ranking function is based on the intuition that a relevant and informative path consists of relevant and useful bridging nodes and edges and we utilize following signals to capture this intuition in the relevance score of a given path between e_a and e_b.

Specificity: This measure is analogous to the inverse document frequency (IDF) concept used frequently in information retrieval models. There are many popular entities that are connected to a disproportionately large number of other entities in the graph. For example, *USA* is connected to a large number of other entities corresponding to different countries, persons, organizations, etc. A path connected through such highly popular bridging nodes may not be as informative and useful as a path connected through relatively rare bridging nodes. Specificity of a given entity is computed as the inverse of total neighbor count of the given entity and specificity of a given path is, in turn, computed as sum of specificity scores of each bridging node in the path.

Connectivity: This measure tries to capture the contribution of constituent edges to the overall relevancy of a given path. A path consisting of stronger edges is more probable to be useful than a path consisting of weaker edges. We posit that the strength of an edge connecting two entities is directly proportional to *relatedness* of the two entities. Computing relatedness scores between two entities in knowledge graphs is a well addressed problem [1,3] and we use the distributional semantics model (DSM) [2] to compute the relatedness score of two connected entities and hence, their *edge strength*. We generate the DSM vector for each entity over Wikipedia concepts and compute the relevance scores between two entities by calculating cosine scores between their vectors.

Cohesiveness: While the previous two signals were concerned with the contribution of individual bridging nodes and constituent edges to the overall relevancy score of a path, this measure reflects the strength of *linkages* between adjacent edges in the path. Connectivity measure as discussed above provides the strength of individual edges in *isolation* and hence, may not capture the relevancy of different edges in context of others. This measure, therefore, tries to capture the *cohesiveness* of successive edges that form a given path as follows. For a pair of consecutive edges p_i and p_{i+1} connecting entities e_a, e_b, and e_c, respectively, we obtain the composite DSM vector of entity e_a and e_b by adding their individual DSM vectors and then take the cosine of the DSM vector e_c with this composite vector. This way, cohesiveness score provides the relevancy of entity e_c to e_a in context of adjacent bridging entity e_b. The overall cohesiveness score of the path is then computed by summing over the cohesiveness score of each consecutive pair of edges in the path.

Finally, the overall relevance score of a given path is obtained by taking a product of all the above three scores.

3   Evaluation

We use a semantic graph constructed from text of all articles in Wikipedia by automatically extracting the entities and their relations by using IBM's Statistical Information and Relation Extraction (SIRE) toolkit[1]. Even though there exist popular knowledge bases like DBPedia that contain high quality data, we

---

[1] http://ibmlaser.mybluemix.net/siredemo.html.

chose to construct a semantic graph using automated means as such a graph will be closer to many real world scenarios where domain specific data is used and high quality curated graphs are often not available. Our graph contains more than 30 millions entities and 192 million distinct statements in comparison to 4.5 million entities and 70 million statements in DBpedia.

Path ranking in knowledge graphs is a relatively new problem and datasets used in previous related research [4,7] for explaining relationships mainly focus on popular paths and ignore rare and inconspicuous relationships. Further, there may not be one single relevant path (or explanation) as the two entities could be related in multiple different ways. In this work, we perform a preliminary qualitative evaluation through a user study involving three human assessors. We asked each of the three volunteers to query our system with three entity pairs of their interest and showed them top 20 paths ranked by their relevance scores as described above. The evaluators were then asked to rate each path using the following criterion – (0) non-relevant path, (1) relevant path, somewhat informative (2) relevant and highly informative path. There are a total of 180 paths (9 pairs times 20 paths) to be evaluated and each evaluator provided judgments for 60 paths. The evaluators rated around 15 % of the paths non-relevant, 75 % as relevant and 10 % as relevant paths with discovery. As an example of results, the top ranked path of length 3 between entities "Abu Bakr Al-Baghdadi" and "Qasem Soleimani" is connected through entities "Islamic States" and "Amirli" and corresponds to the fact (taken from Wikipedia) that "...the fight over Amirli in eastern Iraq has been one of the most important battles against ISIS. The response to ISIS's push against the town was likely formulated by Qassem Suleimani, the head of the Iranian Revolutionary Guards Corps' Qods Force..." Another example of path illustrating the potential of proposed approach in unravelling interesting and hidden facts is the top ranked path between an Indian actor "Aamir Khan" and Hollywood director "Christopher Nolan" which reveals the fact that "Aamir Khan's movie "Ghajini" was the remake of Hollywood movie "Momento" directed by "Christopher Nolan."

4   Conclusion and Future Work

We presented a path ranking mechanism to explain relatedness of two unconnected entities in a knowledge graph and to uncover hidden connections between the two entities. We used specificity, connectivity and cohesiveness features to measure the quality of a path and performed a small scale, preliminary evaluation of the proposed approach. The results from this preliminary study are encouraging and provide some support to the proposed approach's ability to find high quality paths between entities. However, a more rigorous qualitative and quantitative evaluation is required to confirm these initial findings and this will be the focus of our future work.

References

1. Aggarwal, N., Asooja, K., Ziad, H., Buitelaar, P.: Who are the American Vegans related to Brad Pitt? Exploring related entities. In: Proceedings of the 24th International Conference on World Wide Web Companion (2015)
2. Aggarwal, N., Buitelaar, P.: Wikipedia-based distributional semantics for entity relatedness. In: 2014 AAAI Fall Symposium Series (2014)
3. Blanco, R., Cambazoglu, B.B., Mika, P., Torzec, N.: Entity recommendations in web search. In: Alani, H., et al. (eds.) ISWC 2013, Part II. LNCS, vol. 8219, pp. 33–48. Springer, Heidelberg (2013)
4. Fang, L., Sarma, A.D., Yu, C., Bohannon, P.: REX: explaining relationships between entity pairs. Proc. VLDB Endow. **5**(3), 241–252 (2011)
5. Heim, P., Lohmann, S., Stegemann, T.: Interactive relationship discovery via the semantic web. In: Aroyo, L., Antoniou, G., Hyvönen, E., Teije, A., Stuckenschmidt, H., Cabral, L., Tudorache, T. (eds.) ESWC 2010. LNCS, vol. 6088, pp. 303–317. Springer, Heidelberg (2010). doi:10.1007/978-3-642-13486-9_21
6. Nagarajan, M., et al.: Predicting future scientific discoveries based on a networked analysis of the past literature. In KDD. ACM (2015)
7. Pirrò, G.: Explaining and suggesting relatedness in knowledge graphs. In: Arenas, M., et al. (eds.) ISWC 2015. LNCS, vol. 9366, pp. 622–639. Springer, Heidelberg (2015). doi:10.1007/978-3-319-25007-6_36

Iterative TempoWordNet

Mohammed Hasanuzzaman$^{(\boxtimes)}$, Gaël Dias, Stéphane Ferrari, and Yann Mathet

Normandie University, UNICAEN, GREYC UMR 6072, Caen, France
{mohammed.hasanuzzaman,gael.dias,stephane.ferrari,yann.mathet}@unicaen.fr

Abstract. TempoWordNet (TWn) has recently been proposed as an extension of WordNet, where each synset is augmented with its temporal connotation: past, present, future or atemporal. However, recent uses of TWn show contrastive results and motivate the construction of a more reliable resource. For that purpose, we propose an iterative strategy that temporally extends glosses based on TWn^t to obtain a potentially more reliable TWn^{t+1}. Intrinsic and extrinsic evaluation results show improvements when compared to previous versions of TWn.

1 Introduction

There is considerable academic and commercial interest in processing time information in text whether that information is expressed explicitly or implicitly. Recognizing such information can significantly improve the functionality of Natural Language Processing and Information Retrieval applications.

Whereas most of the prior Computational Linguistics and Text Mining temporal studies have focused on temporal expressions and events, there has been a lack of work looking at the temporal orientation of word senses. Towards this direction, [1] developed TempoWordNet (TWn), an extension of WordNet [5], where each synset is augmented with its temporal connotation (past, present, future or atemporal). The authors show that improvements can be reached for temporal sentence classification and temporal query intent classification when unigrams are temporally expanded using the new constructed TWn.

In order to propose a more reliable resource, [3] recently defined two new propagation strategies, Probabilistic and Hybrid, respectively leading to two different TempoWordNet resources called TWnP and TWnH. Although some improvements are evidenced in their experiments, no conclusive remarks can be reached as TWnP evidences highest results in terms of temporal sentence classification but human judgment tends to prefer TWnH.

In this paper, we propose an iterative strategy to build an accurate TWn both in terms of human judgment and classification results. Our underlying idea is simple. Taking into account that (1) synsets are temporally classified based on their gloss content and (2) temporal sentence classification is boosted by TWn, temporally expanding glosses based on a given TempoWordNet at step t (TWn^t) may allow to obtain a more accurate TempoWordNet at step $t + 1$ (TWn^{t+1}) when propagating temporal connotations. Our methodology is intrinsically and extrinsically evaluated in a similar way as [3]. Results show steady improvements of the current iterative TWn when compared to previous versions of TWn.

© Springer International Publishing AG 2016
H. Sack et al. (Eds.): ESWC 2016 Satellite Events, LNCS 9989, pp. 40–45, 2016.
DOI: 10.1007/978-3-319-47602-5_9

2   Methodology

To build a more reliable temporal lexical resource, we propose to rely on two previous findings. First, [1,3] evidenced that synset time-tagging can be achieved by gloss classification with some success. The underlying idea is that the definition of a given concept embodies its potential temporal dimension. Second, [1,3] also mention that temporal sentence classification can be improved when temporal unigrams are expanded with their synonyms stored in a TempoWordNet version.

From these two assumptions, a straightforward enhancement maybe proposed to improve the reliability of the temporal expansion process within WordNet. Indeed, glosses are sentences defining the concept at hand. As a consequence, temporally expanding glosses based on some TempoWordNet version may improve the performance of the classifier used to propagate the temporal connotations and as a consequence meliorate the intrinsic quality of the obtained temporal resource.

Note that this process can be iterated. If a better TempoWordNet can be obtained at some step t, better gloss expansion may be expected and as a consequence a more accurate temporal resource may be obtained at step $t+1$, which in turn may be reused for gloss expansion, and so on and so forth. This iterative strategy is defined in Algorithm 1, where the process stops when a stopping criterion is satisfied.

Algorithm 1. Iterative TempoWordNet algorithm

1: $sl \leftarrow$ list of temporal seed synsets
2: $ps \leftarrow$ propagation strategy
3: $i \leftarrow 0$
4: $\mathrm{Wn}^0 \leftarrow \mathrm{Wn}$
5: **repeat**
6:     $\mathrm{TWn}^i \leftarrow \mathrm{PropagateTime}(\mathrm{Wn}^i, sl, ps)$
7:     $\mathrm{Wn}^{i+1} \leftarrow \mathrm{ExpandGloss}(\mathrm{Wn}, \mathrm{TWn}^i)$
8:     $i \leftarrow i+1$
9: **until** stopping criterion
10: return TWn^i

Let sl be a set of manually time-tagged seed synsets, ps a propagation strategy (e.g. hybrid) and a working version of WordNet Wn^0 (e.g. WordNet 3.0). The iterative construction of TempoWordNet is as follows. The first TempoWordNet TWn^0 is constructed in a similar manner as proposed in [3]. Based on sl, an initial two-class (temporal vs atemporal) classifier is learned over a 10-fold cross validation process, where synsets are represented by their gloss constituents (i.e. unigrams) weighted by gloss frequency. The propagation of the temporal connotations is then run based the learned classifier and the given propagation strategy ps. New expanding synsets are included in the initial seeds list and the exact same learning process iterates N times so to ensure convergence in terms of classification accuracy. At the end of this process, the synsets of the working version of WordNet Wn^0 are either tagged as temporal or atemporal depending on classification probability, thus giving rise to two distinct partitions Wn^0_t (temporal Wn^0) and $Wn^0_{\bar{t}}$ (atemporal Wn^0). In order to fine tune Wn^0_t into past, present and future synsets, the exact same procedure is run exclusively based on the past, present and future synsets from sl over Wn^0_t. When classification convergence is reached[1], all synsets from Wn^0_t are time-tagged as past, present or future giving rise to Wn^0_{ppf}. Note that a synset can neither be exclusively past, present nor future (e.g. "sunday") although it is temporal. In this case, near equal class probabilities are evidenced. At the end of the construction process, $TWn^0 = Wn^0_{\bar{t}} \cup Wn^0_{ppf}$.

Once TWn^0 has been constructed, it can be used to temporally expand Wn glosses giving rise to the second working version of WordNet Wn^1. In particular, within each gloss of Wn, all temporal words from TWn^0 are searched for. If one temporal concept is found in the gloss, then its synonyms are added to the gloss thus enlarging the possible lexical overlap between temporal glosses. So, each gloss is now represented by its unigrams plus the synonyms of its temporal constituents as a bag of words and is noted Wn^1. Note that word sense disambiguation is performed based on the implementation of the Lesk algorithm of the NLTK toolkit[2]. This process refers to line 7 of Algorithm 1.

So, from the new Wn^1, the next TempoWordNet version TWn^1 can be processed, which in turn can give rise to a new WordNet working version Wn^2, which will lead to TWn^2 and so on and so forth. This iterative process stops when some stopping criterion is reached. Many ideas can be presented but within the scope of this paper, we propose that the final TempoWordNet version is obtained when the difference between TWn^i and TWn^{i-1} is marginal in terms of temporal sets, i.e. $TWn^i \setminus TWn^{i-1} \leq \epsilon$, which means that the set of distinct temporal synsets converges.

Table 1. Comparative features of different TempoWordNet versions.

TempoWordNet version	TWnL	TWnP	TWnH^0	TWnH^1	TWnH^2	TWnH^3
# temporal synsets	21213	53001	17174	2020	2804	2832
# past synsets	1734	2851	2547	305	120	1308
# present synsets	16144	19762	842	1247	2181	765
# future synsets	3335	30388	13785	468	503	759
# atemporal synsets	96402	64614	100441	115639	114855	114827
$TWn^i \setminus TWn^{i-1}$	-	-	-	15154	784	28
Fixed-marginal κ	0.507	0.520	0.420	0.625	0.616	0.599
Free-marginal κ	0.520	0.520	0.440	0.850	0.700	0.774

---

[1] N iterations are performed.
[2] http://www.nltk.org/. Last accessed: 15.08.2015.

Table 2. F_1-measure results for temporal sentence, tweet and query intent classification with different TempoWordNet versions performed on 10-fold cross validation with SVM.

TempoWordNet version	without TWn	TWnL	TWnP	TWnH^0	TWnH^1	TWnH^2	TWnH^3
Sentence classification	64.8	66.7	69.3	68.6	68.4	69.7	71.4
Tweet classification	39.7	49.1	51.5	49.8	51.9	52.5	53.1
Query intent classification	75.3	78.0	78.8	75.9	78.3	79.0	80.1

3   Evaluation

Our methodology is evaluated both intrinsically and extrinsically. The underlying idea being that a reliable resource must evidence high quality time-tagging as well as improved performance for some application. As for experimental setups, we used (1) the *sl* seeds list provided by [1], (2) the *ps* hybrid propagation strategy[3] proposed in [3], (3) version 3.0 of WordNet[4] for Wn and (4) $\epsilon = 100$. With respect to the learning procedures, the SVM implementation of Weka[5] was used with parameters tuning and convergence was ensured by iterating the temporal propagation $N = 50$ times. Note that all experimental results as well as produced resources are freely available[6] for the sake of reproducibility.

3.1   Intrinsic Evaluation

In order to assess human judgment about the temporal parts of TempoWordNet, inter-rater agreement (free-marginal and fixed-marginal multirater kappa[7]) with multiple raters is performed. Three annotators are presented with 50 temporal synsets and respective glosses, and must decide upon their correct classification i.e. temporal or atemporal. Note that past, present and future connotations are only indicative of the temporal orientation of the synset but cannot be taken as a strict class. Indeed, there are many temporal synsets, which are neither past, present nor future (e.g. "monthly"). Results are reported in Table 1 and assess moderate agreement for previous versions of TempoWordNet (TWnL [1], TWnP [3] and TWnH^0 [3]) while substantial agreement is obtained for the successive iterative versions. Table 1 also presents figures about the distribution of temporal synsets of each TempoWordNet version. Interestingly, the iterative versions tend to time-tag a much smaller proportion of synsets when compared to previous ones[8].

---

[3] Note that lexical propagation does not spread over a wide range of concepts [1] and probabilistic propagation shows semantic shift problems [3].

[4] https://wordnet.princeton.edu/. Last accessed: 17.04.2016.

[5] http://www.cs.waikato.ac.nz/ml/weka/. Last accessed: 15.08.2015.

[6] https://tempowordnet.greyc.fr/. Last accessed: 17.04.2016.

[7] http://justusrandolph.net/kappa/. Last accessed: 17.04.2016.

[8] Note that the number of past, present and future synsets is based on the highest probability given by the temporal classifier, which does not necessarily imply that the synset belongs to the given class (e.g. almost equal probabilities can be evidenced).

3.2 Extrinsic Evaluation

In order to produce comparative results with prior works, we test our methodology on the balanced data set produced in [1], which consists of 1038 sentences equally distributed as past, present and future. Moreover, we propose to extend experiments on a corpus of 300 temporal tweets[9]. This corpus contains 100 past, 100 present and 100 future tweets, which have been time-tagged by annotators of the crowdflower[10] platform[11]. For both experiments, each sentence/tweet is represented by a feature vector where each attribute is either a unigram[12] or a synonym of any temporal word contained in the sentence/tweet and its value is tf.idf.

In order to strengthen comparative evaluation, we also propose to tackle the TQIC task of NTCIR-11 Temporalia [4], where a given search engine query must be tagged as past, recency, future or atemporal. For that purpose, we use the 400 queries provided by the organisers, which are equally distributed by temporal class. In particular, a query is represented in the same way as sentences and tweets[13] plus the additional time-gapped feature proposed in [2]. Comparative classification results are reported in Table 2. For all experiments, $TWnH^3$ produces highest classification results with respectively 2.1 %, 1.6 % and 1.3 % improvements for sentence, tweet and query temporal classification over the second best (non-iterative) TempoWordNet version. Note that all improvements are statistically relevant and steadily occur between $TWnH^0$, $TWnH^1$, $TWnH^2$ and $TWnH^3$.

4 Conclusions

We proposed an iterative strategy to produce a reliable temporal lexical resource called TempoWordNet. The underlying idea is that based on the temporal expansion of glosses with some version of TempoWordNet at step t, a more accurate resource can be obtained at step $t + 1$. Intrinsic and extrinsic evaluations evidence improved results when compared to recent versions of TempoWordNet. However, deeper experiments must be performed with respect to (1) the stopping criterion of the iterative process, (2) the hard (past, present, future) classification of temporal synsets and (3) the integration of TempoWordNet into temporal classifcation tools.

---

[9] https://tempowordnet.greyc.fr/. Last accessed: 17.04.2016.
[10] http://www.crowdflower.com/. Last accessed: 17.04.2016.
[11] Annotation details are out of the scope of this paper.
[12] Stopwords removal is performed so to better access the benefits of TempoWordNet.
[13] Stopwords are not removed as queries are small.

References

1. Dias, G., Hasanuzzaman, M., Ferrari, S., Mathet, Y.: TempoWordNet for sentence time tagging. In: Companion Publication of the 23rd International Conference on World Wide Web Companion (WWW), pp. 833–838 (2014)
2. Hasanuzzaman, M., Dias, G., Ferrari, S.: Hultech at the NTCIR-11 temporalia task: ensemble learning for temporal query intent classification. In: NTCIR-11 Conference (NTCIR), pp. 478–482 (2014)
3. Hasanuzzaman, M., Dias, G., Ferrari, S., Mathet, Y.: Propagation strategies for building temporal ontologies. In: 14th Conference of the European Chapter of the Association for Computational Linguistics (EACL), pp. 6–11 (2014)
4. Joho, H., Jatowt, A., Blanco, R., Naka, H., Yamamoto, S.: Overview of NTCIR-11 temporal information access (temporalia) task. In: NTCIR-11 Conference (NTCIR), pp. 429–437 (2014)
5. Miller, G.A.: WordNet: a lexical database for English. Commun. ACM **38**(11), 39–41 (1995)

An Ontology to Semantically Declare and Describe Functions

Ben De Meester[⊠], Anastasia Dimou, Ruben Verborgh, and Erik Mannens

Data Science Lab – iMinds, Ghent University, Ghent, Belgium
{ben.demeester,anastasia.dimou,ruben.verborgh,erik.mannens}@ugent.be

Abstract. Applications built on top of the Semantic Web are emerging as a novel solution in different areas, such as decision making and route planning. However, to connect results of these solutions – i.e., the semantically annotated data – with real-world applications, this semantic data needs to be connected to actionable events. A lot of work has been done (both semantically as non-semantically) to describe and define Web services, but there is still a gap on a more abstract level, i.e., describing interfaces independent of the technology used. In this paper, we present a data model, specification, and ontology to semantically declare and describe functions independently of the used technology. This way, we can declare and use actionable events in semantic applications, without restricting ourselves to programming language-dependent implementations. The ontology allows for extensions, and is proposed as a possible solution for semantic applications in various domains.

Keywords: Application · Function · Ontology · Semantic web

1 Introduction

Semantic applications are emerging as a solution to data-driven problems. For instance, the ORCA project aims at automatically assigning nurses to patient calls in a hospital based on their context [1]. Linked Connections define a way to publish raw transit data, to be used for intermodal route planning [3]. In projects like the aforementioned, it is common that the Semantic Web is not solely used as a means to publish data, but also as a catalyst to execute other actions, e.g., calling a real-world nurse, or executing a route planning algorithm. Usually, this implies querying the RDF results for possible actions, parsing the data using custom parsing rules, and executing the actual function. E.g., in the ORCA project, a semantic reasoning system derives the most suitable nurse. The resulting turtle data is queried, resulting in the triple `ex:call1 ex:assign ex:nurseA`. How to convert this triple to the execution of the function `callNurse(nurseA, call1)` is not semantically defined, but is defined ad hoc per use case.

© Springer International Publishing AG 2016
H. Sack et al. (Eds.): ESWC 2016 Satellite Events, LNCS 9989, pp. 46–49, 2016.
DOI: 10.1007/978-3-319-47602-5_10

There exist many specifications handling Web services, both non-semantically (e.g., WSDL and WADL) and semantically (e.g., OWL-S and Hydra)[1]. These specifications target different facets (e.g., HTTP-based vs SOAP-based access, defining RESTful APIs, etc.), but have in common that they *define Web services*. Thus, they clearly specify, e.g., which HTTP method to invoke with which parameter to correctly call the Web service. The big drawback of these specifications is thus that they are very coupled with the technology stack. However, not all actions can be executed using Web APIs, either because of performance or practicality reasons. For example, the nurse call system is a near real-time system, which implies that unnecessary HTTP connections should be avoided.

In this paper, we present a general vocabulary as a data model, specification, and ontology to semantically declare and describe functions. Instead of *defining* technology-specifics, i.e., hard-coding executable actions, the functions are *described* independently of the technology that implements them. Complementary to the current specifications that describe *how* to execute a certain service, this vocabulary describes *what* a functions does. By semantically defining these functions, we provide a uniform and unambiguous solution, and thus, we close the gap between semantic data and any real-world action, enabling semantic applications to be used in real-world scenarios.

2   The Function Ontology

The Function model allows users to declare that a certain function exists, and associates this function with certain problems and the algorithms it implements. Furthermore, it allows for the description of *executions* of functions. Execution descriptions of functions assign values to the input parameters, e.g., `function int sum(int a, int b)` is a description of a function, whilst `sum(2, 4)` is a description of the execution of the `sum` function. The Function model does not describe the internals of a function, as this depends on the used technology. E.g., the aforementioned `sum` function can be interpreted for implementation in JavaScript as `return a + b`, in PHP as `return $a + $b;`, using a Web service, or via another technology.

The Function Ontology[2] consists of a set of base classes. A problem, algorithm or function can be defined as an instance of those base classes. As new problems and algorithmic solution arise daily, it would not be beneficial to include all possible problems and algorithms in the Function Ontology. Problems, functions, and algorithms can be classified using the SKOS model [4]. For example, a `sumProblem` can be classified as `ex:sumProblem skos:broader ex:mathProblem`. These classifications can be reused across domains, independent of the used technologies.

---

[1] See   https://www.w3.org/TR/wsdl20/,   https://www.w3.org/Submission/wadl/, https://www.w3.org/Submission/OWL-S/, and http://hydra-cg.com/spec/latest/ core/, respectively.

[2] Published on http://semweb.mmlab.be/ns/function, with accompanying specification at http://users.ugent.be/~bjdmeest/function/.

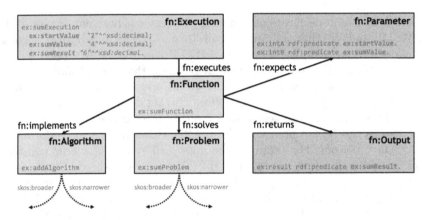

Fig. 1. The Function ontology, with an example of how reification is used to connect named parameters and output values to the function execution.

The Function Ontology (see Fig. 1) consists of the following classes:

- *Function* The declared function (e.g., `function sum`)
- *Problem* A problem that a function solves, e.g., adding two numbers.
- *Algorithm* A declaration of an algorithm. We separate the algorithm concept, as there are no one-to-one mappings between problems, functions, and algorithms [2].
- *Parameter* A parameter of a function (e.g., `function sum` has two parameters, a and b). `rdfs:range` can be used to describe the type of that parameter, and reification (as opposed to, e.g., using RDF lists) is used to create named connections between a function and its parameters (cfr. the parameter and execution declaration in Fig. 1).
- *Output* An output of a function, e.g., `function sum(a, b)` has `sumResult` as output. Typing and reification paradigms are similar as with parameters.
- *Execution* An execution assigns actual input values to function parameters, and holds the output values. E.g., `sum(2, 3)` is an execution of `int function sum(a,b)`.

3   Example and Conclusions

These base classes do not define any specific problem, function or algorithm, but *allow to declare and describe* any specific problem, function, or algorithm. For example, if John Doe needs a description of a *sum* function – given it does not already exist – he can publish his own descriptions as shown in Listing 1. Software systems supporting this description are thus able to compute the execution of that function. The deployment/implementation of the described functions is independent of the programming language and the access interface.

```
1   johndoe:sumProblem   a fn:Problem  ; skos:broader johndoe:mathProblem.
2   johndoe:sumAlgorithm a fn:Algorithm.
3
4   johndoe:sumFunction  a fn:Function;
5       dcterms:title "The sum function"^^xsd:string;
6       dcterms:description "This function can do the sum of two integers."^^xsd:string;
7       fn:solves        johndoe:sumProblem;
8       fn:implements    johndoe:sumAlgorithm;
9       fn:expects  [ rdf:predicate johndoe:startValue; fn:required "true"^^xsd:boolean ];
10      fn:expects  [ rdf:predicate johndoe:sumValue  ; fn:required "true"^^xsd:boolean ];
11      fn:returns  [ rdf:predicate johndoe:sumResult ; fn:required "true"^^xsd:boolean ].
12
13  johndoe:startValue rdfs:range xsd:integer.
14  johndoe:sumValue   rdfs:range xsd:integer.
15  johndoe:sumResult  rdfs:range xsd:integer.
16
17  johndoe:sumExecution  a fn:Execution;
18      fn:executes           ex:sumFunction;
19      johndoe:startValue "2"^^xsd:integer;
20      johndoe:sumValue   "4"^^xsd:integer.
```

Listing 1. Sum function declaration

This example shows how the Function Ontology allows for semantic systems to connect to non-semantic processing functions. By remaining technology-agnostic, both small and large-scale RDF processing actions can be defined, without solely depending on Web services as execution platform. As such, it will be evaluated in the frame of, among others, ORCA, Linked Connections[3], and the COMBUST project[4]. Future work includes the creation of an upper level reference function model and a specification of the technical integration process.

References

1. Arndt, D., et al.: Ontology reasoning using rules in an eHealth context. In: Bassiliades, N., Gottlob, G., Sadri, F., Paschke, A., Roman, D. (eds.) RuleML 2015. LNCS, vol. 9202, pp. 465–472. Springer, Heidelberg (2015). doi:10.1007/978-3-319-21542-6_31
2. Bratas, C., Maglavera, s.N., Quaresma, P.: A framework to describe problems and algorithms in medical informatics via ontologies. In: 4th European Symposium on Biomedical Engineering, pp. 1–4. University of Patras, Patras, Greece, June 2004. http://www.di.uevora.pt/~pq/papers/patras.pdf
3. Colpaert, P., Llaves, A., Verborgh, R., Corcho, O., Mannens, E., Van de Walle, R.: Intermodal public transit routing using Linked Connections. In: Proceedings of the 14th International Semantic Web Conference: Posters and Demos, October 2015. http://ceur-ws.org/Vol-1486/paper_28.pdf
4. Miles, A., Bechhofer, S.: SKOS simple knowledge organization system reference. In: W3C Recommendation, W3C, August 2009. https://www.w3.org/TR/skos-reference/. Accessed 7 Mar 2016

---

[3] http://linkedconnections.org/.
[4] http://www.iminds.be/nl/projecten/2015/03/11/combust.

Context Sensitive Entity Linking of Search Queries in Enterprise Knowledge Graphs

Sumit Bhatia[✉] and Anshu Jain

IBM Watson, Almaden Research Centre, San Jose, CA, USA
{sumit.bhatia,anshu.n.jain}@us.ibm.com

Abstract. Fast and correct identification of named entities in queries is crucial for query understanding and to map the query to information in structured knowledge base. Most of the existing work have focused on utilizing search logs and manually curated knowledge bases for entity linking and often involve complex graph operations and are generally slow. We describe a simple, yet fast and accurate, probabilistic entity-linking algorithm used in enterprise settings where automatically constructed, domain specific Knowledge Graphs are used. In addition to the linked graph structure, textual evidence from the domain specific corpus is also utilized to improve the performance.

1 Introduction

With increasing popularity of virtual assistants like SIRI and Google Now, users are interacting with search systems by asking natural language questions that often contain named entity mentions. Further, a large fraction of queries contain a named entity and searchers tend to use more question-queries for complex information needs [2]. Hence, *fast* and *correct* identification of named entities in user queries is crucial for query understanding and to map the query to information in structured knowledge base. Entity linking in search queries utilizes information derived from query logs and open knowledge bases such as DBPedia and Freebase. Such techniques, however, are not suited for enterprise and domain specific search systems such as legal, medical, healthcare, etc. due to very small user bases resulting in small query logs and absence of rich domain specific knowledge bases. Recently, there have been development of systems for automatic construction of semantic knowledge bases for domain specific corpora [3] and systems that use such domain specific knowledge bases [8]. We describe the method used for entity disambiguation and linking as implemented in one such system, *Watson Discovery Advisor*. It offers users a search interface to search for the indexed information and uses the underlying knowledge base to enhance search results and provide additional entity-centric data exploration capabilities. The system *automatically* constructs a structured knowledge base by identifying entities and their relationships from input text corpora using the method described by Castelli et al. [3]. Thus, for each relationship discovered by the system, the corresponding mention text provides additional contextual information

© Springer International Publishing AG 2016
H. Sack et al. (Eds.): ESWC 2016 Satellite Events, LNCS 9989, pp. 50–54, 2016.
DOI: 10.1007/978-3-319-47602-5_11

about the entities and relationships present in that mention. We posit that the *dense graph structure* discovered from the corpus, as well as the *additional context provided by the associated mention text* can be utilized together for linking entity name mentions in search queries to corresponding entities in the graph. Our proposed entity linking algorithm is intuitive, relies on a theoretical sound probabilistic framework, is fast and scalable with an average response time of $\approx$**100 ms**. Figure 1 shows the working of proposed algorithm in action where top ranked suggestions for named mentions `Sergey` and `Larry` are showed. As will be described in detail in next Section, note that the algorithm is making these suggestions by utilizing the terms in questions (search, algorithm) as well as relationships between all target entities for mentions "Sergey" and "Larry" in the graph. The algorithm figures out that entities "Sergey Brin" and "Larry Page" have strong evidences from their textual content as well as these two entities are strongly connected in the graph, and hence they are suggested as most probable relevant entities in the context of question.

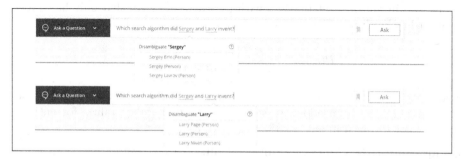

Fig. 1. Entity suggestions produced by proposed approach using text and entity context in search query.

2 Proposed Approach

Let $Q = \{C, T\}$ be the input query where T is the ambiguous token, and $C = \{E_c, W_c\}$ is the context under which we have to disambiguate T. The context is provided by the words ($W_c = \{w_{c1}, w_{c2}, \ldots, w_{cl}\}$) in the query and the set of unambiguous entities $E_c = \{e_{c1}, e_{c2}, \ldots, e_{cm}\}$. Note that initially, this entity set can be empty if there are no unambiguous entity mentions in the query and in such cases, only textual information is considered. The task is to map the ambiguous token T to one of the possible target entities. Let $E_T = \{e_{T1}, e_{T2}, \ldots, e_{Tm}\}$ be the set of target entities for T. A ranked list of target entities can be constructed by computing $P(e_{Ti}|C)$, i.e., the probability that the user is interested in entity e_{Ti} given the context C. Using Bayes' theorem, we can write $P(e_{Ti}|C)$ as follows.

$$P(e_{Ti}|C) = \frac{P(e_{Ti})P(C|e_{Ti})}{P(C)} \propto P(C|e_{Ti}) \tag{1}$$

Since we are only interested in relative ordering of the target entities, we can ignore the denominator $P(C)$ as its value will be same for all the target entities. Likewise, assuming all the entities to be equally probable in absence of any context, $P(e_{T_i})$ can be ignored for ranking purposes. Assuming conditional independence for context terms as well as entities in context, we have:

$$P(e_{Ti}|C) \propto P(W_c|e_{Ti}) \times P(E_c|e_{Ti}) = \underbrace{\prod_{w_c \in W_c} P(w_c|e_{Ti})}_{\text{text context}} \times \underbrace{\prod_{e_c \in E_c} P(e_c|e_{Ti})}_{\text{entity context}} \quad (2)$$

Computing Entity Context Contribution: The *entity context* factor in Eq. 2 corresponds to the evidence for target entity given E_c, the set of entities forming the context. For each individual entity e_c forming the context, we need to compute $P(e_c|e_{Ti})$, i.e., the probability of observing e_c after observing the target entity e_{Ti}. Intuitively, there is a higher chance of observing an entity that is involved in multiple relationship with e_{Ti} than an entity that only has a few relationships with e_{Ti}. Thus, we can estimate $P(e_c|e_{Ti})$ as follows:

$$P(e_c|e_{Ti}) = \frac{relCount(e_c, e_{Ti}) + 1}{relCount(e_c) + |E|} \quad (3)$$

Note that the factor of 1 in numerator and $|E|$ (size of entity set E) in the denominator have been added to smoothen the probability values for entities that are not involved in any relationship with e_{Ti}.

Computing Text Context Contribution: The *text context* factor in Eq. 2 corresponds to the evidence for target entity given W_c, the terms present in the input query. For each individual query term w_c, we need to compute $P(w_c|e_{Ti})$, i.e., the probability of observing w_c given e_{Ti}. This probability can be estimated by using the *mention language model* of e_{Ti} as follows.

$$P(w_c|E_{Ti}) = P(w_c|M_{Ti}) = \frac{\text{no. of times } w_c \text{ appears in mentions of } E_{Ti} + 1}{|M_{Ti}| + N} \quad (4)$$

Here, N is the size of the vocabulary. Since entities are discovered automatically from text, these mentions provide important context information as illustrated in Sect. 1.

3   Evaluation

We use a semantic graph constructed from text of all articles in Wikipedia by automatically extracting the entities and their relations by using IBM's Statistical Information and Relation Extraction (SIRE) toolkit[1]. Even though there exist popular knowledge bases like DBPedia that contain high quality data, we

---

[1] http://ibmlaser.mybluemix.net/siredemo.html.

chose to construct a semantic graph using automated means as such a graph will be closer to many practical real world scenarios where high quality curated graphs are often not available and one has to resort to automatic methods of constructing knowledge bases. Our graph contains more than 30 millions entities and 192 million distinct relationships in comparison to 4.5 million entities and 70 million relationships in DBpedia. For evaluating the proposed approach, we use the KORE50 [5] dataset that contains 50 short sentences with highly ambiguous entity mentions. This widely used dataset is considered amongst the hardest dataset for entity disambiguation. Average sentence length (after stop word removal) is 6.88 words per sentence and each sentence has 2.96 entity mentions on an average. Every mention has an average of 631 candidates to disambiguate in YAGO knowledge base [9]. However, it varies for different knowledge bases. Our automatically constructed knowledge base has *2,261 candidates per mention* to disambiguate illustrating the difficulty in entity linking due to high noise in automatically constructed knowledge bases when compared with manually curated/cleaned knowledge bases such as DBpedia. The results of our proposed approach and various other state-of-the-art methods for entity linking on the same dataset are tabulated in Table 1. We note that the performance of our proposed approach is comparable or better than the other approaches, despite dealing with much noisier data. Further, average response time for proposed approach is about 100 ms, as we utilize the signals from mention text and relationship information about entities instead of performing complex and time consuming graph operations as in other methods, while not sacrificing on the accuracy.

Table 1. Entity disambiguation accuracy

Method	Precision	Method	Precision
Joint-DiSER-TopN [1]	0.72	DBpedia Spotlight [7]	0.35
AIDA-2012 [6]	0.57	**Proposed Method Accuracy @ Rank 1**	0.52
AIDA-2013 [5]	0.64	**Proposed Method Accuracy @ Rank 5**	0.65
Wikifier [4]	0.41	**Proposed Method Accuracy @ Rank 10**	0.74

4   Conclusions

In this paper, we addressed the problem of mapping entity mentions in natural language search queries to corresponding entities in an automatically constructed knowledge graph. We proposed an approach that utilizes the dense graph structure as well as additional context provided by the mention text. Comparative evaluation on a standard dataset with state-of-the-art approaches shows the strengths of our proposed approach in achieving high accuracy with super fast response times. The proposed approach is currently deployed in an enterprise semantic search system called Watson Discovery Advisor and our future work will focus on developing the approach further to utilize user click-feedback for improving the quality of entity suggestions.

References

1. Aggarwal, N., Buitelaar, P.: Wikipedia-based distributional semantics for entity relatedness. In: 2014 AAAI Fall Symposium Series (2014)
2. Aula, A., Khan, R.M., Guan, Z.: How does search behavior change as search becomes more difficult? In: CHI 2010, pp. 35–44 (2010)
3. Castelli, V., Raghavan, H., Florian, R., Han, D.-J., Luo, X., Roukos., S.: Distilling and exploring nuggets from a corpus. In: SIGIR, pp. 1006–1006 (2012)
4. Cheng, X., Roth, D.: Relational inference for wikification. In: EMNLP, pp. 1787–1796 (2013)
5. Hoffart, J., Seufert, S., Nguyen, D.B., Theobald, M., Weikum, G.: KORE: keyphrase overlap relatedness for entity disambiguation. In: CIKM, pp. 545–554 (2012)
6. Hoffart, J., Yosef, M.A., Bordino, I., Fürstenau, H., Pinkal, M., Spaniol, M., Taneva, B., Thater, S., Weikum, G.: Robust disambiguation of named entities in text. In: EMNLP, pp. 782–792 (2011)
7. Mendes, P.N., Jakob, M., García-Silva, A., Bizer, C.: DBpedia spotlight: shedding light on the web of documents. In: Proceedings of the 7th International Conference on Semantic Systems, pp. 1–8. ACM (2011)
8. Nagarajan, M. et al.: Predicting future scientific discoveries based on a networked analysis of the past literature. In: KDD 2015, pp. 2019–2028 (2015)
9. Suchanek, F.M., Kasneci, G., Weikum, G.: Yago: a core of semantic knowledge. In: WWW, pp. 697–706 (2007)

Incorporating Functions in Mappings to Facilitate the Uplift of CSV Files into RDF

Ademar Crotti Junior[✉], Christophe Debruyne, and Declan O'Sullivan

ADAPT Centre for Digital Content Technology Research, Knowledge and Data Engineering Group, School of Computer Science and Statistics, Trinity College Dublin, Dublin 2, Ireland
{crottija,debruync,declan.osullivan}@scss.tcd.ie

Abstract. Many solutions have been developed to convert non-RDF data to RDF. A common task during this conversion is applying data manipulation functions to obtain the desired output. Depending on the data format of the source to be transformed, one can rely on the underlying technology, such as RDBMS for relational databases or XQuery for XML, to manipulate data - to a certain extent - while generating RDF. For CSV files, however, there is no such underlying technology. Instead, one has to resort to more elaborate Extract, Transform and Load (ETL) processes, which can render the generation of RDF more complex (in terms of number of steps), and therefore also less traceable and transparent. One solution to this problem is the declaration and inclusion of functions in mappings of non-RDF data to RDF. In this paper, we propose a method to incorporate functions into mapping languages and demonstrate its viability in Digital Humanities use case.

Keywords: Linked Data · Mapping · Data manipulation

1 Introduction

The CSV file format is a convenient and popular way to exchange structured data, but the semantics of the information captured in such files are not explicit. RDF, on the other hand, provides one means to add meaning to data that software can process. The process of converting data in any non-RDF format (e.g., CSV and relational databases) into RDF is called *uplift*.

There are scenarios where it is necessary to manipulate data during the uplift process. Depending on the source data format, this can be very straightforward. With uplift languages for relational databases such as R2RML[1], for example, one can rely on the underlying RDBMS to support some data manipulation tasks (e.g., string concatenation). The same is true for XSPARQL [5], where one can use XQuery to manipulate the data contained in XML. Sometimes, however, relying on the underlying technology is not sufficient [4]. For CSV datasets there is no such equivalent, standardized underlying technology. When data in CSV files has to be manipulated to generate RDF, one needs to resort to data pre- or post-processing. This increases complexity in terms of number

---

[1] https://www.w3.org/TR/r2rml/.

© Springer International Publishing AG 2016
H. Sack et al. (Eds.): ESWC 2016 Satellite Events, LNCS 9989, pp. 55–59, 2016.
DOI: 10.1007/978-3-319-47602-5_12

of steps necessary to generate RDF, and renders the whole data processing "pipeline" less transparent. A solution to tackle this problem is to capture these manipulations as functions in the mappings.

We propose a method to incorporate functions into mapping languages that draws inspiration from, and generalizes ideas presented in [4]. To demonstrate our method, we extend RML's vocabulary and engine to include notions for function calls and parameter bindings. The main contributions of this paper can be summarized as follows: (i) a method to incorporate functions in a mapping language; (ii) an implementation of the method extending RML; and (iii) a demonstration of functions incorporated into mappings applied to a real world dataset.

2 Related Work

One approach to support uplift is to use an annotation language to relate non-RDF data to RDF (i.e., "mappings"), for which an engine is built. Examples of this approach include R2RML and R2RML-F [4] for relational databases; SML [3] for relational databases and CSV; and RML [1] and KR2RML [7] for an even wider array of non-RDF data formats. These mapping languages usually have access to functionality provided by the underlying technology of the non-RDF data source. For CSV files, these do not exist and one has to apply data pre- or post-processing techniques, which raises problems as explained in Sect. 1.

To the best of our knowledge, KR2RML is the only tool to support data manipulation functions inside a mapping language that does not rely on the underlying technology. Though they provide an editor in which you can load data and input mappings to create functions in Python to manipulate that data, once those functions are stored several problems can be observed. First, a lot of the structured information containing the function is captured as a string. This thus requires both parsing the file and that string. Secondly, the mapping becomes rather complex, which makes it more difficult for users to create similar mappings with other tools. Their editor, however, does facilitate the mapping creation process for their mapping language. In [4] an extension to R2RML called R2RML-F is proposed. R2RML-F adds supports for capturing domain knowledge inside the mapping language for relational databases. Unlike KR2RML, functions in R2RML-F are captured as resources *referred to* by mappings with the RDF data model, allowing functions to be reused in different mappings.

3 Incorporating Functions into Mapping Languages

In this section, we describe how we adopt ideas presented in [4] to develop a more generic, usable and amenable approach to incorporate functions into mapping languages. These functions can be used to capture both domain knowledge (e.g., transforming units) and other – more syntactic – data manipulation tasks (e.g., transforming values to create valid URIs). Function names are unique and each function must have one function name and one function body. A function body defines a function with a return statement; parameters are optional.

Our proof-of-concept extends RML's vocabulary and engine[2] by introducing construct for describing functions, function calls and parameter bindings. Listing 1 defines a function. This function has one string as a parameter and returns a URL concatenated with its camel case version. Although this function executes a simple string transformation, functions in this method are generic and capable of complex data transformations. Furthermore, functions work with any data format and can be reused in the mapping. Listing 2 demonstrates how the function is called. In R2RML – and by consequence RML – a Term Map generates an RDF term (see [1]). In our implementation, we introduced a new Term Map called a *Function Valued* Term Map that generates RDF terms based on the application of a function. The parameters are also Term Maps that are evaluated before the results are passed as arguments.

```
<#Camelize>
 rrf:functionName "camelize" ;
 rrf:functionBody """ function camelize (str) { var camelCaseString =
str.toLowerCase().replace( /[-_]+/g, ' ').replace( /[^\\w\\s]/g, ' ').replace( /
(.)/g, function($1) { return $1.toUpperCase(); }).replace( / /g, '' );
 return "http://dacura.cs.tcd.ie/data/seshat/" + camelCaseString; } """ ; .
```
Listing 1: Declaring a function

```
<#Variable>
 rml:logicalSource [ rml:source "data.csv"; rml:referenceFormulation ql:CSV ];
 rr:subjectMap [ rr:termType rr:BlankNode; ];
 rr:predicateObjectMap [
  rr:predicateMap [ rrf:functionCall [ rrf:function <#Camelize> ;
    rrf:parameterBindings ( [ rml:reference "Variable" ] );]; ];
  rr:objectMap [ rr:parentTriplesMap <#Value> ] ].
```
Listing 2: Calling a function in a PredicateMap

4 Demonstration

The dataset used to demonstrate our approach comes from the Seshat: Global History Databank [6]. This project is developing a knowledge base to describe human history that is created by hand via a wiki where contributors are expected to adhere to certain conventions to structure the facts. The Seshat dataset is currently made available for analysis as CSV files by scraping the wiki pages. A current development within the project is to gather the data into an OWL knowledge base, but predicates from the dataset differ from the predicates defined in the OWL ontology. Thus, there is a need to develop an approach to transform CSV values into the URIs of the ontology's predicates.

For example, in the dataset one predicate is labeled as "Capital", but in the ontology the predicate is *seshat:capital*. Other examples include "Language" and "Supracultural entity". Table 1 shows a fragment of the data where the values for the column "Variable" are transformed into predicates using the mapping from Listings 1 and 2. Example RDF output is shown in Fig. 1.

---

[2] Available at https://github.com/CNGL-repo/RMLProcessor.

Table 1. Excerpt of the CSV file shown as a table.

NGA	Polity	Section	Variable	Value from	Fact type	Value note
Latium	ItRomPr	General variables	Capital	Rome	Simple	Simple
Latium	ItRomPr	General variables	Language	Latin	Simple	Simple
Latium	ItRomPr	General variables	Supracultural entity	Greco-Roman	Simple	Simple

```
<http://dacura.cs.tcd.ie/data/seshat/ItRomPr> <http://dacura.cs.tcd.ie/data/seshat#hasVariable> _:d3SRTs6uGh .
_:d3SRTs6uGh <http://dacura.cs.tcd.ie/data/seshat/capital> _:d8fF2wuiam .
_:d8fF2wuiam <http://www.w3.org/1999/02/22-rdf-syntax-ns#type> <http://dacura.cs.tcd.ie/data/seshat#NameVariable> .
_:d8fF2wuiam <http://dacura.cs.tcd.ie/data/seshat#definiteDataValue> "Rome" .
<http://dacura.cs.tcd.ie/data/seshat/ItRomPr> <http://dacura.cs.tcd.ie/data/seshat#hasVariable> _:genid344T94t1y4CS .
_:genid344T94t1y4CS <http://dacura.cs.tcd.ie/data/seshat/language> _:qjeWSDRDcd .
_:qjeWSDRDcd <http://www.w3.org/1999/02/22-rdf-syntax-ns#type> <http://dacura.cs.tcd.ie/data/seshat#NameVariable> .
_:qjeWSDRDcd <http://dacura.cs.tcd.ie/data/seshat#definiteDataValue> "Latin" .
<http://dacura.cs.tcd.ie/data/seshat/ItRomPr> <http://dacura.cs.tcd.ie/data/seshat#hasVariable> _:genid388sbqxWXT4T .
_:genid388sbqxWXT4T <http://dacura.cs.tcd.ie/data/seshat/supraculturalEntity> _:yeEQUyN1yW .
_:yeEQUyN1yW <http://www.w3.org/1999/02/22-rdf-syntax-ns#type> <http://dacura.cs.tcd.ie/data/seshat#NameVariable> .
_:yeEQUyN1yW <http://dacura.cs.tcd.ie/data/seshat#definiteDataValue> "Greco-Roman" .
```

Fig. 1. RDF output.

5 Conclusions and Future Work

Most tools to convert data to RDF rely on underlying technology for data manipulation, but there is no manipulation language for CSV datasets. Moreover, when data manipulation is needed for CSV datasets one depends on pre- or post-processing techniques, which adds complexity to the uplift process. One solution is to incorporate functions in mappings, but the state-of-the-art does not offer a feasible way to do so. We tackled this problem by presenting a more amenable method to incorporate functions into mapping languages. We demonstrated our approach by extending RML's vocabulary and engine and applied it on a real world dataset.

Future work includes more use cases and experiments to compare performance and expressiveness of our approach and other mapping languages. Since functions can be considered software agents, one can also generate provenance information referring to these functions. Inspiration can be drawn from [2], who proposed a method for creating provenance information while generating RDF.

Acknowledgements. This study is supported by: (i) CNPQ, National Counsel of Technological and Scientific Development – Brazil; (ii) the Science Foundation Ireland ADAPT Centre for Digital Content Technology (Grant 13/RC/2106); (iii) John Templeton Foundation grant to the Evolution Institute [https://evolution-institute.org/project/seshat/]; (iv) the European Union Horizon 2020 ALIGNED [www.aligned-project.eu] (Grant 644055).

References

1. Dimou, A., Vander Sande, M., Colpaert, P., Verborgh, R., Mannens, E., Van de Walle, R.: RML: a generic language for integrated RDF mappings of heterogeneous data. In: Workshop on Linked Data on the Web (2014)
2. Dimou, A., De Nies, T., Verborgh, R., Mannens, E., and Van de Walle, R.: Automated metadata generation for Linked Data generation and publishing workflows. In: Workshop on Linked Data on the Web (2016)
3. Stadler, C., Unbehauen, J., Westphal, P., Sherif, M.A., Lehmann, J.: Simplified RDB2RDF mapping. In: Workshop on Linked Data on the Web (2015)
4. Debruyne, C., O'Sullivan, D.: R2RML-F: towards sharing and executing domain logic in R2RML mappings. In: Workshop on Linked Data on the Web (2016)
5. Bischof, S., Decker, S., Krennwallner, T., Lopes, N., Polleres, A.: Mapping between RDF and XML with XSPARQL. J. Data Semant. **1**(3), 147–185 (2012)
6. Turchin, P., Brennan, R., Currie, T., Feeney, K., Francois, P., Hoyer, D., Manning, J., Marciniak, A., Mullins, D., Palmisano, A., et al.: Seshat: the global history data-bank. Cliodynamics J. Quant. Hist. Cult. Evol. **6**, 77–107 (2015)
7. Slepicka, J., Yin, C., Szekely, P., Knoblock, C.: KR2RML: an alternative interpretation of R2RML for heterogeneous sources. In: Proceedings of the 6th International Workshop on Consuming Linked Data (2015)

Automatic Extraction of Axioms from Wikipedia Using SPARQL

Lara Haidar-Ahmad[1(✉)], Amal Zouaq[1,2], and Michel Gagnon[1]

[1] Department of Computer and Software Engineering, Ecole Polytechnique de Montreal,
Montreal, Canada
{lara.haidar-ahmad,michel.gagnon}@polymtl.ca
[2] School of Electrical Engineering and Computer Science, University of Ottawa, Ottawa, Canada
azouaq@uottawa.ca

Abstract. Building rich axiomatic ontologies automatically is a step towards the realization of the Semantic Web. In this paper, we describe an automatic approach to extract complex classes' axioms from Wikipedia definitions based on recurring syntactic structures. The objective is to enrich DBpedia concept descriptions with formal definitions. We leverage RDF to build a sentence representation and SPARQL to model patterns and their transformations, thus easing the querying of syntactic structures and the reusability of the extracted patterns. Our preliminary evaluation shows that we obtain satisfying results, which will be further improved.

1 Introduction

Building rich ontologies with reasoning capabilities is a difficult task, which can be time consuming. It requires both the knowledge of domain experts and the experience of ontology engineers. This is one of the main reasons why current Semantic Web and linked data rely mostly on lightweight ontologies. The automatization of axiom extraction is a step towards creating richer domain concept descriptions [4] and building a Semantic Web that goes beyond explicit knowledge for query answering. Ontology learning, i.e. the automatic extraction of ontologies from text, can help automatize the extraction of primitive, named and complex classes. Few state of the art approaches were developed to achieve this goal, mostly pattern-based approaches [2, 3]. To our knowledge, LExO [3] is the most advanced system for complex class extraction. This paper describes our approach to extract defined and primitive class axioms from Wikipedia concept definitions using SPARQL. The main contribution of this work is (i) the utilization of SPARQL graph matching capabilities to model patterns for axiom extraction (ii) the description of SPARQL patterns for complex class extractions from definitions and (iii) The enrichment of DBpedia concept descriptions using OWL axioms and defined classes. We also briefly compare our preliminary results with those of LEXO.

© Springer International Publishing AG 2016
H. Sack et al. (Eds.): ESWC 2016 Satellite Events, LNCS 9989, pp. 60–64, 2016.
DOI: 10.1007/978-3-319-47602-5_13

2   Methodology

We rely on a pattern-based approach to detect syntactic constructs that denote complex class axioms. These axioms are extracted from Wikipedia definitions.

Definition Representation and General Pipeline: We process definition sentences and first construct an RDF graph that represents the dependency structure of the defini-tion and the words' part of speech and positions in the sentence. This step makes the subsequent step of pattern matching using SPARQL requests easier. For every word, we specify its label, its part of speech, its position in the sentence and its grammatical relations with the other words based on the output of the Stanford parser [1]. Figure 1 presents an example of the RDF graph of a definition. For this example, we use the definition of the Wikipedia concept *Vehicle* from our dataset, which is "Vehicles are non-living means of transportation".

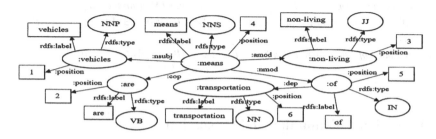

Fig. 1. The RDF representation of the definition of *vehicles*.

Based on this RDF representation, we execute a pipeline of SPARQL requests on the obtained RDF graphs (see Fig. 2).

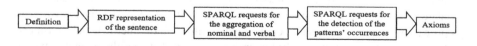

Fig. 2. Pipeline.

First, we execute SPARQL aggregation requests to extract complex expressions such as nominal and verbal groups and define subclass axioms. For instance, for the sentence *vehicles are non-living means of transportation*, we obtain the following expressions: *vehicles, non-living means of transportations* and *means of transportations*. We also extract the axiom *subClassOf (Non-living means of transportation, Means of transpor-tation)*. Finally, we execute a set of SPARQL axiom queries to identify occurrences of patterns that can be mapped to OWL complex class definitions.

SPARQL Pattern Representation: Based on a randomly chosen set of 110 definitions from Wikipedia and their sentence representation, we identified several recurring syntactic structures manually and built their corresponding SPARQL patterns. Next, we mapped patterns to complex class axioms using SPARQL CONSTRUCT. Table 1 presents the most common patterns that we identified in our dataset, in addition to their corresponding axioms. Each pattern is modeled using a single SPARQL request. This mechanism provides simple ways to enrich our approach with patterns that we do not support yet.

Table 1. Most frequent patterns and their respective axioms.

Frequent patterns for the definitions of concepts	Corresponding axioms
(1) SUBJ copula COMP	SUBJ $\subseteq$ COMP
Vehicles are non-living means of transportation	*Vehicles $\subseteq$ NonLivingMeansOfTransportations*
(2) SUBJ copula COMP that VERB OBJ	SUBJ $\equiv$ (COMP $\cap$ $\exists$VERB.OBJ)
A number is an abstract entity that represents a count or measurement	*Number $\equiv$ (AbstractEntity $\cap$ $\exists$represents. (Count $\cup$ Measurement))*
(3) SUBJ copula COMP VERB preposition NOUN	SUBJ $\equiv$ (COMP $\cap$ $\exists$VERB_prep_NOUN)
A lake is a body of water surrounded by land	*Lake $\equiv$ (BodyOfWater $\cap$ $\exists$surroundedBy.Land)*

3 Preliminary Evaluation and Discussion

We compared the generated axioms with a manually-built gold standard containing 20 definitions chosen randomly from our initial dataset[1]. We assessed the correctness of the axioms using standard precision and recall by focusing on named classes, predicates and complete axioms (see Table 2). Complete axioms metrics are calculated by counting the number of classes, predicates and logical operators matched with the ones in the gold standard. We obtain a macro precision and recall of 0.86/0.59 respectively. We also propose an axiom evaluation based on the Levenstein similarity metric which considers each axiom as a string. The higher the Levenstein similarity between the generated axiom and the reference, the most similar the axioms are. We tested multiple similarity levels as shown in Table 3. We notice that we usually generate the right axioms for (i) small sentences (ii) sentences with a simple grammatical structure and (iii) longer sentences which have no grammatical ambiguities. We also notice that false positives are rarely generated, and the errors in our results are usually caused by incomplete axioms. This is explained by the limited number of implemented patterns (10 patterns).

---

[1] The dataset and gold standard are available at: http://westlab.herokuapp.com/axiomfactory/dataESWC16.

Table 2. Evaluation results.

	Classes		Predicates		Complete axioms	
	Precision	Recall	Precision	Recall	Precision	Recall
Macro	0.87	0.66	0.94	0.54	0.86	0.59
Micro	0.86	0.61	0.76	0.36	0.78	0.48

Table 3. Axioms' precision based on Levenstein similarity with the gold standard.

Similarity level	0.70	0.80	0.90	1.00
Levenstein precision	0.55	0.50	0.35	0.30

While LExO [3] adopted a similar approach to ours, they did not rely on standard Semantic Web languages such as SPARQL for their patterns and did not take into account the aggregation of nominal and verbal groups, or the extraction of taxonomical relations. For example, given the definition *A minister or a secretary is a politician who holds significant public office in a national or regional government*, LExO generates *(Minister ⊔ Secretary) ≡ (Politician ⊓ ∃holds.((Office ⊓ Significant ⊓ Public) ⊓ ∃in. (Government ⊓ (National ⊔ Regional))))*. In contrast, our system generates the axiom *Minister ≡ Secretary ≡ (Politician ⊓ ∃ holds.(SignificantPublicOffice ⊓ ∃in.(NationalGovernment ⊔ RegionalGovernment)))*, and in addition, it generates a taxonomy where, *SignificantPublicOffice* is a subclass of *PublicOffice*, and *NationalGovernment* and *RegionalGovernment* are subclasses of *Government*.

4   Conclusion and Future Work

The paper describes an approach to extract OWL axioms with the aim to *logically define* DBpedia concepts from Wikipedia definitions using SPARQL requests. We are currently working on the implementation of our pipeline as a Web service, which has not been proposed yet in the state of the art. More importantly, one original contribution of this paper is the reliance on Semantic Web languages (RDF, SPARQL) to model sentences, patterns and axioms, thus easing the reusability and enrichment of the defined patterns.

Acknowledgement.   This research has been funded by the NSERC Discovery Grant Program.

References

1. De Marneffe, M.-C., Manning, C. D.: The Stanford typed dependencies representation. In: Proceedings of the Workshop on Cross-Framework and Cross-Domain Parser Evaluation. ACL (2008)
2. Bühmann, L., Fleischhacker, D., Lehmann, J., Melo, A., Völker, J.: Inductive lexical learning of class expressions. In: Janowicz, K., Schlobach, S., Lambrix, P., Hyvönen, E. (eds.) EKAW 2014. LNCS, vol. 8876, pp. 42–53. Springer, Switzerland (2014)

3. Völker, J., Haase, P., Hitzler, P.: Learning expressive ontologies. In: Proceedings of the Conference on ontology Learning and Population, pp. 45–69. IOS Press (2008)
4. Font, L., Zouaq, A., Gagnon, M.: Assessing the quality of domain concepts descriptions in DBpedia. In: SITIS 2015, pp. 254–261 (2015)

Towards a Platform for Curation Technologies: Enriching Text Collections with a Semantic-Web Layer

Peter Bourgonje, Julian Moreno-Schneider, Jan Nehring, Georg Rehm$^{(\boxtimes)}$, Felix Sasaki, and Ankit Srivastava

Language Technology Lab, DFKI GmbH, Alt-Moabit 91c, 10559 Berlin, Germany
georg.rehm@dfki.de

Abstract. In an attempt to put a Semantic Web-layer that provides linguistic analysis and discourse information on top of digital content, we develop a platform for digital curation technologies. The platform offers language-, knowledge- and data-aware services as a flexible set of workflows and pipelines for the efficient processing of various types of digital content. The platform is intended to enable human experts (knowledge workers) to get a grasp and understand the contents of large document collections in an efficient way so that they can curate, process and further analyse the collection according to their sector-specific needs.

Keywords: Digital Curation · Linguistic Linked Data · NLP

1 Introduction

The target audience of our platform are knowledge workers who conduct research in specific domains with the goal of, for example, preparing museum exhibitions or writing news articles. Typically they only have limited time available to accomplish their tasks, ranging from several hours to one or two weeks at most. Owing to the diversity of tasks, the domains are often new to them. The output of their work is typically used in online or traditional media (e.g., newspapers, agencies, tv stations) or by a museum for an exhibition. In the project Digitale Kuratierungstechnologien[1] (DKT, Digital Curation Technologies, [7]), we aim at automating specific parts of the workflows, which consist of looking for information related to and relevant for the domain, learning the key concepts, selecting the most relevant parts and preparing the information to be used. We build a platform that integrates various Natural Language Processing (NLP), Information Retrieval (IR) and Machine Translation (MT) components to retrieve information and recombine them to produce output that improves the curation processes and makes them more efficient. We work with heterogeneous data sets from public and non-public sources. The output will be in the form of a semantically enriched hypertext graph, stored and accessed using linked data

---

[1] http://digitale-kuratierung.de.

H. Sack et al. (Eds.): ESWC 2016 Satellite Events, LNCS 9989, pp. 65–68, 2016.
DOI: 10.1007/978-3-319-47602-5_14

technologies. Our goal is to enable knowledge workers to explore and curate document collections easier and more efficiently [6].

2   Related Work

Our digital curation platform integrates individual components by linking content with metadata [2]. The components include open-source tools for NLP tasks such as Named Entity Recognition (NER), Information Extraction (IE), IR and MT. The platform uses an architecture developed in the project FREME [8]. A related platform focusing on the localisation industry as the main use case is described by [1]. Our platform targets multiple other industry sectors, has different digital content formats in focus and deploys a different approach to representing curated information.

3   Semantic Layer

For the purpose of this paper we focus on written text documents. Eventually we will also be able to handle the conversion of non-textual data into text (e.g., transcripts for audio, subtitles for video, etc.). On top of text data we generate a Semantic Layer (SL) that contains semantic annotations. The SL creates an interlinked representation connected to external information sources. It is produced by a set of tools that communicate using the NLP Interchange Format (NIF).[2] It operates in a pipelined workflow where the output of each service is used as input for the next one. The SL can be used for exploratory search. The user query is sent through the same pipeline used to generate the SL over the whole document collection. This allows us to search the index for the plain words in the query, but also any entities or temporal expressions that were recognized. The components of the pipelined workflow are:

– **NLP:** This component consists of NER combining a model approach and a dictionary approach. It works with three types of entities: persons, locations and organizations. Any entities found in the input are annotated using NIF. After NER, we also perform Entity Linking using DBPedia Spotlight [4] to retrieve the relevant (DBPedia) URI for entities recognized with the model and on the URI directly taken from the dictionary (the dictionary specifies a key – the entity – and a value, i.e., a URI in some ontology) for entities recognized with the dictionary. Subsequently the NLP workflow performs a temporal expression analysis. This module consists of a language-specific regular expression grammar and currently supports German and English. The expressions are normalized to a machine readable format and added to the NIF model.

---

[2] http://persistence.uni-leipzig.org/nlp2rdf/.

- **Information extraction:** We use Lucene[3] to create an index for our document collection that enables text-based IR. In addition to indexing the text content, entities and temporal expressions have their own specific fields in order to allow search in the SL as well. Indexing entities also allows us to disambiguate based on entity clustering (planned for the next phase in this two-year project).
- **Semantic Storage:** The semantic information generated during the NLP processes is stored in the triple store Sesame.[4] We use an ontology relating the semantic information extracted from the documents. It relies on Schema.org to describe entities and contains documents and concepts, where the concepts are divided into locations, organizations, persons and temporal expressions.
- **Multilingual component:** This component is based on Moses[5] enhanced with pre-/post-processing modules to leverage the information obtained from preceding steps (e.g., NER, temporal analysis). The MT system is capable of translating both segments (sentences, subtitles) and documents enabling knowledge workers to retrieve information and to present the semantically-enriched output in several languages (English, German, Spanish, Arabic). Preliminary experiments show as much as a 5 % improvement in the overall MT system performance for multiple language pairs and domains.

4   Experiments

Our goal is to reduce the time knowledge workers invest in their sector-specific curation processes. A proper evaluation would require us to measure the time it takes knowledge workers to get from input to output with and without utilization of our platform. This is rather difficult to measure and to quantify. We are in an early stage of the project and do not have access to suitable data for such an evaluation yet. As the project progresses, we will acquire real-world data (to be provided by the industrial partners involved in the project), annotate the data to construct a gold standard so that the platform can be evaluated as a whole. For now, we can offer isolated evaluations of individual components. For evaluating the German version of the temporal expression analyzer, we use the German WikiWars corpus [3]. This corpus is a collection of 22 documents sourced from Wikipedia pages about military conflicts and contains 2.240 temporal expressions. Evaluating against this corpus, we can report an f-score of *0.83*. However, we developed against this same corpus and since we are mainly interested in coverage of our regular expressions, the corpus was not divided into training and test sets. We consider our f-score an acceptable baseline and will continue to improve this during the project. For evaluating the German version of the NER module, we selected the German wikiNER corpus [5]. This corpus contains NER annotations in CoNLL format. For this we can report f-scores of *0.78*, *0.87* and *0.76* for locations, persons and organizations, respectively. These numbers will serve as baselines for future work as well.

---

[3] https://lucene.apache.org/core/.
[4] http://rdf4j.org/.
[5] http://www.statmt.org/moses/.

5   Conclusion and Future Work

This article addresses the issue of combining NLP, IR and MT procedures into a system that enables knowledge workers to explore a collection of documents in an intuitive and efficient way. Our focus is on combining the individual components and linking the output of the methods, rather than trying to improve upon the output of individual state-of-the-art procedures. In this early stage of the project, we can aggregate the information contained in multiple documents and present this in a way that allows the knowledge worker to see what is inside. For the future we plan to work on making our tools easily adaptable to new domains. This poses a challenge since we expect to deal with domains for which only limited amounts of training data are available. We also plan to exploit the linked open data framework more by plugging in new datasets. Future applications are related to the project goals: *text summarization* of documents will help the curation process and *semantic story-telling* will assist in text generation processes, relating individual document components at a semantic level.

Acknowledgments. "Digitale Kuratierungstechnologien" is supported by the German Federal Ministry of Education and Research, Unternehmen Region, WK-P (No. 03WKP45).

References

1. Lewis, D., Brennan, R., Finn, L., Jones, D., Meehan, A., O'sullivan, D., Hellmann, S., Sasaki, F.: Global intelligent content: active curation of language resources using linked data. In: Proceedings of LREC 2014, Reykjavik (2014)
2. Lewis, D., Gómez-Pérez, A., Hellman, S., Sasaki, F.: The role of linked data for content annotation and translation. In: Proceedings of the 2014 European Data Forum. EDF 2014 (2014). http://2014.data-forum.eu
3. Mazur, P., Dale, R.: WikiWars: a new corpus for research on temporal expressions. In: Proceedings of the 2010 Conference on Empirical Methods in Natural Language Processing. EMNLP 2010, pp. 913–922. Association for Computational Linguistics, Stroudsburg (2010)
4. Mendes, P.N., Jakob, M., García-Silva, A., Bizer, C.: DBpedia spotlight: shedding light on the web of documents. In: Proceedings of the 7th International Conference on Semantic Systems. I-Semantics 2011, pp. 1–8. ACM, New York (2011)
5. Nothman, J., Ringland, N., Radford, W., Murphy, T., Curran, J.R.: Learning multilingual named entity recognition from Wikipedia. Artif. Intell. **194**, 151–175 (2012)
6. Rehm, G.: Hypertextsorten: Definition - Struktur - Klassifikation. Ph.D. thesis, Institutfür Germanistik, Angewandte Sprachwissenschaft und Computerlinguistik, Justus-Liebig-Universität Giessen (2005)
7. Rehm, G., Sasaki, F.: Semantische Technologien und Standards für das mehrsprachige Europa. In: Ege, B., Humm, B., Reibold, A. (eds.) Corporate Semantic Web, pp. 247–257. Springer, Heidelberg (2015)
8. Sasaki, F., Gornostay, T., Dojchinovski, M., Osella, M., Mannens, E., Stoitsis, G., Richie, P., Declerck, T., Koidl, K.: Introducing freme: deploying linguistic linked data. In: Proceedings of the 4th Workshop of the Multilingual Semantic Web. MSW 2015 (2015)

Towards Entity Summarisation
on Structured Web Markup

Ran Yu$^{(\boxtimes)}$, Ujwal Gadiraju, Xiaofei Zhu, Besnik Fetahu, and Stefan Dietze

L3S Research Center, Leibniz Universität Hannover, Hannover, Germany
{yu,gadiraju,zhu,fetahu,dietze}@L3S.de

Abstract. Embedded markup based on Microdata, RDFa, and Microformats have become prevalent on the Web and constitute an unprecedented source of data. However, statements extracted from markup are fundamentally different to traditional RDF graphs: entity descriptions are flat, facts are highly redundant and granular, and co-references are very frequent yet explicit links are missing. Therefore, carrying out typical entity-centric tasks such as retrieval and summarisation cannot be tackled sufficiently with state of the art methods. We present an entity summarisation approach that overcomes such issues through a combination of entity retrieval and summarisation techniques geared towards the specific challenges associated with embedded markup. We perform a preliminary evaluation on a subset of the Web Data Commons dataset and show improvements over existing entity retrieval baselines. In addition, an investigation into the coverage and complementary of facts from the constructed entity summaries shows potential for aiding tasks such as knowledge base population.

Keywords: Entity summarisation · Web Data Commons · Fact Selection

1   Introduction

Markup annotations embedded in HTML pages have become prevalent on the Web, building on standards such as RDFa[1], Microdata[2] and Microformats[3], and driven by initiatives such as schema.org by Google, Yahoo!, Bing and Yandex.

The Web Data Commons [2], a recent initiative investigating a Web crawl of 2.01 billion HTML pages from over 15 million pay-level-domains (PLDs) found that 30 % of all pages contain some form of embedded markup already, resulting in a corpus of 20.48 billion RDF quads[4]. Considering the upward trend of adoption - the proportion of pages containing markup increased from 5.76 % to 30 % between 2010 and 2014 - and the still comparably limited nature of the investigated Web crawl, the scale of the data suggests potential for a range of tasks, such as entity retrieval, knowledge base population, or entity summarisation.

---

[1] RDFa W3C recommendation: http://www.w3.org/TR/xhtml-rdfa-primer/.
[2] http://www.w3.org/TR/microdata.
[3] http://microformats.org.
[4] http://www.webdatacommons.org.

© Springer International Publishing AG 2016
H. Sack et al. (Eds.): ESWC 2016 Satellite Events, LNCS 9989, pp. 69–73, 2016.
DOI: 10.1007/978-3-319-47602-5_15

However, facts extracted from embedded markup have different characteristics when compared to traditional knowledge graphs and Linked Data. Co-references are very frequent (for instance, in the WDC2013 corpus, 18,000 entity descriptions of type `Product` are returned for query 'Iphone 6'), but are not linked through explicit statements. In contrast to traditional strongly connected RDF graphs, RDF markup statements mostly consist of isolated nodes and small subgraphs. In addition, extracted RDF markup statements are highly redundant and often limited to a small set of predicates, such as `schema:name`. Moreover, data extracted from markup contains a wide variety of syntactical and semantic errors, ranging from typos to the frequent misuse of vocabulary terms.

These distinctive characteristics highlight the challenges when aiming to summarise entities sourced from embedded markup. Initial works such as the *Glimmer* search engine[5] have applied traditional *entity retrieval* techniques [1] to embedded markup (WDC corpus). However, given the large amount of flat and highly redundant entity descriptions, practical use of search results obtained in that way is limited. One major issue with such approaches are the unresolved entity co-references and entity fact redundancies. Therefore, applying *entity summarisation* techniques in order to obtain a homogeneous entity summary, assembled from all extracted facts seems the most promising approach in order to answer entity-centric queries. In this work we present an entity summarisation approach building on established entity retrieval approaches for obtaining a set of candidate entity descriptions and combined with algorithms for selecting and ranking distinct facts based on the clustering of candidate facts.

2   Approach

An entity summary consists of a set of facts, i.e. ⟨predicate, object⟩ pairs. Given an entity-centric query, we generate the corresponding entity summary through a two-step approach: (i) entity retrieval, and (ii) fact selection. Figure 1 shows the summarisation process for query `Forrest Gump, type:(Movie)`.

Entity Retrieval. A prerequisite to construct entity summaries is the entity retrieval process. In our case, entities consist of a collection of facts ⟨predicate, object⟩. For this purpose, we build a standard IR index, where instead of documents we consider entity descriptions (set of facts associated with an entity). Next, we retrieve entity descriptions based on the BM25 retrieval model.

A necessary step to further improve the entity retrieval step is to resolve the *object properties* in such entity descriptions, i.e., ⟨s_1, `schema:actor`, s_2⟩. We rely on a simple heuristic, where if such object values correspond to another entity in the WDC dataset, we replace it with it literal label `schema:name` in s_2.

Clustering-Based Fact Selection (CBFS). In this step, we select facts from the entity descriptions retrieved in the previous step. Our approach is restricted to facts associated with predicates from *schema.org*.

---

[5] http://glimmer.research.yahoo.com/.

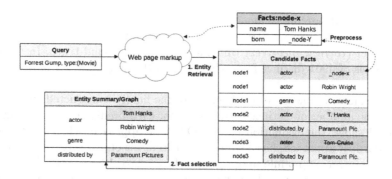

Fig. 1. Overview of the entity summary pipeline

Clustering. One major issue to address in the fact selection process, is canonicalizing different surface forms corresponding to a specific entity, e.g., `Tom Hanks` and `T. Hanks` are equivalent surface forms representing the same entity. To find duplicates and near duplicates, we first cluster the associated values at the predicate level into n clusters $(c_1, c_2, \cdots, c_n) \in C$. In this way, based on string similarity metrics we can canonicalize equivalent surface forms. For the clustering process we employ the X-means algorithm [3], which automatically determines the amount of clusters.

Fact Cardinality. Another challenge we address is the cardinality of predicates. Depending on the predicate, the number of correct statements varies. For example, `schema:actor` is associated with multiple values, whereas `schema:duration` normally has only one valid statement.

Fact Selection. Assuming that false entity facts have lower frequency, we eliminate irrelevant facts by choosing facts that are closer to the cluster's centroid and further meet the following criteria:

$$|c_j| > \beta \cdot \max(|c_k|), c_k \in C \qquad (1)$$

where $|c_j|$ denotes the size of cluster c_j, and β is a parameter used to adjust the number of facts. In our experiments, β is empirically set at 0.5.

3   Experimental Setup and Evaluation

Dataset and Queries. We use a subset of the WDC 2014 dataset for experiments (entities of type `movie`), which contains 77 million RDF quads. We randomly select entities of type *movie* from Wikipedia as queries. Next, we keep only those queries that have at least 50 retrieved entities and have a high BM25 score (higher than 8). This is to ensure the correctness of entity summaries. Finally, we are left with 26 queries for use in our evaluation.

Performance. We consider BM25 as the baseline, and take the top 50 distinct facts from the retrieved entities, where the facts are ranked based on the associated rank to the original entity descriptions.

We use crowdsourcing to identify the precision of the retrieved facts for each entity-centric query. Five different crowd workers provide us with binary labels for each fact (*correct, incorrect*) which are used as a ground truth. In the end, we get labels of 4901 distinct fact, which is also the candidate set of our CBFS approach.

Furthermore, we measure the diversity of the finally constructed entity summary, as the number of distinct *correct* facts for each predicate in our summary.

Table 1 shows the evaluation results. Our final approach, when compared to the standard IR approach, has a gain of 5.5 % in terms of precision. Additionally, due to the clustering module in our approach, we have 12.5 % more facts at the top 50 cutoff when compared to the baseline. This however, is intuitive given that the standard IR does not perform any de-duplication process.

Table 1. Performance of the proposed entity summarization approach

Approach	Fact	Correct	P	Distinct	$Dist\%$
CBFS	1075	895	**0.833**	876	**97.9**
BM25@50	1124	877	0.778	749	85.4

Coverage Gain. To evaluate the potential of our approach for aiding knowledge base augmentation tasks, we measure the coverage gain by comparing our results to DBpedia. We use *coverage gain (CG)*, i.e. the percentage of facts detected that are not available in DBpedia. We manually compare the fact selection results with corresponding DBpedia resources to determine the *CG*.

We found that 57 % of the facts detected by our approach do not exist in DBpedia. Some of the facts correspond to new predicates. We also calculate the extra coverage considering only the predicates that exist in DBpedia, and the *CG* is 33.4 %. This suggests that WDC dataset provides richer and diverse information in comparison to DBpedia and our approach as well as markup data in general are able to complement the knowledge available in DBpedia.

4   Conclusion and Future Work

We have introduced an approach to entity summarisation on Web markup where experimental results suggest potential for exploiting Web markup as a novel resource for tasks such as entity summarisation or knowledge base population. While our current approach is based on a simplistic pipeline involving a range of heuristics, as part of ongoing work we are working on a supervised approach for fact selection. In addition, more focused experiments aim at investigating the performance of our approach for specific knowledge base population tasks.

References

1. Blanco, R., Mika, P., Vigna, S.: Effective and efficient entity search in RDF data. In: Aroyo, L., Welty, C., Alani, H., Taylor, J., Bernstein, A., Kagal, L., Noy, N., Blomqvist, E. (eds.) ISWC 2011, Part I. LNCS, vol. 7031, pp. 83–97. Springer, Heidelberg (2011)
2. Meusel, R., Petrovski, P., Bizer, C.: The WebDataCommons microdata, RDFa and microformat dataset series. In: Mika, P., Tudorache, T., Bernstein, A., Welty, C., Knoblock, C., Vrandečić, D., Groth, P., Noy, N., Janowicz, K., Goble, C. (eds.) ISWC 2014, Part I. LNCS, vol. 8796, pp. 277–292. Springer, Heidelberg (2014)
3. Pelleg, D., Moore, A.W. et al.: X-means: extending k-means with efficient estimation of the number of clusters. In: ICML, pp. 727–734 (2000)

Taking Advantage of Discursive Properties for Validating Hierarchical Semantic Relations from Parallel Enumerative Structures

Mouna Kamel$^{(\boxtimes)}$ and Cassia Trojahn

Institut de Recherche en Informatique de Toulouse, Toulouse, France
{mouna.kamel,cassia.trojahn}@irit.fr

Abstract. This paper presents an approach for automatically validating candidate hierarchical relations extracted from parallel enumerative structures. It relies on the discursive properties of these structures and on the combination of resources of different nature, a semantic network and a distributional resource. The results show an accuracy of between 0.50 and 0.67, with a gain of 0.11 when combining the two resources.

1 Introduction

Relation extraction is a key task in ontology learning from texts. The identification of candidate relations has been the subject of large body of literature and many approaches have been proposed (linguistic, statistical or hybrid approaches, based or not on learning methods). However, this is an error-prone step (imprecise lexico-syntactic patterns, accuracy of learning techniques under 100 %, chaining of NLP tools in pre-processing steps, etc.). Validating candidate relations is a crucial step before integrating them into semantic resources.

This paper concerns the validation of candidate hierarchical relations, the backbone of ontologies. While manual validation is a time-consuming task requiring domain expert judges, automatic ones rely on external semantic resources (such as WordNet, BabelNet), which are usually non domain-specific, or gold standards, which may suffer of imperfections or low domain coverage. The proposal here relies on the extraction of hierarchical semantic relations from parallel enumerative structures (called hereafter PES) [4]. This choice is motivated by the following reasons: (1) PES often carry hierarchical relations; (2) they are frequent in corpora, especially in scientific or encyclopedic texts (rich sources of semantic relations); and (3) they have well-established discursive properties bringing up a semantic unit within the structure. The originality of our approach lies in the discourse properties of PES for disambiguating candidate relations and in the combination of two complementary external resources, a semantic network and a distributional resource. While the semantic network allows for validating the candidate relations with a good level of precision, the distributional resource, which does not specify the nature of the relation but offers a good coverage, allows for emerging new relations, which may enrich the network itself. Although evaluated for the French language, the approach remains reproducible for any other language.

© Springer International Publishing AG 2016
H. Sack et al. (Eds.): ESWC 2016 Satellite Events, LNCS 9989, pp. 74–78, 2016.
DOI: 10.1007/978-3-319-47602-5_16

2   Parallel Enumerative Structures

An enumerative structure is a textual structure expressing hierarchical knowledge through different components: a primer, a list of items (at least two) constituting the enumeration, and possibly a conclusion. Different typologies have been proposed [3,5]. Here, we consider enumeratives structures for which the enumeration items are functionally equivalent (from a syntactic and rhetoric point of view) (Fig. 1). From a discursive point of view, the items are independent in a given context: they are in turn connected by a multi-nuclear rhetoric relation (or coordination), the first item being linked to the primer by a nuclear-satellite relation (or subordination) (Fig. 2). According to the RST (Rhetorical Structure Theory) [2], if "DU_j (where DU corresponds to Discourse Unit) is subordinated to DU_i, hence each DU_k coordinated with DU_j is subordinate to DU_i". Thereby, N nuclear-satellite relations between DU_0 and DU_i, for $i=1,...,N$ (if N is the number of items in the ES) can be inferred. These N relations can be specialised in N semantic relations $R(H, h_i)_{i=1,...,N}$ of same nature, where H correspond to a term of DU_0, and h_i to a term of DU_i. From Fig. 2, three relations can be identified: $R(disease, Cholera)$, $R(disease, Colorectal\ cancer)$, and $R(disease, Diverticulitis)$.

```
There are a number of diseases affecting the gastrointestinal system:
- Cholera
- Colorectal cancer
- Diverticulitis
```

Fig. 1. Example of PES.

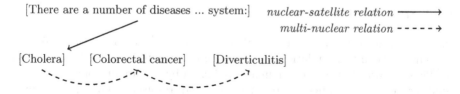

Fig. 2. Discursive representation of the PES of Fig. 1 according to the RST

3   Proposed Approach

The validation principle exploits the discourse properties of PES to jointly validate the relations $R(H, h_i)$ $(i = 1, .., N)$ where R is the hypernym relation:

1. if $R(H, h_i)$ corresponds to an entry in the semantic network SN, $R(H, h_i)$ is validated.
2. if $R(H, h_i)$ has no entry in SN, but an entry corresponding to $R(H, h_j)$ exists in SN and h_i is a neighbour of h_j in the distributional resource DR, then $R(H, h_i)$ is validated.

From SN, we retrieve $Synsets(H)$, the synsets of H, and $SuperHyperyms_{SN}^{k}$ (h_i), the hypernym synsets of h_i of rank k (k being the maximum length of the path from h_i to one of its hypernym synsets in SN, based on a depth-first search strategy). From DR we retrieve $p(h_i, h_j)$, the semantic proximity between h_i and h_j. This process is described in Algorithm 1.

Algorithm 1. Algorithm for validating a set of relations from a PES

$V \leftarrow \emptyset$, $\overline{V} \leftarrow \bigcup_{i=1}^{N} R_i$ // V set of validated relations, $\overline{V}$ set of non validated relations
for each relation $R_i(H, h_i) \in \overline{V}$ **do**
   if $SuperHyperyms_{SN}^{k}(h_i) \cap Synsets(H) \neq \emptyset$ **then**
     // H is a hypernym of h_i
     $validate(R_i(H, h_i)) \leftarrow 1$ // R_i is validated
     $V = V \bigcup \{R_i\}$
     $\overline{V} = \overline{V} - \{R_i\}$
   end if
end for
if $V \neq \emptyset$ et $\overline{V} \neq \emptyset$ **then**
   //at least one relation has been validated and one has not been yet
   for each relation $R_j(H, h_j) \in \overline{V}$ **do**
     $validate(R_j(H, h_j)) = \frac{\sum_{R_i \in V} p(h_i, h_j)}{|V|}$
     //proximity between h_j and the hyponyms h_i from the validated relations
   end for
end if

---

4   Experimentation

Data set and resources. The evaluation data set[1] is composed of 67 PES involving 262 candidate relations, automatically extracted from Wikipedia pages [4]. These relations have been manually validated by two annotators in a double-blind process. 27 conflicts were identified and resolved. 206 relations were assessed as correct and 56 as incorrect. This set constitutes our *gold standard*. With respect to the resources, we have used the multilingual semantic network *BabelNet* [6] and the distributional resource *Voisins de Wikipédia* [1]. They have been chosen because they support French language and they are built from the same corpus as the one used for constructing the evaluation data set.

Results and discussion. Two sets of candidate relations were considered (Table 1): S, the whole set of true positive relations from the *gold standard* (206 relations) and S_{BN}, the subset of S for which H exists in *BabelNet* (116 relations). For both sets, 76 out of 78 relations were correctly validated by the system. 12 out of 76 have been correctly validated thanks to the distributional resource, what corresponds to an improvement of the performance up to 11%.

---

[1] Available at https://www.irit.fr/~Cassia.Trojahn/PES.zip.

Table 1. Overall results of the validation process combining both SN and DR. (+) corresponds to the specific gain of using DR.

	Precision $(+_{DR})$	Recall $(+_{DR})$	FMeasure $(+_{DR})$	Accuracy $(+_{DR})$
S	.97 (+0.0)	.37 (+.06)	.54 (+.07)	.50 (+.05)
s_{BN}	.97 (+0.0)	.66 (+.11)	.78 (+0.8)	.67 (+.11)

In terms of recall, we have a lower performance (76 relations out of 206 for S but 76 out of 116 for S_{BN}). In terms of accuracy, 130 relations have been validated (out of 262) for the set S and 88 relations (out of 131) for the set S_{BN}.

Although the precision is quite high, we could identify the reasons for the noisy cases. It is due to the fact that we are using BabelSynsets which group terms of similar meaning. For instance, for the candidate relation $R(country, Horn\ of\ Africa)$, the BabelSynset $bn{:}00028934n = \{land,\ dry\ land,\ earth,\ ground,\ terra\ firma\}$ belongs to the intersection of the sets $SuperHyperyms^3_{BN}(Horn\ of\ Africa)$ and $Synsets(country)$. With respect to the low recall, we observed two main phenomena. First, 62 hypernyms (from S) have no entries in *BabelNet*. In this case, no relation within the PES could be validated. Second, considering $k = 3$ (empirically chosen) as maximum length of the path from h_i to one of its hypernyms seems to be insufficient. We could also observe that the distributional resource allows for identifying missing entries in the semantic network. For example, the relation $R(chromosomal\ abnormality, insertion)$ was validated due to the fact that *insertion* and *deletion* are semantically near in the distributional resource. Although the entries in this resource overwhelmingly correspond to single words and 40 % of our hyponyms correspond to compounds, we improved the performance up to 11 % when combining both resources. Distributional resources supporting compounds may further improve our results.

5   Conclusions and Future Work

This paper proposed an approach for automatically validating semantic relations, relying on discursive properties and combining a semantic network and a distributional resource. As future work, we plan to exploit alternative resources (in particular, distributional resources with compounds), analyse the trade-off between depth-first and breath-first search strategies and their computational complexity, exploiting larger semantic networks or combining several resources together. We intent as well to extend our approach to validate other semantic relations like meronymy, synonymy and antonym.

Acknowledgement. Cassia Trojahn is partially supported by the French FUI SparkinData project.

References

1. Adam, C., Fabre, C., Muller, P.: Évaluer et améliorer une ressource distributionnelle: protocole d'annotation de liens sémantiques. TAL **54**(1), 71–97 (2013)
2. Asher, N.: Reference to abstract objects in discourse: a philosophical semantics for natural language metaphysics. In: SLAP, vol. 50. Kluwer (1993)
3. Christophe, L.: Représentation et composition des structures visuelles et rhétoriques du textes. Approche pour la génération de textes formatés. PhD thesis (2000)
4. Fauconnier, J.P., Kamel, M.: Discovering hypernymy relations using text layout. In: Joint Conference on Lexical and Computational Semantics, Denver, pp. 249–258. ACL (2015)
5. Hovy, E., Arens, Y.: Readings in intelligent user interfaces. In: Automatic Generation of Formatted Text, pp. 256–262. Morgan Kaufmann Publishers (1998)
6. Navigli, R., Ponzetto, S.P.: BabelNet: the automatic construction, evaluation and application of a wide-coverage multilingual semantic network. Artif. Intell. **193**, 217–250 (2012)

Separating Wheat from the Chaff –
A Relationship Ranking Algorithm

Sumit Bhatia[(✉)], Alok Goel, Elizabeth Bowen, and Anshu Jain

IBM Watson, Almaden Research Centre, San Jose, CA, USA
{sumit.bhatia,alok.goel,elizabeth.bowen,anshu.n.jain}@us.ibm.com

Abstract. We address the problem of ranking relationships in an automatically constructed knowledge graph. We propose a probabilistic ranking mechanism that utilizes entity popularity, entity affinity, and support from text corpora for the relationships. Results obtained from preliminary experiments on a standard dataset are encouraging and show that our proposed ranking mechanism can find more informative and useful relationships compared to a frequency based approach.

1 Introduction

We are transitioning from the era of Big Data to Big Knowledge, and semantic knowledge bases such as knowledge graphs play an important role in this transition. A knowledge graph consists of real world entities (or concepts) as nodes. A given node may be connected with many other nodes and there could be multiple edges between two given nodes, with each edge representing a real world fact. Thus, a knowledge graph is an essential source for getting important facts about real world entities and how these entities are related to each other. Knowledge Graphs can be constructed either manually (facts authored by humans) or automatically (facts extracted from text using Machine Learning tools). Manually curated knowledge graphs such as DBPedia have little or no noisy facts as they are carefully authored, but they require very large human efforts. This problem is further exacerbated in enterprise domains and custom domains such as life sciences, finance, intelligence, etc. where domain expertise is also crucial to add good quality facts in the graph. As a result, efforts have been made for development of systems for automatic construction of semantic knowledge bases for domain specific corpora [1] and systems that use such domain specific knowledge bases [4] are gaining prominence. In such systems, a machine learning based annotator is used to extract entities and relationships from domain specific corpus. As a result, in such knowledge bases, for each discovered relationship, *associated text mentions* from the corpus are also available and number of times a relationship is observed in the corpus can be used as a proxy for that relationship's *evidence or strength*. Such information may or may not be available with manually curated knowledge graphs.

Typically, a given entity may be involved in many relationships and we argue that all such relationships are not equally important and informative and thus,

© Springer International Publishing AG 2016
H. Sack et al. (Eds.): ESWC 2016 Satellite Events, LNCS 9989, pp. 79–83, 2016.
DOI: 10.1007/978-3-319-47602-5_17

there is a need for methods that can identify the most relevant and meaningful relationships for a given entity. There have been some efforts to rank the entities and relationships in knowledge bases. Li et al. [3] propose an entity-relationship structured query mechanism for ranking Wikipedia entities and their relationships. However, their method is dependent on the hyperlinks structure derived from Wikipedia and thus, can not be applied for graphs constructed from generic corpora. Instead of finding most important relationships, Zhang et al. [5] propose an alternative way to cluster similar relationships together and then allowing the users to further explore clusters of interest. In this paper, we address the problem of *ranking relationships for an entity in an automatically constructed knowledge graph*. We propose a probabilistic framework that judges the relevance of a relationship by utilizing various measures such as entity popularity, strength of evidence for a relationship, and affinity between input and target entities. We use a semantic graph constructed from text of all articles in Wikipedia by automatically extracting the entities and their relations by using IBM's Statistical Information and Relation Extraction (SIRE) toolkit[1]. Even though there exist popular knowledge bases like DBPedia that contain high quality data, we chose to construct a semantic graph using automated means as such a graph will be closer to many practical real world scenarios where high quality curated graphs are often not available and one has to resort to automatic methods of constructing knowledge bases. Our graph contains more than 30 millions entities and 192 million distinct relationships in comparison to 4.5 million entities and 70 million relationships in DBpedia.

2 Proposed Relationship Ranking Algorithm

Let us consider a Knowledge Graph $\mathcal{G} = \{E, R\}$, where, $E = \{e_1, e_2, \ldots, e_n\}$ is the set of nodes (or entities) and $R = \{r_1, r_2, \ldots, r_m\}$ is the set of edges (or relationships). Further, let $w : R \rightarrow \mathbb{R}_+$ is a weight function that gives weight of any edge in the graph. This function can be defined in various ways to measure the importance of an edge. We chose the frequency of occurrence of the given fact in the text corpus (mention count) as the weight of the edge corresponding to that relation in the graph. Given an input entity e, and a set $R_e \in R$ of all the relations involving e, we want to produce an ordered list of all the elements of R_e, ordered by their importance/relevance. The task of selecting a relationship involving input entity e can be decomposed in two steps – first selecting a target entity e_t, and then selecting an edge that connects these two entities. For example, for input *Barack Obama*, we first select a target entity, say *United States*, and then decide which of the two relationships out of *citizenOf* and *presidentOf* should be picked. Mathematically,

$$P(r, e_t|e) = P(e_t|e)P(r|e_t, e) = P(r|e, e_t)\frac{P(e_t)P(e|e_t)}{P(e)} \qquad (1)$$

---

[1] http://ibmlaser.mybluemix.net/siredemo.html.

In the above equation, $P(e)$ can be ignored for ranking purposes since this factor will remain same for all target entities and relationships. The above equation can then be written as follows:

$$P(r, e_t | e) \propto \underbrace{P(e_t)}_{\text{Entity Prior}} \times \underbrace{P(e|e_t)}_{\text{Entity Affinity}} \times \underbrace{P(r|e, e_t)}_{\text{Relationship Strength}} \tag{2}$$

The above equation represents the ranking function that can be used to rank all relationships of a given entity. We now discuss the three components in the above equation contributing to the overall relevance score of a given relationship.

Entity Prior: This component captures the intuition that in absence of any other information, the input entity has a higher chance of having a relationship with a popular entity in the graph as compared to a rare entity in the graph. It can be computed as follows:

$$P(e_t) \propto relCount(e_t) \tag{3}$$

where, $relCount(e_t)$ is the number of relationships entity e_t is involved in.

Entity Affinity: A target entity that has most of its relationships (or most of its strongest relationships) with the input entity is more important as compared to a target entity that has very few (or very weak relationships) with the input entity. For example, in our knowledge graph, both "Florida" and "France" have almost equal number of relationships in the graph, however, "Florida" has a larger fraction of its relationships with "USA" and some of its strongest relationships are with "USA". Hence, compared to "France", "Florida" is a more specific entity to "USA". Mathematically, it can be expressed as follows:

$$P(e|e_t) = \frac{\sum_{r_i \in R(e, e_t)} w(r_i) \times r_i}{\sum_{r_i \in R(e_t)} w(r_i) \times r_i} \tag{4}$$

where, $R(e_t)$ is the set of all relationships e_t is involved in and $R(e, e_t)$ is the set of all relationships between e and e_t.

Relationship Strength: While the previous two components were concerned with capturing the relevance of target entities for a given input entity, this component tries to measure the relative importance of different relationship types once we have selected the target entity. Given two entities, there could be multiple relationships between them. For example, "Barack Obama" is connected to "USA" with multiple relationships such as *presidentOf, citizenOf, livesAt, bornAt,* etc. In absence of any other information, we hypothesize that a relationship having more support/evidence from the corpus is more important than a relationship that has little supporting evidence. Mathematically,

$$P(r|e, e_t) = \frac{mentionCount(r, e, e_t)}{\sum_{r \in R_{e, e_t}} mentionCount(r, e, e_t)} \tag{5}$$

where, $mentionCount(r, e, e_t)$ represents the number of times relationship r connecting e and e_t was mentioned in the text corpus, and R_{e, e_t} is the set of all relationships between entities e and e_t.

3   Evaluation

For evaluating our proposed relationship ranking approach, we use the set of entities provided in the KORE entity relatedness dataset [2]. This dataset provides 21 seed entities from various domains. However, the dataset does not provide a ranked list of important relationships for the seed entities. Hence, we took help of two human evaluators to assess the quality of results produced by proposed ranking approach. For each input entity, we generated top 10 relationships ranked by our proposed ranking function and also top 10 most popular relationships as a baseline. Each evaluator was asked to rate the resulting relationships using a three point scale – 0 for an incorrect/noisy relationship, 1 for a correct but not useful relationship, and 2 for a correct and highly interesting relationship. As an example, "Brad Pitt" is *spouseOf* "Angelina Jolie" is a much more useful and informative relationship when compared to a generic relationship "Brad Pitt" is a *partOfMany* "Actors", even though both relationships are correct. One evaluator provided judgments for relationships for 11 entities and one provided for 10 entities. The results are tabulated in Table 1. We observe that while almost all the relationships produced by both the approaches were correct (corresponding to scores of 1 and 2), the proposed approach was much better at finding more informative relationships (70.48 % relationships with score 2 compared to only 47.14 % for popularity based ranking). The results of this preliminary evaluation are encouraging and provide strength to our hypothesis that not all facts about an entity are equally important and hence, the need for an appropriate relationship ranking algorithm.

Table 1. Results for relationship ranking as provided by human evaluators.

Method	Score 0	Score 1	Score 2
Popularity	8 (3.8 %)	103 (49.05 %)	99 (47.14 %)
Proposed approach	10 (4.76 %)	52 (24.76 %)	148 (70.48 %)

4   Conclusions and Future Work

We discussed the problem of ranking facts about a given entity in a knowledge graph and proposed a probabilistic framework to rank relationships/facts. Results of a preliminary evaluation study are encouraging and our future work will focus on enhancing the evaluation, both qualitatively and quantitatively. One major limitation of proposed approach is its inability to find facts that are customized to a user's requirements. For example, for the input entity *Barack Obama*, a user researching about *presidential elections* will be interested in different facts than a user interested in his *education history*. Therefore, our future research work will focus on *context sensitive* ranking of relationships so that users can get facts that are most important to their information needs.

References

1. Castelli, V., Raghavan, H., Florian, R., Han, D.-J., Luo, X., Roukos, S.: Distilling and exploring nuggets from a corpus. In: SIGIR, p. 1006 (2012)
2. Hoffart, J., Seufert, S., Nguyen, D.B., Theobald, M., Weikum, G.: Kore: Keyphrase overlap relatedness for entity disambiguation. In: CIKM 2012, pp. 545–554 (2012)
3. Li, X., Li, C., Yu, C.: Entity-relationship queries over wikipedia. ACM Trans. Intell. Syst. Technol. **3**(4), 70:1–70:20 (2012)
4. Nagarajan, M. et al.: Predicting future scientific discoveries based on a networked analysis of the past literature. In: KDD, pp. 2019–2028 (2015)
5. Zhang, Y., Cheng, G., Qu, Y.: JIST 2013, Selected Papers, chapter Towards ExploratoryRelationship Search: A Clustering-Based Approach, pp. 277–293 (2014)

Semantic Web Conference Ontology - A Refactoring Solution

Andrea Giovanni Nuzzolese[1], Anna Lisa Gentile[2(✉)], Valentina Presutti[1], and Aldo Gangemi[1,3]

[1] Semantic Technology Lab, ISTC-CNR, Roma, Italy
andrea.nuzzolese@istc.cnr.it, {valentina.presutti,aldo.gangemi}@cnr.it
[2] Data and Web Science Group, University of Mannheim, Mannheim, Germany
annalisa@informatik.uni-mannheim.de
[3] LIPN, Université Paris 13, Sorbone Cité, UMR CNRS, Paris, France

Abstract. The Semantic Web Dog Food (SWDF) is the reference linked dataset of Semantic Web community about papers, people, organisations, and events related to its academic conferences. In this paper we analyse the existing problems, of generating, representing and maintaining Linked Data for the SWDF. Accordingly, we discuss a refactoring of the Semantic Web Conference Ontology by adopting best ontology design practices (e.g., Ontology Design Patterns, ontology reuse and interlinking). We regenerate metadata for a set of conferences already existing in SWDF, using cLODg (conference Linked Open Data generator), an Open Source workflow which adopts the proposed refactoring.

1 Introduction

A good practise in the Semantic Web community is to encourage the publication of Linked Data about scientific conferences in the field, as a way of "eating our own dog food" [5]. The main example is the *Semantic Web Dog Food*[1] (SWDF), a corpus that collects Linked Data about papers, people, organisations, and events related to academic conferences. Currently, all main Semantic Web conferences and related events publish their data as Linked Data on SWDF, but for many other conferences, events and publication venues information is still not available in a structured and linked form. On the other hand the growth of available content with respect to the early times of SWDF poses data management issues and reveals design problems which where not foreseen when the dataset was at its initial stage.

There are several challenges to pursue the maintenance of a healthy and sustainable SWDF for the future: (i) the availability of appropriate vocabularies to express the current state of the data; (ii) the shared knowledge of such vocabularies; (iii) the availability of tools to ease the task of data acquisition, conversion, integration, augmentation, verification and finally publication; (iv) the ongoing maintenance of the dataset.

---

[1] SWDF: http://data.semanticweb.org.

© Springer International Publishing AG 2016
H. Sack et al. (Eds.): ESWC 2016 Satellite Events, LNCS 9989, pp. 84–87, 2016.
DOI: 10.1007/978-3-319-47602-5_18

In this work we focus on (i) and (ii) and use cLODg[2] (conference Linked Open Data generator) [4] to tackle (iii) (so far cLODg has been used to gather and publish metadata for ESWC2014 and ESWC2015 and it will be utilised for ESWC2016). As main contribution we identify issues of intentional nature on the SWDF dataset and we propose a refactoring solution for the Semantic Web Conference (SWC) ontology[3].

2   State of the Art

The first considerable effort to offer comprehensive semantic descriptions of conference events is represented by the metadata projects at ESWC 2006 and ISWC 2006 conferences [6], with the SWC Ontology being the vocabulary of choice to represent such data. Increasing number of initiatives are pursuing the publication about conferences data as Linked Data, mainly promoted by publishers such as Springer[4] or Elsevier[5] amongst many others. For example, the knowledge management of scholarly products is an emerging research area in the Semantic Web field known as Semantic Publishing [8]. Semantic Publishing aims at providing access to semantic enhanced scholarly products with the aim of enabling a variety of semantically oriented tasks, such as knowledge discovery, knowledge exploration and data integration. Despite these continuous efforts, it has been argued that lots of information about academic conferences is still missing or spread across several sources in a largely chaotic and non-structured way and a viable solution is a strong cooperation between researchers and publishers [1]. In this work we provide an analysis of existing modelling issues in the SWDF in order to provide a reference ontology and foster its adoption to close this gap.

3   Towards a Sustainable SWDF

The SWDF uses the SWC ontology as the reference ontology for modelling data about academic conferences. The SWC ontology combines existing widely accepted vocabularies (i.e., FOAF, SIOC and Dublin Core) and relies on the SWRC (Semantic Web for Research Communities) ontology for modelling entities of academic conferences, such as accepted papers, authors, their affiliations, talks and other events, the organizing committee and all other roles involved. Namely, the core types of SWC ontology are `foaf:Person` for describing people, `foaf:Organization` for describing the organisations (e.g., universities, research institutions, etc.) people are affiliated to, `foaf:Organization`, `swc:Artefact` for describing documents (e.g., papers, proceedings, etc.), `swc:OrganisedEvent` for describing any event related to an academic conference and `swc:Role` for describing the roles held by people at a conference. In the following we briefly list the main intentional issues affecting the data in the SWDF.

---

[2] https://github.com/AnLiGentile/cLODg.
[3] http://data.semanticweb.org/ns/swc/swc_2009-05-09.html.
[4] http://lod.springer.com/wiki/bin/view/Linked+Open+Data/About.
[5] http://data.elsevier.com/documentation/index.html.

Affiliations. Namely, the SWC ontology uses (i) the object property `swrc:affiliation` from the SWRC ontology in order to represent affiliations of people to organisations and (ii) the property `foaf:member` for expressing membership relations between organisations and people. Though this representation is very intuitive, it ignores the temporal dimension (i.e., the time when a given affiliation was held by an actor) that is relevant to interpret affiliations correctly. For example, it would be not possible to provide a correct answer to a simple competency question, such as "What was the affiliation of a person who authored a certain paper?".

Roles held by people at a conference. Roles such as program chair, track chair, etc. are model by using a commonly adopted ontology pattern based on the reification of a n-ary relation. Namely, the n-ary relation is identified by individuals of the class `swc:Role`. In fact, these individuals represents actual roles and relates people to events. The SWC ontology contains a very basic set of role classes (i.e., `swc:Chair`, `swc:Delegate`, `swc:Presenter` and `swc:ProgrammeCommitteeMember`) represented as sub-classes of `swc:Role`. This choice allows to instantiate the small set of different Role classes and cover the roles at specific events. For example, instead of sub-classing the `swc:Chair` class with MainChair, WorkshopChair, TutorialChair, etc., the different types of chairs should simply be instances of the generic Chair class and be labelled appropriately (e.g., `eswc2014:general-chair`[6]). The problem is that the individuals representing roles are defined locally to each conference. This means that for each conference, there is, for example, a different individual for representing the role "general chair". Hence, it is very complex to query the dataset in order to retrieve all the general chairs of the various editions of ESWC. More in detail, this comes from the erroneous reification of the n-ary relation on individuals of the class `swc:Role`, instead of using individuals of a different class representing the description of a role assignment situation.

We propose a refactoring of the SWC ontology, exploiting Ontology Design Patterns (ODP) [2]. We choose the Time indexed person role[7] ontology design pattern for modelling affiliations and assignment of roles to people. It is based on the reification of a n-ary relation, whose individuals are instances of the class `timeindexedpersonrole:TimeIndexedPersonRole`, representing the fact that a certain situation (e.g., a person affiliated to an organisation) occurs at a certain time interval. The classes `conf:Affiliation` and `conf:RoleAssignments` are defined as specializations of the class `timeindexedpersonrole:TimeIndexedSituation`[8]. According to this representation we are able to represent affiliations as n-ary relations having (i) a person (i.e., the agent holding the affiliation), (ii) an organisation and (ii) an associated time. The time can be an interval consisting of the conference dates or the instant when the paper was submitted. This allows us

---

[6] The prefix `eswc2014:` stands for the namespace http://data.semanticweb.org/conference/eswc/2014/.

[7] http://ontologydesignpatterns.org/wiki/Submissions:Time_indexed_person_role.

[8] http://ontologydesignpatterns.org/wiki/Submissions:TimeIndexedSituation.

to represent cases in which a person moves to another organisation in the time interval between the paper submission and the conference event.

Additionally, we aligned the ontology to (i) Dolce D0[9] classes that define the top level of the ontology and to (ii) FaBIO [7]. The defined ontology is available on-line for download[10] and the final data generation is performed using the cLODg Open Source workflow that uses our proposed refactored data model.

4 Conclusions and Future Work

This paper identifies current issues in the SWC ontology and proposes a refactoring solution. Given the new proposed data model we regenerate a subset of the SDWF dataset using the cLODg workflow [3] (the current cleaned dataset is provided on http://www.scholarlydata.org/). As future work we plan to collaborate for the cleansing of the official SWDF on http://data.semanticweb.org/ and to add more advanced alternative implementations for each step of cLODg and provide data maintenance services.

References

1. Bryl, V., Birukou, A., Eckert, K., Kessler, M.: What is in the proceedings? Combining publishers and researchers perspectives. In: Proceedings of SePublica 2014, Anissaras, 25 May 2014
2. Gangemi, A., Presutti, V.: Ontology design patterns. In: Staab, S., Studer, R. (eds.) Handbook on Ontologies, 2nd edn, pp. 221–243. Springer, Heidelberg (2009). doi:10. 1007/978-3-540-92673-3_10
3. Gentile, A.L., Acosta, M., Costabello, L., Nuzzolese, A.G., Presutti, V., Recupero, D.R., Live, C.: Accessible and sociable conference semantic data. In: Proceedings of WWW 2015 (Companion Volume), pp. 1007–1012. ACM (2015)
4. Gentile, A.L., Nuzzolese, A.G.: cLODg - conference linked open data generator. In: Proceedings of the ISWC 2015 Posters & Demonstrations Track co-located with the 14th International Semantic Web Conference (ISWC-2015), Bethlehem, 11 October 2015
5. Harrison, W.: Eating your own dog food. Industrial, Organizational Psychology, 5–7 June 2011
6. Möller, K., Heath, T., Handschuh, S., Domingue, J.: Recipes for semantic web dog food — the ESWC and ISWC metadata projects. In: Aberer, K., et al. (eds.) ASWC 2007 and ISWC 2007. LNCS, vol. 4825, pp. 802–815. Springer, Heidelberg (2007). doi:10.1007/978-3-540-76298-0_58
7. Peroni, S., Shotton, D.: FaBiO and CiTO: ontologies for describing bibliographic resources and citations. J. Web Semant. **17**, 33–43 (2012)
8. Shotton, D.: Semantic publishing: the coming revolution in scientific journal publishing. Learn. Publ. **22**(2), 85–94 (2009)

---

[9] http://www.ontologydesignpatterns.org/ont/dul/d0.owl.

[10] http://www.scholarlydata.org/ontology/conference-ontology.owl.

Semantic Context Consolidation and Rule Learning for Optimized Transport Assignments in Hospitals

Femke Ongenae[✉], Pieter Bonte, Jeroen Schaballie, Bert Vankeirsbilck, and Filip De Turck

IBCN Research Group, INTEC Department, Ghent University,
iGent Building, Technology Park 15, 9052 Zwijnaarde, Belgium
Femke.Ongenae@intec.ugent.be

Abstract. The increase of ICT infrastructure in hospitals offer opportunities for cost reduction by optimizing workflows, while maintaining quality of care. This work-in-progress poster details the AORTA system, which is a semantic platform to optimize transportation task scheduling and execution in hospitals. It provides a dynamic scheduler with an up-to-date view about the current context by performing semantic reasoning on the information provided by the available software tools and smart devices. Additionally, it learns semantic rules based on historical data in order to avoid future delays in transportation time.

1 Introduction

Healthcare is under high financial pressure and many hospitals struggle to balance budgets while maintaining quality. Hence, they are investigating ways to optimize care delivery processes. One area of interest is the organization of logistic services, which may account for more than 30 % of hospital costs [4]. Logistic operations in hospitals require more resources than reasonable, both in terms of staff responsible for scheduling and dispatching the activities and in terms of executing them. Moreover, not all tasks are performed by logistic personnel. It is estimated that nurses will spend on average 10 % of their time performing logistic tasks [2]. One area where there are huge opportunities in terms of gain in efficiency and cost reduction is the transport logistics of patients and equipment.

The advent of Internet of Things and intelligent decision support systems enable the design of advanced software systems to automatically assign the most suitable staff member to a transport based on all the available information about the context (e.g. location of staff and patients & how crowded the hospital is), the staff (e.g. competences), the patient (e.g. physical condition) and the specific transportation task (e.g. pick-up location & priority). In the AORTA project[1], we aim to build such as software system to enable flexible transport task scheduling and execution through the use of smart devices, semantic context models, dynamic scheduling algorithms and self-learning models.

---

[1] www.iminds.be/en/projects/2015/03/10/aorta.

H. Sack et al. (Eds.): ESWC 2016 Satellite Events, LNCS 9989, pp. 88–92, 2016.
DOI: 10.1007/978-3-319-47602-5_19

2   Architecture

The overall architecture of the designed AORTA system is visualized in Fig. 1.

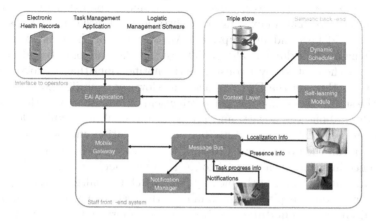

Fig. 1. The architecture of the designed AORTA system

Smart devices, wearables and sensors are introduced into the hospital environment to provide the software system with more contextual information. These devices generate dynamic data that indicate, e.g., the location of the staff or the status of a transport task. The **Message Bus** intercepts all these events and forwards them to interested services, which can subscribe to particular types of data through this bus. The bus also receives the notifications from the intelligent back-end software, e.g., the Dynamic Scheduler. The **Notification Manager** subscribes to these notifications and makes sure they are formatted in such a manner that they can be optimally shown on the device used by the staff member. The Message Bus then forwards this formatted notification to the mobile applications where they are visualized to the staff.

The **Mobile Gateway** forms the link between the events coming from the devices and the Enterprise Application Integration (EAI) Application. It is responsible for filtering the events generated by the mobile devices that are of interest to the back-end to schedule and execute the transportation tasks. It also makes sure that the decisions and notifications of the intelligent back-end are communicated to the front-end devices.

The **EAI Application** is responsible for integrating and linking the different existing software systems at the hospital. These existing software tools manage information such as the electronic health records of the patients, the transportation tasks being requested by the different hospital departments and information about the staff members and the logistics. The EAI Application is responsible for extracting the data relevant for scheduling and executing the transportation tasks from these software systems and providing it to the Context Layer.

The **Context Layer** consolidates and links the heterogeneous context data provided by the EAI Application by using an ontology. This semantic data is stored in a triple store. Reasoning is performed on this triple store to derive information that can be provided to the Dynamic Scheduler, e.g., the location & current tasks of the staff, the needed transportation mode, the average walking speed within particular areas, the general commotion within the hospital and known busy periods. Moreover, based on the available context data, the reasoning can also derive that an existing transportation task needs to be re-assigned, e.g., detected delays and interruptions. An overview of the task assignments by the Dynamic Scheduler is also maintained in the ontology. Based on these assignments, the context layer constructs the notifications that are forwarded to the end-users through the EAI Application. How these notifications should be sent to the staff is determined by performing semantic reasoning on the available context data about the devices and profile of the staff, e.g., which device is a staff member currently using or his/her current location and occupation. This information can then be exploited to optimally decide which information should be contained in the notification sent to the staff member, e.g., on a mobile device an optimal route to follow to the pick-up point can be visualized and dynamically updated, while on a display on the wall only the assignment itself is shown.

The **Dynamic Scheduler** receives the transportation requests from the Context Layer and constructs an optimal schedule such that all the requests can be handled in a timely manner with an optimal use of resources. To achieve this optimal rostering, the scheduler will request the dynamic context information from the Context Layer, e.g., the locations, availability, competences, work load & average walking speed of the staff, busy areas and possible causes of delay. It constantly maintains an overall optimal schedule and updates this schedules as new requests and status updates of on-going transports come in. When a staff member indicates that a transport has been finished, the Context Layer will communicate this to the Dynamic Scheduler, which will then assign a new task to this staff member based on this overall optimized schedule.

The **Self-learning Module** studies the scheduled assignments and accompanying executed transportation tasks to derive the reasons why some transports did not arrive on time. It extracts semantic rules, in the form of OWL axioms and adds them to the ontology. For example, the module could learn that certain transports during the visiting hour on Friday are often late and more time should be reserved for them. The incorporation of the knowledge modeled in the ontology, allows to learn more accurate rules. Furthermore, learning semantic rules allows to understand and validate the learned results. This module enables the generic system to adapt itself automatically to the workflows and particular context of the hospital in which it is deployed. Moreover, it gives insights to the staff on particular bottlenecks in their workflows or organization.

3   Implementation

The communication between the various components is achieved by exchanging JSON messages by using HTTP REST APIs. The Message Bus is implemented with RabbitMQ[2]. The EAI Application was implemented by extending the software of Xperthis[3].

An ontology was constructed by extending the Task Model Ontology[4], the Ambient-aware Continuous Care Ontology[5] and the Amigo Location Ontology[6]. Two triple stores are currently supported, i.e., Stardog[7] & RDFox[8]. Semantic reasoning is implemented by defining rules in these stores. The Context Layer communicates with the Dynamic Scheduler by translating the incoming JSON requests to SPARQL queries on the stores and forwarding the results.

The organization of transports in hospitals fits within the problem class of dynamic pick-up and delivery problems (DPDPs), which focus on goods requiring pick-up from and delivery to specific locations by a fleet of vehicles [1]. However, the constraints imposed by hospital policies and management (e.g., certain corridors that can only be used to transport goods), the relatively high request arrival rates and short transport times make it a challenging variant. In close collaboration with the hospital staff, a weighted objective function of key performance indicators, e.g., tardiness, travel time and avoiding unoccupied staff, was defined. The Dynamic Scheduler implements an optimization algorithms that iteratively tries to minimize the objective function, while upholding the defined constraints. As such, a trade-off can be made between the time available to perform the scheduling and the achieved algorithm performance.

Inductive Logic Programming was used to learn the semantic rules by employing DL-Learner [3]. As there are various reasons that cause transports to be delayed, semantic clustering is first performed on the set of late transports. Each set is then fed to DL-Learner separately and the results are merged afterward.

4 Conclusions and Next Steps

With this poster, we hope to get some initial feedback on the overall concept of using semantic technologies in order to optimize transportation task scheduling and execution within hospitals. We are currently collecting anonymized data sets from two hospitals during a period of three months, containing a realistic overview of the requested transports, their associated context and how they were scheduled and executed. This data will be used to evaluate the designed system.

Acknowledgment. This research was partly funded by the AORTA project, co-funded by the IWT, iMinds, Xperthis, Televic Healthcare, AZMM and ZNA.

---

[2] www.rabbitmq.com/.
[3] www.xperthis.be/nl/integratie.
[4] www.semanticdesktop.org/ontologies/2008/05/20/tmo/.
[5] users.intec.ugent.be/pieter.bonte/ontology/accio.htm.
[6] gforge.inria.fr/projects/amigo/.
[7] stardog.com.
[8] www.cs.ox.ac.uk/isg/tools/RDFox/.

References

1. Berbeglia, G., et al.: Dynamic pickup and delivery problems. Eur. J. Oper. Res. **202**(1), 8–15 (2010)
2. Landry, S., Philippe, R.: How logistics can service healthcare. Supply Chain Forum **5**(2), 24–30 (2004)
3. Lehmann, J.: DL-Learner: learning concepts in description logics. J. Mach. Learn. Res. **10**, 2639–2642 (2009)
4. Hastreiter, S., et al.: Benchmarking logistics services in German hospitals: a research status quo. In: ICSSSM, pp. 803–808 (2013)

Demo Papers

LinkedPipes ETL: Evolved Linked Data Preparation

Jakub Klímek[1,2,3(✉)], Petr Škoda[1,2], and Martin Nečaský[1]

[1] Faculty of Mathematics and Physics, Charles University in Prague,
Malostranské nám. 25, 118 00 Praha 1, Czech Republic
{klimek,helmich,necasky}@ksi.mff.cuni.cz
[2] University of Economics, Prague,
Nám. W. Churchilla 4, 130 67 Praha 3, Czech Republic
[3] Faculty of Information Technology, Czech Technical University in Prague,
Thákurova 9, 160 00 Praha 6, Czech Republic

Abstract. As Linked Data gains traction, the proper support for its publication and consumption is more important than ever. Even though there is a multitude of tools for preparation of Linked Data, they are still either quite limited, difficult to use or not compliant with recent W3C Recommendations. In this demonstration paper, we present LinkedPipes ETL, a lightweight, Linked Data preparation tool. It is focused mainly on smooth user experience including mobile devices, ease of integration based on full API coverage and universal usage thanks to its library of components. We build on our experience gained by development and use of UnifiedViews, our previous Linked Data ETL tool, and present four use cases in which our new tool excels in comparison.

Keywords: Linked Data · RDF · ETL · Transformation

1 Introduction

There is already a multitude of tools for Linked Data preparation. The goal of each such tool should be to ease as much as possible the shift from internal or open data in relational databases or Excel, CSV, XML and JSON files to Linked Data. What is often overlooked is the fact, that the target user of such tool is not only a computer scientist willing to use a system with unfriendly interface and sketchy documentation and a steep learning curve. One of the typical users is a government employee in charge of data in legacy systems who just learned about Linked Data and is enthusiastic to publish it. For him, the tool needs to be easy to install, well documented, user friendly, extensible and ideally ready to

This work was supported in part by the Czech Science Foundation (GAČR), grant number 16-09713, in part by the project SVV-2016-260331 and in part by the H2020 project no. 645833 (OpenBudgets.eu).

H. Sack et al. (Eds.): ESWC 2016 Satellite Events, LNCS 9989, pp. 95–100, 2016.
DOI: 10.1007/978-3-319-47602-5_20

be integrated into the systems of his government office. This is actually our real-world experience gained by supporting the Czech Social Security Administration in publishing their Linked Data[1].

In this paper, we present LinkedPipes ETL (LP-ETL), our Linked Data preparation tool, which we created based on our experience with development, usage and support of UnifiedViews [1], our previous Linked Data ETL tool, and the feedback we got for it. With LP-ETL we focus on APIs and configurations based on Linked Data principles, so that the tool is usable by and ready to be integrated with other software. We design the user interface based on Google Material Design[2], which includes responsiveness, native support for mobile devices and overall ease of use. We make the typical use cases as easy as possible and we do not bother the user with unnecessary interactions.

The rest of the paper is structured as follows. In Sect. 2 we survey related work. In Sect. 3 we introduce the basic ETL concepts and our tool. In Sect. 4 we present the qualities of LP-ETL on four use cases covering its typical usage. In Sect. 5 we indicate our future plans with the tool and we conclude.

2   Related Work

ETL (Extract Transform Load) is a designation for a data processing task where data is first gathered (extracted), then transformed and at the end, loaded to a local database or file system. LP-ETL is based on our experience gained from *UnifiedViews* (UV), which is also an RDF-enabled ETL tool. UV has various problems which we address in LP-ETL. From the programmer perspective, the biggest flaw is an incomplete API, which only supports execution and monitoring of existing pipelines, which makes it impossible to integrate into a system that would create the pipelines through the API. From the user perspective, the biggest flaw was the frontend application, which was unfriendly and prone to crashes, lags and loss of data caused e.g. by connection interruptions. Another ETL solution is the *LinDa Workbench* [2]. It allows the users to do simple tabular to RDF transformations from CSV/Excel files and relational DBs and run analyses and visualizations on top of the transformed RDF data. However, it does not support further data linking, cleansing, publishing or cataloging and it is not documented how it can be extended with additional plugins. A cloud-based open data publishing platform *DataGraft*[3] is similar to the LinDa Workbench. It also focuses on tabular data to RDF transformation and has a user-friendly interface. At the end of the transformation, the user gets a data page and a SPARQL endpoint for querying his data. However, it also does not support linking, cleansing nor cataloging of the data and it is not documented how additional plugins can be created. In addition, a cloud based solution is often not acceptable for government offices and their data because of trust issues and they prefer a software that they can install or integrate into their own, secured environment.

---

[1] https://data.cssz.cz.
[2] https://design.google.com.
[3] https://datagraft.net/.

3 LinkedPipes ETL (LP-ETL)

LinkedPipes ETL[4] runs using Java and Node.js and it can be compiled from source using Git and Maven. It contains a backend, which runs the data transformations and exposes the APIs, and a frontend, which is a web application and provides a pipeline editor and an execution monitor to the user. A pipeline is a repeatable data transformation process consisting of configurable components, each responsible for an atomic data transformation task, such as a SPARQL query of CSV to RDF transformation.

4 Demonstrated Use Cases

The individual use cases of LinkedPipes ETL will be demonstrated on the live demo instance[5]. The aim is to show how LP-ETL helps with the creation and debugging of typical ETL pipelines with Linked Data output. To demonstrate the complexity of typical ETL pipelines let us discuss 182 datasets published by the COMSODE project team[6]. For each of the datasets, a UV pipeline was developed. In total, 89 different components were used in these pipelines. These are responsible for gathering input data, its transformation to RDF including alignment with various ontologies and linkage to third-party datasets and for cleansing of datasets. Each pipeline contained an average number of 23 component instances. The execution time of the pipelines ranged from several minutes in the case of transformation of small XLS files with codelists to RDF representation in SKOS, to several days in the case of transformation of the Czech Registry of Addresses to its Linked Data representation including alignment with various ontologies and linkage to external data sources. This shows that real-world ETL pipelines are complex, cover many processing steps and may run for a very long time. Therefore, the support for smarter workflow for creating ETL pipelines and their advanced debugging demonstrated by our use-cases is imperative.

4.1 Smarter Workflow for Creating ETL Pipelines

In this use case, we will demonstrate the process of creating a simple ETL pipeline (see Fig. 1), from downloading the data sources, transforming them using SPARQL and other techniques and loading them to a triplestore. This will be demonstrated on the CSV open data of the Czech Trade Inspection Authority[7]. The development of the pipelines in LinkedPipes ETL is governed by an evolved workflow. In each step where a user adds another component to the pipeline, the components are recommended according to various criteria. The first criterion is *data unit compatibility*, which is checked and only components that consume what the previous component produces will be offered. Another criterion is the

---

[4] http://etl.linkedpipes.com.

[5] http://demo.etl.linkedpipes.com.

[6] http://data.comsode.eu/.

[7] http://www.coi.cz/userdata/files/dokumenty-ke-stazeni/open-data/kontroly.csv.

probability of the appearance of the component after the last component in the pipeline, to which it will be connected. This probability is based on existing pipelines in the instance. The last criterion is *full text search and tag based search*. Each component has a name and a set of tags assigned to it. The user can start typing and the offered components get filtered both according to their names and the tags. The usefulness of tags can be demonstrated on the Decompress component that unpacks ZIP and BZIP2 files. When a user searches for a component that could unzip his input files, he may type *unzip* or *zip*, which will not suggest the Decompress component simply based on its name. This is were tags help, the Decompress component can have tags for each of the supported formats (i.e. *zip*, *bzip2*) and for the actions (*unzip*, *unbzip*, *decompress*, *unpack*) which increases the chances to find the required component even without the knowledge of its name.

4.2 Advanced Debugging Support

In a typical case of development of an ETL pipeline, errors happen. They can be caused for example by external services failing or misconfiguration of components. In these situations, having a proper support for debugging is crucial. Imagine an ETL pipeline that runs for a week, or a complex pipeline consisting of a hundred components, both of which happen in the real world. At any time, the pipeline may fail while already having done some substantial work. In UnifiedViews, when this happened, there were only two options. The user could either rerun the pipeline or search the server for the RDF data.

In LP-ETL we developed a new *debug from* functionality, which allows the user to resume a failed pipeline from an arbitrary point of its previous execution after it is fixed. This saves some effort, especially for long running pipelines, since the user can fix the component that was misconfigured and continue execution from that point. In addition, LinkedPipes ETL implements the *debug to* functionality, which allows the user to run a part of a pipeline from its start to a certain point. A typical use case is when following the *progressive enrichment* Linked Data pattern[8]. First, a complete pipeline with simple transformation of input data is developed, including data gathering, loading and cataloging. Later, the data transformation is enhanced e.g. by improving a SPARQL query in one component, leaving the rest of the pipeline intact. This improved debugging functionality will be demonstrated on the same pipeline as the use case in Subsect. 4.1. There is a pipeline which failed on the *RDF to file* component due to misconfiguration. We fix it and restart the pipeline from this component. Next, the pipeline fails again (see Fig. 1) on the other *RDF to file*, again due to misconfiguration. The dark green bordered components are done in the first execution, the light green bordered components done in the current one and the red component is the one which failed.

---

[8] http://patterns.dataincubator.org/book/progressive-enrichment.html.

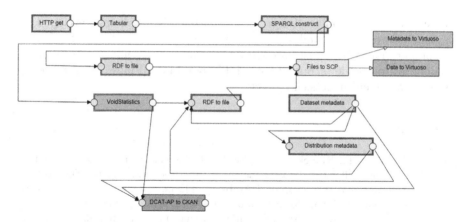

Fig. 1. Pipeline in LinkedPipes ETL fails again, fresh and old executions distinguished (Color figure online)

4.3 Full API Coverage and RDF Configuration

LinkedPipes ETL can be integrated in larger platforms and extended by alternate user interfaces. It was designed with this in mind from the beginning. All of the components have REST API interfaces and use RDF for configuration of components and pipelines. The same interface is used by LinkedPipes ETL itself and this means that it can be also used by any other software. This will be demonstrated by showing how a pipeline can be uploaded, executed and its state monitored from the command line using *curl*.

4.4 Sharing Pipelines and Fragments

LinkedPipes ETL eases sharing of pipelines and pipeline fragments. Since pipeline definitions are simple RDF files in JSON-LD serialization, they can be shared on the web or on GitHub for reuse. We will demonstrate this on the pipeline from Subsects. 4.1 to 4.2 where we will download it to demonstrate the debugging capabilities and the fully working version from the LinkedPipes ETL web documentation.

5 Future Work and Conclusions

We are currently developing *wizards* for typical ETL tasks, e.g. conversion of a CSV source to Linked Data similar to DataGraft. This task in its basic form requires only the location of the source CSV file and the ID of the target dataset, that can be used for the target folder structure, file names and IRIs. The pipeline consisting of the Tabular component, which implements the CSV on the Web W3C Recommendation[9], RDF and files conversion, SPARQL queries and loaders can then be generated based on this input and the settings of an LP-ETL

---

[9] https://www.w3.org/TR/csv2rdf/.

instance. An *import tool for UnifiedViews pipelines* is also being developed. In this paper we presented LinkedPipes ETL, a lightweight ETL tool for Linked Data preparation. We demonstrated its added value on four use cases, comparing it mainly to UnifiedViews, our previous ETL tool. LinkedPipes ETL is currently in use by the OpenBudgets.eu project[10].

References

1. Knap, T., Škoda, P., Klímek, J., Nečaký, M.: UnifiedViews: towards ETL tool for simple yet powerfull RDF data management. In: Proceedings of the Dateso 2015 Annual International Workshop on DAtabases, TExts, Specifications and Objects, Nepřívěc u Sobotky, Jičín, Czech Republic, 14 April 2015, pp. 111–120 (2015)
2. Thellmann, K., Orlandi, F., Auer, S.: LinDA - visualising and exploring linked data. In: Proceedings of the Posters and Demos Track of 10th International Conference on Semantic Systems - SEMANTiCS 2014, Leipzig, 9 (2014)

---

[10] http://openbudgets.eu.

DataGraft: Simplifying Open Data Publishing

Dumitru Roman[1]([✉]), Marin Dimitrov[2], Nikolay Nikolov[1], Antoine Putlier[1],
Dina Sukhobok[1], Brian Elvesæter[1], Arne Berre[1], Xianglin Ye[1],
Alex Simov[2], and Yavor Petkov[2]

[1] SINTEF, Forskningsveien 1a, 0373 Oslo, Norway
{dumitru.roman,nikolay.nikolov,antoine.putlier,dina.sukhobok,
brian.elvesaeter,arne.berre,xianglin.ye}@sintef.no
[2] Ontotext AD, Tsarigradsko Shosse 47A, 1784 Sofia, Bulgaria
{marin.dimitrov,alex.simov,yavor.petkov}@ontotext.com

Abstract. In this demonstrator we introduce DataGraft – a platform for Open Data management. DataGraft provides data transformation, publishing and hosting capabilities that aim to simplify the data publishing lifecycle for data workers (i.e., Open Data publishers, Linked Data developers, data scientists). This demonstrator highlights the key features of DataGraft by exemplifying a data transformation and publishing use case with property-related data.

1 Introduction

In the recent years, various government organisations around the world have committed to making data accessible under open licenses and, in most cases, in reusable formats. Unfortunately, due to the high cost and domain-specific expertise required for publishing and maintaining Open Data, this approach is still not adopted by the majority of government institutions.

DataGraft started with the goal to alleviate some of these obstacles by providing new tools for faster and lower-cost publication of Open Data. The lifecycle for the creation and provisioning of (Linked) Open Data typically involves raw data *cleaning, transformation,* and *preparation* (most often from tabular formats), *mapping* to standard linked data vocabularies and *generating a semantic RDF graph.* The resulting semantic graph is then *stored in a triple store*, where applications can easily access and query the data. Conceptually, this process is rather straightforward; however, such an integrated workflow is not commonly implemented. Instead, publishing and consuming (Linked) Open Data remains a tedious task due to a variety of reasons:

1. The *technical complexity* of preparing Open Data for publication is high – toolkits are poorly integrated and require expert knowledge, particularly for publishing of linked data;
2. There is *considerable cost* for publishing data and providing reliable access to it. In the absence of clear monetisation channels and cost recovery incentives, the relative investment costs can easily become excessively high for many organisations;

© Springer International Publishing AG 2016
H. Sack et al. (Eds.): ESWC 2016 Satellite Events, LNCS 9989, pp. 101–106, 2016.
DOI: 10.1007/978-3-319-47602-5_21

3. The *poorly maintained and fragmented supply* of Open Data reduces the reuse of data: datasets are often provided through disconnected outlets; sequential releases of the same dataset are often inconsistently formatted and structured.

What is needed is an integrated platform for effective and efficient data publication and reuse. At the very core, this means automating the Open Data publication process to a significant extent – in order to increase the speed and lower its cost.

2   The DataGraft Platform

DataGraft[1] was developed as a platform for data workers to manage their data in a simple, effective, and efficient way, supporting the data publication and access process discussed in Sect. 1. Its key features and benefits are:

- *Interactive design of data transformations*: transformations that provide instant feedback to publishers on how data changes speed-up the transformation process and improve the quality of the outcome;
- *Repeatable data transformations*: data transformation/publication processes often need to be repeatedly executed as new data arrives (e.g., publishing monthly budget reports). Executable and repeatable transformations are a key requirement for a lower-cost data publication process;
- *Shareable and reusable data transformations*: Capabilities to share, reuse and extend existing data transformations created by other developers further improves the speed and lowers the cost of the data publication;
- *Reliable data access*: provisioning data reliably is another key aspect for the third party data services and applications utilising Open Data.

The key enablers of DataGraft are shown in Fig. 1. *Grafterizer* is a front-end framework for data cleaning and transformation. It builds on Grafter[2], which is a framework of reusable components designed to support complex and reliable data transformations. *Grafter* provides a domain-specific language (DSL), which allows the specification of data transformations.

Another key enabler is the *semantic graph database-as-a-service* (DBaaS) triple store [1], which is used for accessing the Linked Data on the platform. With this DBaaS solution, publishers do not have to deal with typical administrative tasks such as installation, upgrades, provisioning and deployment, back-ups, etc. The utilization of cloud resources by the DBaaS depends on the utilisation of the DataGraft platform, and resources are elastically provisioned or released to match the current usage levels.

Finally, the *portal* integrates the previously discussed components together in a web-based interface designed to ensure a natural flow of the supported data processing and publication workflow. The entire process of publishing data is reduced to a simple wizard-like interface, where publishers can simply drop their data and enter some basic

---

[1] http://datagraft.net/.
[2] http://grafter.org/.

metadata. The portal also provides a module that helps visualize data from the semantic graph database (triple store). Currently, the platform provides a number of visualization widgets, including tables, line charts, bar charts, pie charts, scatter charts, bubble charts and maps (using the Google Maps widget).

The key capabilities of DataGraft are also accessible via RESTful services, so that they can be easily incorporated into 3rd party applications and data publishing workflows.

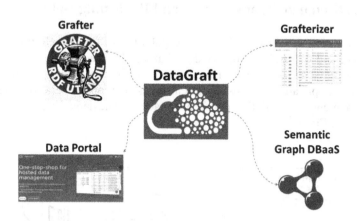

Fig. 1. DataGraft key enablers

Related Work. In the current state-of-the art there are several software tool ecosystem solutions that provide support for publication of Linked Data (data extraction, RDF-isation, storage, querying). Examples of such are the *Linked Data Stack*[3] and the *LinDA project*[4]. Whereas they may come functionally close to the features supported by Data-Graft, neither solution is provided "as-a-service", thus leaving the burden of deploying the services and managing the infrastructure around the toolsets.

The *COMSODE project*[5] provides a set of software tools and methodology for Open Data processing and publishing. The *COMSODE* tools are focused on specifying, monitoring and debugging data workflows on Linked Data. Data workflow specification addresses an aspect that is orthogonal to DataGraft's transformation approach, which is focused on lower level operations such as cleaning and RDF-isation of the actual data. Additionally, similar to *LinDA* and the *Linked Data Stack*, *COMSODE* tools are not provided as-a-service.

OpenRefine[6], with its RDF plugin implements an approach with similar capabilities to DataGraft when it comes to data cleaning, transformation, and RDF-isation. However,

---

[3] http://stack.linkeddata.org.
[4] http://linda-project.eu.
[5] http://www.comsode.eu.
[6] http://openrefine.org.

OpenRefine is unsuitable for use in a service offering context, such as the one DataGraft was built for. Additionally, the processing engine itself is not suitable for robust ETL processes, as it is inefficient with larger data volumes – it implements a multi-pass approach to individual operations, and is thus memory-intensive. Nevertheless, *Open-Refine* currently provides some powerful RDF mapping features.

3   Demo Scenario: Transforming and Publishing Data

The demonstration scenario highlights the capabilities of the DataGraft platform by transforming and publishing property data that will be used by the State of Estate (SoE) service – a registration and reporting portal for state-owned properties in Norway. The application will use different types of property-related datasets for overall data integration with the support of DataGraft. The demonstration scenario is summarised in Fig. 2.

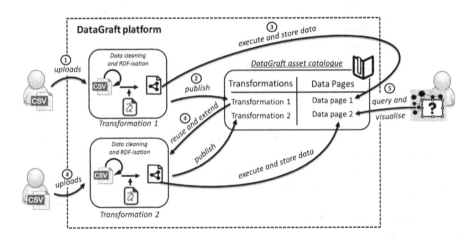

Fig. 2. Demo scenario

The usage scenario will demonstrate the following core aspects of DataGraft:

1. Interactive specification of tabular data transformations and mapping of tabular data to graph data (RDF);
2. Publication of data transformations on the DataGraft asset catalogue;
3. Execution and storage of transformed data on the semantic graph DBaaS;
4. Sharing, reusing and extending user-generated content;
5. Querying data from the live endpoint and visualising query results (Fig. 3).

```
Query
 1   select ?title ?lat ?lng where {
 2     ?s <http://www.statsbygg.no/specific-attributes/hasLat> ?lat .
 3     ?s <http://www.statsbygg.no/specific-attributes/hasLon> ?lng .
 4     ?s <http://www.statsbygg.no/specific-attributes/PMANSVARSSTEDKODE> ?regionCode .
 5     ?s <http://www.statsbygg.no/specific-attributes/PMANSVARSSTED> ?sbRegion .
 6     ?s <http://prodatamarket.eu/vocabs#hasNumber> ?sbIdentifier .
 7     ?s <http://prodatamarket.eu/vocabs#hasName> ?buildingName .
 8     ?s <http://prodatamarket.eu/vocabs#hasAddress> _:bn .
 9     _:bn <http://prodatamarket.eu/vocabs#hasZipCode> ?zipCode .
10     _:bn <http://prodatamarket.eu/vocabs#hasPostLocation> ?postLocation .
11     _:bn <http://prodatamarket.eu/vocabs#hasDistrict> ?municipality .
12     _:bn <http://prodatamarket.eu/vocabs#hasDistrict> "OSLO" .
13     _:bn <http://prodatamarket.eu/vocabs#hasAddress> ?address .
```

Statsbygg owned buildings in Oslo

Information on buildings owned by Statsbygg. Includes basic information (e.g. address, area) and accessibility information.

Fig. 3. Data query and visualization in DataGraft

A visitor of the demonstration will learn how to:

- Use DataGraft to simplify the tasks of data transformation and data publishing;
- Create data transformations with minimal effort through DataGraft's portal/GUI (for tabular data cleaning/transformation and mapping to graph data);
- Share and reuse data transformations already published in DataGraft;
- Run data transformations and host/publish the resulting data on DataGraft's reliable, cloud-based semantic graph database;
- Query data hosted/published on DataGraft;
- Work with the transformations and data catalogues in DataGraft;
- Use DataGraft for real life applications (publishing property data).

DataGraft is available via http://datagraft.net/ and further details can be found in [2].

Acknowledgements. This work was partly funded by the European Commission within the following research projects: *DaPaaS* (FP7 610988), *SmartOpenData* (FP7 603824), *InfraRisk* (FP7 603960), and *proDataMarket* (H2020 644497).

References

1. Dimitrov, M., Simov, A., Petkov, Y.: Low-cost open data as-a-service in the cloud. In: Proceedings of the 2nd Semantic Web Developers Workshop (SemDev 2015), Part of ESWC 2015, 31 May 2015, Portoroz, Slovenia (2015)
2. Roman, D., Nikolov, N., Putlier, A., Sukhobok, D., Elvesæter, B., Berre, A., Ye, X., Dimitrov, M., Simov, A., Zarev, M., Moynihan, R., Roberts, B., Berlocher, I., Kim, S., Lee, T., Smith, A., Heath, T.: DataGraft: One-Stop-Shop for Open Data Management. Semant. Web J. (SWJ) (2016, to appear). http://www.semantic-web-journal.net/system/files/swj1428.pdf

Link++: A Flexible and Customizable Tool for Connecting RDF Data Sources

Ali Masri[1,2]([✉]), Karine Zeitouni[1], Zoubida Kedad[1], and Gabriel Kepeklian[2]

[1] DAVID Laboratory, University of Versailles Saint-Quentin-en-Yvelines,
Versailles, France
{karine.zeitouni,zoubida.kedad}@uvsq.fr
[2] VEDECOM Institute, Versailles, France
{ali.masri,gabriel.kepeklian}@vedecom.fr

Abstract. Existing interlinking tools focus on finding similarity relationships between entities of distinct RDF datasets by generating owl:sameAs links. These approaches address the detection of equivalence relations between entities. However, in some contexts, more complex relations are required, and the links to be defined follow more sophisticated patterns. This paper introduces Link++, an approach that enables the discovery of complex links in a flexible manner. Link++ enables the users to generate rich links by specifying a link pattern as well as rules and functions to discover them. When visiting the demo, attendees will be introduced to all the aspects of the system explaining the required steps to define custom functions, connection patterns and linking rules until finally obtaining custom connections.

Keywords: Linked open data · Data interlinking · Semantic web

1 Introduction

Data interlinking aims to find equivalence relations between entities of different datasets in the Web of data. Roughly speaking, a user defines two data sources and a linking rule which specifies how different entities can be related together. The results are a set of triples in the form $x\ owl{:}sameAs\ y$ which are used to navigate from one data source to another in order to gain richer information.

Existing interlinking tools [1–3,5,6] support this task by providing a platform with a set of similarity functions that can be combined to form a designated rule. The interlinking engine processes the given datasets, applies an interlinking rule and returns the results.

In the same spirit as for data interlinking, our aim is to answer the need for more complex relations in some application domains. We want to enable the possibility of discovering complex relationships carrying more information in the form of properties that would be used for analysis purposes.

For instance, consider two transportation datasets representing a set of bus stops and a set of train stations respectively. Assume that our goal is to link bus

© Springer International Publishing AG 2016
H. Sack et al. (Eds.): ESWC 2016 Satellite Events, LNCS 9989, pp. 107–111, 2016.
DOI: 10.1007/978-3-319-47602-5_22

stops that can be reached from a train station, with the intention of developing a multi-modal trip planner that uses these links to compute trips. Using the existing tools, we are faced with two main limitations.

- The restriction to a predefined set of functions for composing linking rules. In our case, we lack of precise functions to calculate the closeness of a bus stop and a train station. Existing tools support geographical distances to calculate distance similarity which are not always reliable in real life, e.g., two geographically close stops that are separated by a river might be hard to reach from one another, therefore connecting them does not make sense.
- The representation of the generated output. Supporting complex relations requires more complex output patterns. Considering our example, interlinking based on the stations that can be reached from another gives the output *BusStop1 nextTo TrainStation132* which delivers no idea about the semantics of this relation. They are next to each others but how close are they? and what are the transportation modes that we can use? etc.

Our contribution is twofold, first we define connection patterns – a new way of customizing the output of a linking process. Second, we propose a platform that enables dynamic functions insertion and integration within the linking process.

In this work we introduce Link++, a tool that enables users to produce complex links by using their defined functions and connection patterns to support any type of relation between datasets.

The paper is structured as follows: In Sect. 2 we give an overall description of our system. We test the approach via a use case explained in Sect. 3 then we finally conclude in Sect. 4.

2   System Design and Implementation

To enable the discovery of complex links we introduce the notions of customized *Connection Patterns*. A connection pattern defines the content and the format of the link to be generated by defining the properties along with the associated properties. Figure 1 shows an overall view of the designed platform to generate customized connections.

TO begin, a configuration task is executed to define a connection pattern represented by a set of connection properties where each property is calculated by a function. Custom functions can be provided either by the user or by a common pre-coded functions library. In our implementation the functions are represented within a JAVA class and the library dependencies are regular .jar files. The configuration parameters are inputs of the connection generation component where the linking rule is defined. The linking rule is a file that describes the conditions to generate a connection pattern between entities. In short, if a rule between two entities is valid, a connection is instantiated and filled based on the given template (connection pattern). In the connection generation phase the system passes over the data sources and test the validity of a user-defined linking rule. If a rule is valid the connection pattern is evaluated and the resulting connection is

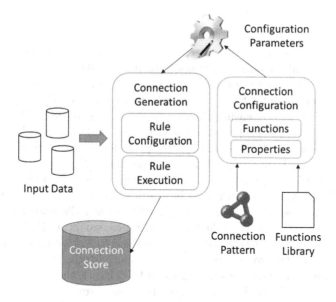

Fig. 1. An approach for flexible and customizable connection generation

stored in a connection store. In our implementation, both the configuration file (connection pattern) and the linking rule are XML files described by a DTD and the connection is generated in Turtle format[1]. Figure 2 shows an example of a connection pattern. An executable version of the system can be found online via the link: https://github.com/alimasri/link-plus-plus.git in addition to a video tutorial on: https://youtu.be/u2gr7Wa4eT4.

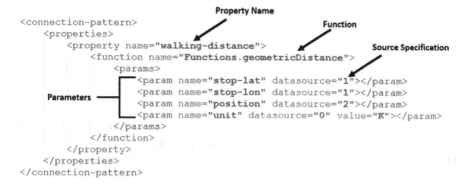

Fig. 2. An example of a connection pattern

---

[1] https://www.w3.org/TR/turtle/.

3 Scenario

The scenario we present in this demo describes how we can use our tool to connect transportation points of transfer to provide data for multimodal trip planning solutions. Transportation companies work in isolation and provide specific planning solutions for their data. We have used our solution to expand the transportation view by creating chains of connections in cities. We have considered two data sources representing SNCF train stations[2] and VELIB[3] bike sharing stations in the Paris area in France. Both data sources are pre-processed and translated into RDF using the DataLift platform [4].

The specified linking rule aims to find for each train station the nearby bike sharing stations, which is defined to be the walking distance since it is more relevant for our case, and it can be calculated via any distance API over the web. In this scenario we choose the Google distance matrix API[4], however users are free to choose any existing API or create their own since this is completely independent from the process. For the connection pattern, we are interested in getting the source and destination (generated by default), the calculated walking distance and the estimated walking time. Both files are provided as inputs to our system and we have successfully generated an output file representing the needed connections as shown in a snapshot of our system in Fig. 3.

Fig. 3. Screen shot of the implemented system Link++

The output file can be used as an input for trip planning algorithms to create multimodal trips and to facilitate and optimize passengers trips. The reliability of the results depends on the chosen functions and this requires a careful selection from the user.

---

[2] http://gtfs.s3.amazonaws.com/sncf_20131211_1451.zip.

[3] http://opendata.paris.fr/explore/dataset/stations-velib-disponibilites-en-temps-reel/.

[4] https://developers.google.com/maps/documentation/distance-matrix/.

4   Conclusion

Current interlinking tools focus on discovering equivalence relationships between datasets. In the same spirit as these tools, we have proposed an approach which discovers more complex relations to support specific application requirements such as linking transportation data sources.

In this paper we introduced Link++, a flexible interlinking tool which provides user defined semantic relationships between entities. The system enables the users to define their own functions and connection patterns thus providing customized and flexible linking capability.

Future work will target the dynamic nature of the created links. For instance, in some application domains (e.g. transportation) the created links can be dynamic and vary in real time which indeed affect their correctness and reliability. Moreover, we are interested in defining a functions library where users can share their functions and connection patterns allowing reusability and knowledge sharing.

Acknowledgments. We thank Mr. Bertrand Leroy for the fruitful discussion.

References

1. Jaffri, A., Glaser, H., Millard, I.C.: Managing URI synonymity to enable consistent reference on the semantic web. In: Proceedings of the Workshop on Identity, Reference, and the Web (IRSW) (2008)
2. Ngomo, A-C.N., Auer, S.: LIMES: a time-efficient approach for large-scale link discovery on the web of data. In: Proceedings of the Twenty-Second International Joint Conference on Artificial Intelligence, IJCAI 2011, vol. 3, pp. 2312–2317. AAAI Press (2011)
3. Raimond, Y., Sutton, C., Sandler, M.B.: Automatic interlinking of music datasets on the semantic web. In: Linked Data on the Web, vol. 369 (2008)
4. Scharffe, F., Atemezing, G., Troncy, R., Gandon, F., Villata, S., Bucher, B., Hamdi, F., Bihanic, L., Képéklian, G., Cotton, F., et al.: Enabling linked data publication with the datalift platform. In: Proceedings of AAAI Workshop on Semantic Cities (2012)
5. Scharffe, F., Liu, Y., Zhou, C., Rdf-ai: an architecture for rdf datasets matching, fusion and interlink. In: Proceedings of IJCAI 2009 Workshop on Identity, Reference, and Knowledge Representation (IR-KR) (2009)
6. Volz, J., Bizer, C., Gaedke, M., Kobilarov, G.: Silk-a link discovery framework for the web of data. In: Linked Data on the Web, vol. 538 (2009)

LinkedPipes Visualization: Simple Useful Linked Data Visualization Use Cases

Jakub Klímek[✉], Jiří Helmich, and Martin Nečaský

Faculty of Mathematics and Physics, Charles University in Prague,
Malostranské nám. 25, 118 00 Praha 1, Czech Republic
{klimek,helmich,necasky}@ksi.mff.cuni.cz

Abstract. There is a need for being able to effectively demonstrate the
benefits of publishing Linked Data. There are already many datasets and
they are no longer limited to research based data sources. Governments
and even companies start publishing Linked Data as well. However, a
tool, which would be able to immediately demonstrate the Linked Data
benefits to those, who still need convincing, was missing. In this paper,
we demonstrate LinkedPipes Visualization, a tool based on our previous
work, the Linked Data Visualization Model. Using this tool, we show
four simple use cases that immediately demonstrate the Linked Data
benefits. We demonstrate the value of providing dereferenceable IRIs and
using vocabularies standardized as W3C Recommendations on use cases
based on SKOS and the RDF Data Cube Vocabulary, providing data
visualizations on one click. LinkedPipes Visualization can be extended to
support other vocabularies through additional visualization components.

Keywords: Linked Data · RDF · Visualization · Discovery

1 Introduction

The research in the area of Linked Data shifts from Linked Data preparation
towards Linked Data consumption and visualization. There are many tools that
allow the users to somehow visualize RDF data, however, the process tends to
be more complicated than necessary. The potential of simplification is thanks
to the Linked Data principles and standardized vocabularies (W3C Recommendations) such as SKOS[1], the RDF Data Cube Vocabulary[2] and others. One of
the promises used to convince data publishers to use Linked Data is, that by
using these standardized vocabularies, users and Linked Data enabled applications around the world will easily understand the published data and will be able
to reuse it. An easy way of showing the publishers the added value of Linked
Data would be to show them their own data used in those various applications

---

This work was supported in part by the Czech Science Foundation (GAČR), grant
number 16-09713 and in part by the project SVV-2016-260331.

[1] https://www.w3.org/TR/skos-reference/.
[2] https://www.w3.org/TR/vocab-data-cube/.

H. Sack et al. (Eds.): ESWC 2016 Satellite Events, LNCS 9989, pp. 112–117, 2016.
DOI: 10.1007/978-3-319-47602-5_23

immediately after publishing it. This is because many data publishers still feel that publishing data is not enough, that a user also needs a visualization and that they are the ones best qualified to develop it. Unfortunately, this is the moment when the Linked Data arguments stop being convincing and shrink to a simple statement of the fact that once the data is published as Linked Data, others can link to it. The cause of this problem is the lack of simple and ready to use applications that could immediately show the benefits of publishing data as Linked Data.

In this paper, we introduce LinkedPipes Visualization (LP-VIZ), a Linked Data visualization tool built on top of our Linked Data Visualization Model (LDVM) [6], which we developed and experimented with earlier. This time, we focus on the demonstration of benefits of usage of standardized vocabularies and the ability to dereference an IRI, to access it to get additional data.

The rest of the paper is structured as follows. In Sect. 2 we survey the related work. In Sect. 3 we briefly mention our Linked Data Visualization Model (LDVM) as the theoretical basis and we introduce its implementation, Linked-Pipes Visualization, which we demonstrate on use cases in Sect. 4. In Sect. 5 we state our future efforts and we conclude.

2   Related Work

There are many RDF visualization approaches, but most of them are not easy to use and require the user to provide unnecessary information about the data that could be automatically discovered from it instead. For their overview, see [6] and [5]. With LinkedPipes Visualization, we focus on providing simple visualizations with minimal user effort and we try to get as much information from the data itself as possible. From this point of view, the most related is LodLive [3], which enables users to quickly browse their Linked Data provided that their IRIs are dereferencable. It also parses labels of properties and classes based on the RDF Schema (RDFS)[3] vocabulary. However, LodLive does not provide vocabulary specific visualizations and even though it provides a map visualization, it does not recognize the Schema.org GeoCoordinates class. Another tool in early prototype stage heading in the direction of analyzing data based on Linked Data vocabularies and principles is LDVizWiz [1], which detects categories of data in a SPARQL endpoint.

3   Linked Data Visualization Model and LinkedPipes Visualization

In our previous work, we defined the Linked Data Visualization Model (LDVM) [2]. It consists of components of four types. The *data sources* represent the source RDF data, the *analyzers* extract data from data sources, the *transformers* manipulate the data, e.g. transform it between formats according

---

[3] https://www.w3.org/TR/rdf-schema/.

to different vocabularies, and, finally, the *visualizers* provide the visualization itself. A key concept of LDVM is the *component compatibility checking*, which allowed us to automatically compose a visualization pipeline out of the components registered in a LDVM instance, based on their capabilities described by RDF data samples, SPARQL input descriptors and the source data.

Based on LDVM, we showed advanced use cases [6] where we were able to automatically compose visualization pipelines combining multiple data sources. In [4] we show how RDF data cubes can be visualized using LDVM and our Data cube visualizer.

LinkedPipes Visualization[4] is our current LDVM implementation. Besides being able to support the advanced use cases from [6], it now supports much simpler and straightforward use cases, on which we can more directly demonstrate the benefits of Linked Data and which we cover in this paper.

4   Demonstrated Use Cases

In this section, we demonstrate the usability of LinkedPipes Visualization on four use cases. We focus mainly on simple demonstrations of Linked Data benefits regarding use of vocabularies and providing dereferencable IRIs.

4.1   Validation

An important feature to have in a data consumption tool is *validation*. This means that instead of seeing an error or no result when trying to visualize data, the user has an option to have his data validated and see what is wrong. So far, we have two validators to choose from, SKOS for taxonomies and Data Cube for statistical data cubes. There already are validators for SKOS[5] and RDF Data Cubes[6]. However, we focus on validation of specific common errors that break our visualizations.

Because we assume that the data is wrong, first we need to know, what the data is expected to be. Therefore, in this use case, we start with selecting the validator we want to use. For the demonstration, we will use the simpler SKOS validator, which checks mainly for missing SKOS concept schemes, links from concepts to the concept schemes via the `skos:inScheme` property and missing target concepts of `skos:broader` and `skos:narrower` links and their transitive variants.

First, the user inputs a broken SKOS concept scheme[7]. Second, the validator detects, that the object of one of the `skos:broader`, one of the `skos:narrower` and one of the `skos:inScheme` properties is missing in the data. This could be caused by omitting an entity or a simple misspell during manual creation of the concept scheme.

4.2 One Click Visualization

One of the promised advantages of using Linked Data is that when the data is modeled according to standardized vocabularies, it can be easily used in tools that support such vocabularies. When applied to visualizations, the user expects that when he models his data correctly, he gets a visualization specific for used vocabularies, easily. From the workflow perspective, this means that the user has to only provide the right data to get a vocabulary-specific visualization, nothing more.

In this use case we demonstrate, how LinkedPipes Visualization can achieve just that, a *one click user data visualization*. This is given that the data is a SKOS concept scheme, a Data Cube or a set of points on a map according to Schema.org[8], for which we currently have visualization components installed in our demo instance[9]. More vocabularies (or their combinations) can be supported by adding additional visualization components with corresponding input descriptors.

First, the user inputs a URL of an RDF file with a SKOS Concept scheme. This can be e.g. some of the newly published Named Authority Lists (NALs)[10] in the European Metadata Registry, such as the EU programmes[11] and clicks on *Visualize*. Second, the concept scheme is detected and visualized using our library of hierarchy visualizers from the D3.js[12] library. For the EU programmes, the Tree visualization is the most appropriate, even though it is really a code list, not a hierarchy.

4.3 Enhancing Visualization with Dereferencing of IRIs

Another one of the key principles of Linked Data states that the data publisher should link to other data to provide context. This is almost mandatory in the case of publishing data cubes, as they consist of observations identified by dimension values and those values typically come from a code list represented as a SKOS concept scheme. This concept scheme is usually shared by dimensions of multiple data cubes, and possibly even other types of datasets, and therefore, it is almost necessarily a different dataset. This is especially true for highly reused concepts such as the ones for sex[13], time periods[14] or geographical locations. Therefore, when the user wants to visualize a data cube, which is using such dimensions, all the values for those dimensions will almost certainly not be available in the same file or even the same SPARQL endpoint as the actual data cube data. Instead, they will be simple links to those other datasets. For conventional data cube

---

[8] http://schema.org.

[9] http://demo.visualization.linkedpipes.com.

[10] http://publications.europa.eu/mdr/authority/.

[11] http://publications.europa.eu/mdr/resource/authority/eu-programme/skos/
eu-programme-skos.rdf.

[12] https://d3js.org/.

[13] http://purl.org/linked-data/sdmx/2009/code#sex.

[14] https://datahub.io/dataset/data-gov-uk-time-intervals.

visualization methods, this usually means that the dimension values will not be displayed correctly, because the human readable labels are not available in the same data source as the observations, and only their IRIs will be visible.

In this use case we demonstrate, how our Data Cube visualizer can handle this situation naturally by *dereferencing the IRIs* of dimension values, for which there are no labels in the original data source and which can be dereferenced. First, the user inputs the data cube[15] and its Data Structure Definition (DSD)[16]. LP-VIZ detects that there in fact is a data cube in the data and displays it. Next, for dimension values, labels are displayed even though they are not in the source data, because they are obtained through dereferencing of their IRIs.

4.4   Embedding Visualizations

In this short use case, we will create a visualization in LinkedPipes Visualization and show how it can be easily embedded into another website as an HTML *iframe*. This can help data publishers to easily develop simple visualizations with little extra cost, again thanks to the usage of Linked Data. We will start with the visualization from Subsect. 4.3 and pre–select some dimension values. Next, we will use the *embed link* to add this visualization as a resource to a CKAN data catalog.

5   Future Work and Conclusions

Currently, we are working on extending our library of visualizers with ones allowing us to demonstrate the Linked Data benefits on a wider range of well–known vocabularies.

In this paper, we demonstrated LinkedPipes Visualization (LP-VIZ), a tool implementing our Linked Data Visualization Model (LDVM). On four use cases we showed how the Linked Data benefits promised to data publishers can be immediately exploited by validating the data, producing vocabulary specific visualizations on one click, enhancing them with dereferenceable IRIs and embedding them in websites.

References

1. Atemezing, G.A., Troncy, R.: Towards a Linked-Data based visualization wizard. In: Hartig, O., Hogan, A., Sequeda, J. (eds.) Proceedings of the 5th International Workshop on Consuming Linked Data (COLD 2014) Co-located with the 13th International Semantic Web Conference (ISWC 2014), Riva del Garda, Italy, 20 October 2014, vol. 1264. CEUR Workshop Proceedings, CEUR-WS.org (2014)
2. Brunetti, J.M., Auer, S., García, R., Klímek, J., Nečaský, M.: Formal Linked Data visualization model. In: Proceedings of the 15th International Conference on Information Integration and Web-based Applications & Services (IIWAS 2013), pp. 309–318 (2013)

---

[15] http://visualization.linkedpipes.com/example/datacube.ttl.
[16] http://visualization.linkedpipes.com/example/dsd.ttl.

3. Camarda, D.V., Mazzini, S., Antonuccio, A.: LodLive, exploring the web of data. In: Presutti, V., Pinto, H.S. (eds.) 8th International Conference on Semantic Systems, I-SEMANTICS 2012, Graz, Austria, 5–7 September 2012, pp. 197–200. ACM (2012)
4. Helmich, J., Klímek, J., Nečaský, M.: Visualizing RDF data cubes using the Linked Data visualization model. In: Presutti, V., Blomqvist, E., Troncy, R., Sack, H., Papadakis, I., Tordai, A. (eds.) ESWC 2014. LNCS, vol. 8798, pp. 368–373. Springer, Heidelberg (2014). doi:10.1007/978-3-319-11955-7_50
5. Klímek, J., Helmich, J., Nečaský, M.: Application of the Linked Data visualization model on real world data from the Czech LOD cloud. In: Bizer, C., Heath, T., Auer, S., Berners-Lee, T. (eds.) Proceedings of the Workshop on Linked Data on the Web co-located with the 23rd International World Wide Web Conference (WWW 2014), Seoul, Korea, 8 April 2014, vol. 1184. CEUR Workshop Proceedings. CEUR-WS.org (2014)
6. Klímek, J., Helmich, J., Nečaský, M.: Use cases for Linked Data visualization model. In: Bizer, C., Auer, S., Berners-Lee, T., Heath, T. (eds.) Proceedings of the Workshop on Linked Data on the Web, LDOW 2015, Co-located with the 24th International World Wide Web Conference (WWW 2015), Florence, Italy, 19 May 2015, vol. 1409. CEUR Workshop Proceedings, CEUR-WS.org (2015)

ADA – Automated Data Architecture

Creating User Journeys Through Content Using Linked Data

Jo Kent[✉]

BBC Radio & Music Multiplatform, London, UK
jo.kent@bbc.co.uk

Abstract. The BBC has a wealth of permanently available programmes across a wide range of subjects with very low usage. We wanted to create a route into these programmes which balanced the need for curated, high quality journeys between programmes and the limited resource available for that curation effort. I will demonstrate ADA, a system created to create consistent, meaningful high-quality links between programmes with limited user input.

1 Introduction

There is a need for content providers to create consistent, high quality onward journeys to available content. Across the industry solutions used range from the heavily internally manually curated approach of Netflix[1], to the user-driven algorithmically determined approach of Spotify[2].

In this demo I will demonstrate ADA (Automated Data Architecture) which uses minimal manual curation and linked data to provide high quality serendipitous onward journeys.

2 Understanding the Problem Space

2.1 Assigning Metadata

The BBC has at least 34,000 permanently available speech radio programmes to which the traffic is low. There is no easy path into all available content. Some programmes have archive navigation, but these are isolated, specialised and heavily curated, and with decreasing team sizes, even these are not sustainable in the long term.

Our news and sport teams have long used linked data to dynamically populate article pages, which are set up using a strict, pre-existing ontology[3]. This constrains the browser

---

[1] Netflix manual tagging process: http://www.techradar.com/news/television/netflix-wants-to-pay-you-to-watch-shows-here-s-why-1256098.
[2] Spotify's algorithm explained: http://qz.com/571007/the-magic-that-makes-spotifys-disc over-weekly-playlists-so-damn-good/.
[3] Sport ontology: http://www.bbc.co.uk/ontologies/sport.

H. Sack et al. (Eds.): ESWC 2016 Satellite Events, LNCS 9989, pp. 118–122, 2016.
DOI: 10.1007/978-3-319-47602-5_24

to a rigid structure which may not match their world view. In any case, such an ontology does not exist across all programmes, the subject matter is too diverse.

Crowdsourcing metadata creation has been used by our R&D department on the World Service archive, this achieved at best 30.3 % precision (36.7 % recall)[4] largely because every person may have a different perception of the subject matter or the meaning of a term [1]. Also it was inconsistent: people only added tags to programmes which interested them, so some programmes have lots of tags and some none at all.

A team of researchers would provide better quality, more consistent data [2], but at a permanently high cost, in staff time. This is unfeasible with fewer staff available.

A fully automated system would not be able to deliver consistent quality standards for an audience facing offer: the automated interpretation of a homograph can give an erroneous or even offensive connection (e.g. Georgia the country as opposed to the American state). Any loss of data quality can cause a loss of trust in our content [3].

A middle ground needs to be found between these levels of automation, without compromising quality. We cannot expect producers to classify consistently, but they do know the precise subject of their programme. Therefore we need a system which only requires them to enter that subject, (e.g. a programme on autism can just be tagged with autism) without the need to classify the concept. Without this classification therefore, we need a system that will automatically supply the links.

2.2 Classification Systems

The ideal classification system would need to be recognisable and therefore trusted by our audiences and also be flexible and maintainable over time, as perceptions change [4].

Maintenance of our own ontology requires a significant staff time overhead, but the use of eternally maintained ontologies means we cannot control when changes happen, and still have to adapt when they do. Given the diversity of subject matter (subjects include the A470 (a road in Wales), Munch's "The Scream", virtue, Canada geese, existentialism and the Battle of Bosworth Field) the task of creating an ontology to cover and group every possible subject would be unfeasibly large. To make it manageable, we would have to make arbitrary choices about classification to make the multidimensional world fit in a two dimensional hierarchical structure. This is increasingly viewed as an outmoded and dictatorial organisational method, compared to open ontologies and collaborative folksonomies [5], and any arbitrary divisions of data are no longer semantic distinctions but simply an organisational tool.

3   Unlocking the Power of Linked Data to Provide Automated Onward Journeys

The most promising linked open data sources were the Wikipedia/Dbpedia and Wikidata datasets. We found that there was no consistent hierarchical navigation or grouping

---

[4] http://www.bbc.co.uk/rd/blog/2014/08/data-generated-by-the-world-service-archive-experiment-draft.

information applied to the datasets. Wikidata has classification such as Library of Congress and Dewey Decimal mappings, but these are inconsistently applied[5]. It also offers classes and subclasses[6], again inconsistently applied, which often simply cut off without reaching the top class of 'Thing'. Dbpedia has classes[7], but again these are only applied to a fraction of instances[8]. Dbpedia has categories for every subject, which offer a skos:broader[9] journey to other categories, however, due to the way it is structured, often the category that was two hops broader was the initial category we started with, which meant that we had simply introduced more categories without any additional clarity (Fig. 1).[10]

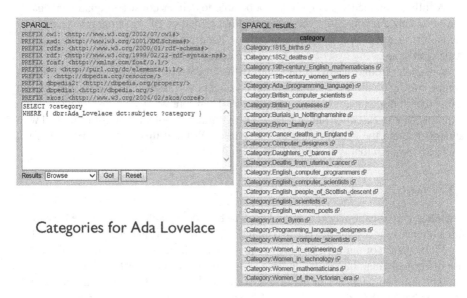

Categories for Ada Lovelace

Fig. 1. Categories in dbpedia

Having found no usable hierarchical or grouping information we looked again at categories in Wikipedia/dbpedia. These have been added by Wikipedia editors, each adding the facts they felt were most salient. Anyone can remove them if they disagree, so they are effectively crowdsourced and peer reviewed. This means they have the recognisable relevance that people will respond to, while being of a high quality.

Asking producers to simply identify the subject for their programme means we can assure the quality of the initial reference, and automatically link to all of the categories (an average of seven per subject), which are matched to others to create user journeys

---

[5] Fewer than 3,500 Wikidata entities have Dewey Classifications attached.

[6] https://www.wikidata.org/wiki/Property:P279.

[7] http://mappings.dbpedia.org/server/ontology/classes/.

[8] Only 182 of 720 (25 %) sampled had types applied.

[9] http://www.w3.org/2009/08/skos-reference/skos.html#broader.

[10] Using the 3.9 dataset, this issue seems to have been improved in the 1/4/2016 release.

that we could not create using manual curation without hours of research. At best a curatorial team might have added tags to Ada Lovelace like 'computer scientist' or 'mathematician' but here we have links to such diverse groups as 'programming language designers' and 'British countesses'. These small, precise categories give a serendipitous feel to the journey and allow the users to learn more about the subjects even as they are navigating between programmes.

By discarding the notion of a hierarchy and instead presenting a graph, the journeys are not constrained in to a single worldview. We know from Lobel and Sadler's work on homophily (i.e., love of the same) that "In a relatively sparse network, diverse preferences present a clear barrier to information transmission. In contrast, in a dense network, preference diversity is beneficial" [6]. So we can see that people respond better to a wide range of links that may not match their world view than to a narrow one. Therefore providing a broad range of linking categories (which have been selected by their peers as relevant) to each subject will present links the user will instinctively have a positive response to (Fig. 2).

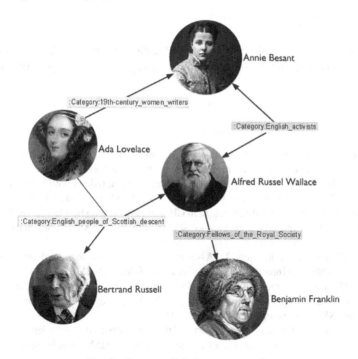

Fig. 2. Category links between people

4   Evaluation

Beginning with our initial sample of 610 programmes, we extracted over 1000 categories, of which 554 were linked to more than one programme, some of them to as many as 12. We only use categories which link to two programmes or more because only those offer an onward journey. We keep the non-matching categories in the ADA triple store so that they can be used as soon as a new programme with a matching category is added. Some maintenance categories such as "World Digital Library related"[11] or "Articles with inconsistent citation formats"[12] were added to a blacklist as these are not useful user journeys. We were then able to examine the quality of the journeys offered. A programme on Roman Satire yielded links to 14 other programmes through five different categories; a greater and more detailed level of linking than in our bespoke archives.

We launched a beta[13] to gauge the audience reaction to the new navigation, and the response has been overwhelmingly positive with a rating of 4.15 (out of 5) stars on BBC Taster[14] and 164 (out of 250) positive verbatim responses through the demo feedback link. Once we rolled out ADA to all of our programmes, we plan to roll this out to other departments in the BBC to bring in news articles and educational literature and then to partner agencies (particularly cultural heritage organisations and learning institutions) to create learning journeys across all of our content by subject, rather than content type.

5   The Demo

In the demo I'll be showing the beta and the API calls that power it. Visitors will be able to see and experiment with the semantic onward user journeys made possible by the use of linked data.

References

1. Feyisetan, O., Luczak-Roesch, M., Simperl, E., Tinati, R., Shadbolt, N.: Towards hybrid NER: a study of content and crowdsourcing-related performance factors. In: Gandon, F., Sabou, M., Sack, H., d'Amato, C., Cudré-Mauroux, P., Zimmermann, A. (eds.) ESWC 2015. LNCS, vol. 9088, pp. 525–540. Springer, Heidelberg (2015). doi:10.1007/978-3-319-18818-8_32
2. Lykke, M., Lund, H., Skov, M.: Metadata in CHAOS: how researchers tag and annotate radio broadcasts. In: Knowledge Organization – making a difference (ISKO UK biennial conference) (2015)
3. Johnson, F., Sbaffi, L., Rowley, J.: Assessing trustworthiness in digital information. In: International Data and Information Management Conference (IDIMC) (2014)
4. Cupar, D.: Diachronic semantics: changes of meaning of words over time and the consequences for keeping classification systems up to date. In: Knowledge Organization – making a difference (ISKO UK biennial conference) (2015)
5. Busch, J.: Web-based content organization and the transformation of traditional classification systems. In: Knowledge Organization – making a difference (ISKO UK biennial conference) (2015)
6. Lobel, I., Sadler, E.: Preferences, homophily, and social learning. In: Operations Research (2014)

---

[11] https://en.wikipedia.org/wiki/Category:World_Digital_Library_related.

[12] https://en.wikipedia.org/wiki/Category:Articles_with_inconsistent_citation_formats.

[13] http://www.bbc.co.uk/inourtimeprototype.

[14] Feedback on BBC Taster overall rating 4.15 stars: http://www.bbc.co.uk/taster/projects/in-our-time.

Graph-Based Editing of Linked Data Mappings Using the RMLEditor

Pieter Heyvaert[(✉)], Anastasia Dimou, Ruben Verborgh, Erik Mannens, and Rik Van de Walle

Data Science Lab, Ghent University - IMinds, Ghent, Belgium
`pheyvaer.heyvaert@ugent.be`

Abstract. Linked Data is in many cases generated from (semi-) structured data. This generation is supported by several tools, a number of which use a mapping language to facilitate the Linked Data generation. However, knowledge of this language and other used technologies is required to use the tools, limiting their adoption by non-Semantic Web experts. We demonstrate the RMLEditor: a graphical user interface that utilizes graphs to easily visualize the mappings that deliver the RDF representation of the original data. The required amount of knowledge of the underlying mapping language and the used technologies is kept to a minimum. The RMLEditor lowers the barriers to create Linked Data by aiming to also facilitate the editing of mappings by non-experts.

1 Introduction

Linked Data [1] is one of the most important aspects that drives the adoption of Semantic Web technologies, as it interlinks data whose semantically enriched representation is available. Most of the current Linked Data stems originally from data in (semi-)structured formats. Mappings languages, such as R2RML [2] and RML [3], specify in a declarative way how Linked Data is generated from such (semi-)structured data. This data is possessed by data specialists, who are not Semantic Web experts or developers. Therefore, data specialists should be able to specify the mappings, modify and extend them at any time, while limiting the awareness of the underlying mapping languages and technologies.

Nevertheless, dedicated environments that support users to intuitively edit mappings were not thoroughly investigated yet, and each have their limitations. First, *step-by-step* wizards, e.g., the fluidOps editor[1], prevailed as an easy-to-reach solution. However, such applications restrict data publishers' editing options, hamper altering parameters in previous steps, and detach mapping definitions from the overall knowledge modeling, since related information is separated in different steps. Second, a number of tools, such as the fluidsOps editor,

---

The described research activities were funded by Ghent University, iMinds, the Institute for the Promotion of Innovation by Science and Technology in Flanders (IWT), the Fund for Scientific Research Flanders (FWO Flanders), and the European Union.

[1] http://www.fluidops.com.

H. Sack et al. (Eds.): ESWC 2016 Satellite Events, LNCS 9989, pp. 123–127, 2016.
DOI: 10.1007/978-3-319-47602-5_25

sheet2RDF[2] and -ontoPro-[3], limit the user to a specific data format that can be used to generate Linked Data. This limitation makes integration of data distributed across data sources in different data formats impossible. Though, other tools, e.g., Karma[4], DataOps[5] and RDF123[6] do support heterogeneous data sources. Nevertheless, third, all aforementioned tools require users to understand the mapping language's syntax, as it is widely used in their graphical user interface (GUI). Therefore, data specialist are only able to create their own mappings with the help of Semantic Web experts or by acquiring the required knowledge of the language themselves.

We propose a demo of the RMLEditor[7], an editing environment for specifying mappings of (semi-)structured data to their RDF representation based on graph visualizations, that does not suffer from the aforementioned limitations. The demo accompanies a paper in the in-use track of the ESWC2016 conference [4]. Participants are able to perform their own mappings, on multiple data sources in different data formats, using a live instance of the RMLEditor, as shown via the following screencast that is available at https://www.youtube.com/watch?v=J7OtSYnZD9I.

2   RMLEditor

A mapping editor is an application that allows users to describe how Linked Data is generated based on (semi-)structured data, including tabular data (e.g., CSV) and hierarchical data (e.g., XML and JSON). It should have a GUI that is understandable and usable by non-Semantic Web experts. To achieve this, a list of seven desired features were defined in previous work [5]. The RMLEditor covers all of them:

1. The RMLEditor is independent of the underlying mapping language, so users are able to create mappings with a limited amount of knowledge of the language's syntax.
2. It allows users to execute the mappings outside of the editor, because the editor is mainly meant to edit the mappings. Users can export the mappings and execute them using different tools.
3. It enables users to map multiple data sources at the same time, as it might occur that data that is required to describe a knowledge domain is spread across multiple sources.
4. It supports data sources in different data formats, e.g., CSV, and XML, as the Linked Data is independent of the data's original format.
5. As multiple schemas (ontologies and vocabularies) can be used to create a mapping, it supports the use of both existing and customs schemas.

---

[2] http://art.uniroma2.it/sheet2rdf/.
[3] http://ontop.inf.unibz.it/components/sample-page/.
[4] http://usc-isi-i2.github.io/karma/.
[5] http://www.fluidops.com.
[6] http://ebiquity.umbc.edu/project/html/id/82/RDF123.
[7] http://rml.io/RMLeditor; licensing options available on request.

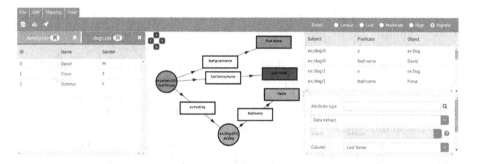

Fig. 1. The RMLEditor with the *Input Panel* on the left, the *Modeling Panel* in the center and the *Results Panel* on the right

6. It allows multiple alternative modeling approaches [6], as certain use cases might benefit from using a specific approach.
7. By supporting non-linear workflows, users are able to keep an overview of the mapping model and its relationships.

The RMLEditor is an application available in the browser. The mappings are visualized in the GUI by using graphs. Users can create a new or upload an existing mapping. Creating, updating and extending the mappings is done by performing the corresponding graph manipulations. The RMLEditor triggers the mapping processor which executes the mappings, exported by the RMLEditor, and generates RDF statements. We chose the RML as the RMLEditor's underlying mapping language. However, any other mapping language could be used instead, if it allows to implement the features, which is the case for RML.

3 Graphical User Interface

One of the most important aspects of a mapping editor is its GUI, as indicated by the aforementioned features. In the RMLEditor, we offer three panels, that implement these features, to the users: *Input Panel*, *Modeling Panel* and *Results Panel* (see Fig. 1).

In the following, we elaborate on how the features are implemented in the RMLEditor. The first feature is implemented via the *ModelingPanel*, as this panel offers a generic representation of the mappings, independent of the underlying mapping language, by using a graph representation. Mappings are created and edited by manipulating the nodes and edges. The second feature is also implemented via this panel, as it allows to export both the graph representation and the RML statements. This allows to execute the mappings outside the RMLEditor. The third feature is implemented via the *Input Panel*, because in this panel users are able to view the different data sources. Each data source can be uniquely identified by its color that is automatically and arbitrarily assigned by the RMLEditor. Additionally, the nodes' and edges' colors depend on the data source that is used in that specific mapping. The fourth feature is implemented by choosing an adequate visualization for each data source based on the data format. The fifth feature is facilitated by allowing users, through manipulations on the graph, to add semantic annotations using multiple schemas.

In previous work [6] we described the different mapping generation approaches. The *data-driven* approach uses the input data sources as the basis to construct the mappings. The classes, properties and datatypes of the schemas are then assigned to the mappings. When users start with the schemas to generate the mappings, the *schema-driven* approach is followed. Next, data fractions from the data sources can be associated to the mappings. The sixth feature is facilitated via the *Input Panel*, *Modeling Panel* and the *Result Panel*, as their functionality and interaction supports these approaches, as explained in more detail in Sect. 4. The latter panel shows the resulting RDF dataset when the mappings defined in the *Modeling Panel* are executed on the data in the *Input Panel*. For each RDF triple of the dataset it shows the subject, predicate and object. The last feature is facilitated by allowing the users to decide which panel they use and at what moment in the mapping process they use it. In linear workflows, this is not the case. Additionally, the panels are aligned next to each other, and when users want to focus on a specific panel, they are able to hide the other panels.

4   Editing Mappings

For Linked Data, an entity is one of the most important aspects. An *entity is something in the world, identified by a unique name (URI)*. Anything can be an entity, including physical things, documents and abstract concepts. A URI for a person named 'John Doe' could be http://www.example.com/john_doe. We assume for this example that every person's name is unique. Therefore, using the name in the URI results in a unique URI for each person. Additionally, the use of this functional attribute ensures that always the same URI is generated between different mapping executions. The classes defined in schemas are used to define the type of an entity, e.g. `foaf:Person` (where `foaf:` is expanded to the full URI of the vocabulary http://xmlns.com/foaf/0.1/). When no URI is provided for an entity, we call it a blank node. Information about an entity is represented by both *attributes* and *relationships*. Examples of attributes are 'John Doe' (name of a person) and 'BE0596.342.234' (VAT number of a company). Relationships connect entities and attributes. For example, the relationship with property `foaf:name` states that the string 'John Doe' (attribute) is the name of http://www.example.com/john_doe (entity).

Specifying entities, attributes and relationships is what composes the creation of mappings. This can be done using the two aforementioned approaches. When following the data-driven approach, users first load all the data in the RMLEditor. Next, they create entities and attributes based on the different data fractions, by interacting with the corresponding elements in the *Input Panel*. If blank nodes are required, they are created via the *Modeling Panel*. Users define the classes of the entities and blank nodes, and the datatypes of the attributes. This is done by clicking on a node, which brings up a panel where the node's details can be edited. Additionally, for the entities they define how the URIs are generated. Users also need to define relationships to connect entities with their attributes or other entities, together with selecting the correct property from the schemas. This is done in the same way as with nodes. The Linked Open Vocabularies[8] can be consulted via the GUI to get suggestions on which classes, properties and datatypes to use. Finally, the mapping is executed, and the resulting RDF triples are visible in the *Results Panel*. This last step can also be performed earlier on when not all data fractions are mapped. This allows users to inspect the RDF triples before the mapping is complete, which makes it possible to fix errors earlier on in the process.

When following the schema-driven approach, users mostly interact with the *Modeling Panel*. Through this panel, they can create entities, blank nodes and attributes, based on one or multiple schemas. Relationships are added, based on the properties defined in the schemas. Up until now no references to data sources are made. Therefore, the relevant data sources are loaded and they appear in the *Input Panel*. Every mapping definition is updated to incorporate the use of the data sources where applicable. Finally, again the mapping is executed, and the resulting RDF triples are available in the *Results Panel*.

Regardless of the approach users apply, the RMLEditor assists (non-)Semantic Web experts in editing their Linked Data mappings, while limiting the amount of knowledge needed of RML or the other used technologies. Additionally, facilitating the editing of mappings further lowers the barriers of obtaining Linked Data and thus stimulates the adoption of Semantic Web technologies.

References

1. Bizer, C., Heath, T., Berners-Lee, T.: Linked data-the story so far. In: Emerging Concepts, Semantic Services, Interoperability and Web Applications (2009)
2. Das, S., Sundara, S., Cyganiak, R.: R2RML: RDB to RDF mapping language. In: Working group recommendation, W3C, September 2012. http://www.w3.org/TR/r2rml/
3. Dimou, A., Sande, M.V., Colpaert, P., Verborgh, R., Mannens, E., Van de Walle, R.: RML: a generic language for integrated RDF mappings of heterogeneous data. In: Workshop on Linked Data on the Web (2014)
4. Heyvaert, P., Dimou, A., Herregodts, A.-L., Verborgh, R., Schuurman, D., Mannens, E., Van de Walle, R.: RMLEditor: a graph-based mapping editor for linked data mappings. In: Sack, H., Blomqvist, E., d'Aquin, M., Ghidini, C., Ponzetto, S.P., Lange, C. (eds.) ESWC 2016. LNCS, vol. 9678, pp. 709–723. Springer, Heidelberg (2016). doi:10.1007/978-3-319-34129-3_43
5. Heyvaert, P., Dimou, A., Verborgh, R., Mannens, E., Van de Walle, R.: Towards a uniform user interface for editing mapping definitions. In: Proceedings of the 4th Workshop on Intelligent Exploration of Semantic Data (2015)
6. Heyvaert, P., Dimou, A., Verborgh, R., Mannens, E., Van de Walle, R.: Approaches for generating mappings to RDF. In: Proceedings of the 14th International Semantic Web Conference: Posters and Demos (2015)

---

[8] http://lov.okfn.org/dataset/lov/.

SPARQL Query Recommendations by Example

Carlo Allocca[(✉)], Alessandro Adamou, Mathieu d'Aquin, and Enrico Motta

Knowledge Media Institute, The Open University, Milton Keynes, UK
{carlo.allocca,alessandro.adamou,mathieu.daquin,enrico.motta}@open.ac.uk

Abstract. In this demo paper, a SPARQL Query Recommendation Tool (called SQUIRE) based on *query reformulation* is presented. Based on three steps, *Generalization*, *Specialization* and *Evaluation*, SQUIRE implements the logic of reformulating a SPARQL query that is satisfiable w.r.t a *source* RDF dataset, into others that are satisfiable w.r.t a *target* RDF dataset. In contrast with existing approaches, SQUIRE aims at recommending queries whose reformulations: (i) reflect as much as possible the same intended meaning, structure, type of results and result size as the original query and (ii) do not require to have a mapping between the two datasets. Based on a set of criteria to measure the similarity between the initial query and the recommended ones, SQUIRE demonstrates the feasibility of the underlying *query reformulation* process, ranks appropriately the recommended queries, and offers a valuable support for query recommendations over an unknown and unmapped target RDF dataset, not only assisting the user in learning the data model and content of an RDF dataset, but also supporting its use without requiring the user to have intrinsic knowledge of the data.

1 Introduction

One of the main aspects that characterises Linked Open Data (LOD) is *Heterogeneity*: it is not hard to find RDF datasets that describe overlapping domains using different vocabularies [11]. A long-standing challenge raised by this state of affairs is related to the *access* and *retrieval* of data. In particular, a common scenario that is playing a central role to accomplish a number of tasks, including integration, enriching and comparing data from several RDF datasets, can be described as follows: given a query Q_o (e.g. `Select distinct ?mod ?title where { ?mod a ou:Module. ?mod dc:title ?title }`[1]) formulated w.r.t a *source* RDF dataset D_s (e.g. D_{ou} = http://data.open.ac.uk/query), we need to reformulate it w.r.t another similar *target* RDF dataset D_t (e.g. D_{su} = http:// sparql.data.southampton.ac.uk). Achieving this goal usually involves quite intensive and time consuming ad-hoc pre-processing [12]. In particular, it requires spending time in exploring and understanding the target RDF dataset's data model and content, and then, iteratively reformulating and testing SPARQL queries until the user reaches a query formulation that is right for his/her

---

[1] Select the names of the modules available at The Open University.

© Springer International Publishing AG 2016
H. Sack et al. (Eds.): ESWC 2016 Satellite Events, LNCS 9989, pp. 128–133, 2016.
DOI: 10.1007/978-3-319-47602-5_26

needs [3]. Reformulating a query over many RDF datasets can be very laborious but, if aided by tool support that recognises similarities and provides prototypical queries that can be tested without the user's prior knowledge of the dataset, the time and effort could be significantly reduced. In this demo paper, we propose a novel approach and a tool (called SQUIRE) that, given a SPARQL query Q_o that is satisfiable w.r.t a *source* RDF dataset (D_s), provides query recommendations by automatically reformulating Q_o into others Q_{r_i} that are satisfiable w.r.t a *target* RDF dataset (D_t). In contrast with existing approaches (see Sect. 2), SQUIRE aims at recommending queries whose reformulations: i) reflect as much as possible the same intended meaning, structure, type of results and result size as the original query and ii) do not require to have an ontology mapping and/or instance matching between the datasets. Based on a set of criteria to measure the similarity between the user-provided query Q_o and the recommended ones Q_{r_i}, we have prototyped our approach. Demo session attendants will have the opportunity to experiment with SQUIRE over real-world SPARQL endpoints, thus demonstrating the feasibility of the underlying *query reformulation* and *query recommendation* processes. The paper structure is as follows: Sect. 2 discusses existing works, Sect. 3 details the SQUIRE's approach and its implementation. Finally, Sect. 4 concludes and points out future research.

2   Related Work

To the best of our knowledge, there is no other study investigating SPARQL query recommendations over unmapped RDF datasets that take user queries into account. On the contrary, several solutions exist to address the issue of SPARQL query rewriting for implementing data integration over linked data. For instance, [7] devised a query rewriting approach that makes full use of schema mapping, whereas [2] relies on an explicit ontology alignment between the source D_s and the target D_t. Similarly, [9] described a method for query approximation where the entities appearing in the query can be generalized w.r.t an given ontology mapping. Moreover, several systems have been proposed, with very good achievements, to support users with no knowledge on SPARQL or RDF to build appropriate queries from scratch. Just to mention a few, Sparklis [4], QUICK [12], QueryMed [10] are designed on a query building process that is based on a guided interactive questions and answers. The authors of RDF-GL [6] designed a method based on a visual query language where a query can be viewed in a natural language-like form and in a graphical form. On the same line, but hiding the SPARQL language syntax, SparqlFilterFlow [5] and SPARQLViz [1] proposed an approach based on visual interface where the queries can be created entirely with graphical elements.

   Closer to our goal, [3] aims at alleviating the effort of understanding the potential use of an RDF dataset by automatically extracting relevant natural language questions that could be formulated and executed over it. Although all the above studies were useful for us as they contribute interesting elements to build on, they are mainly driven by a context where the user is not familiar with the underlying technologies (which is not our case) and having in mind the goal that semantic access and retrieval of data can be made more usable through an appropriate natural language based systems. In contrast, we are focusing on a method to make SPARQL query recommendations by reformulating a user query for accessing and retrieving data from unmapped RDF datasets.

3   Method and Implementation

To achieve our goal, SQUIRE proposes and implements a mechanism based on three steps: *Generalization*, *Specialization* and *Evaluation*. To present each of them, let us consider the case in which we want to build recommendations for the example query Q_{ou} but w.r.t. the Southampton University RDF dataset D_{su}.

Generalization aims at generalizing the entities (classes, properties, individuals and literals) of Q_o that are not present in D_t into variables (marked as *template variables*)[2]. By applying this step, we build what we called the *Generalized Query Template* (GQT). Back to the query Q_{ou}, the GQT is obtained from it by turning the entities ou:Module as a class and dc:title as a datatype property into two template variables, that is ?ct1 and ?dtp1, respectively. The result is shown in the root node of the tree in Fig. 1.

Specialization aims at specializing consistently the obtained GQT by applying two main operations: (a) *Instantiation* (I) instantiates consistently a template variable with a corresponding concrete value that belongs to D_t (e.g.

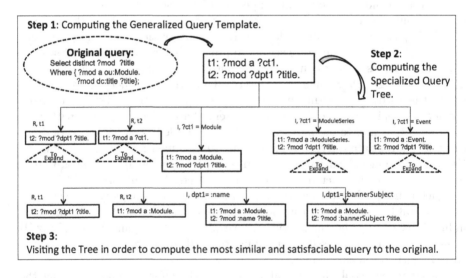

Fig. 1. Part of the *Specialized Query Tree* by applying the two operations.

---

[2] It is a query variable that has been consistently indexed with natural numbers and named according to its type. We used ct, it, opt, dpt and lt, for class, instance, object property, data type property and literal, respectively.

we instantiate ?ct1[?dtp1] over each class [datatype property] of D_{su}); and (b) *Removal* (R) deletes an entire triple pattern from the Q_o's GQT. We called the output of this step *Specialized Query Tree*. Figure 1 shows a part of it for Q_{ou}.

Evaluation. As a result of the previous two steps, each tree node is considered to be a reformulated query that is a candidate for recommendation. However, some of them are more "similar" to the original one than the others. Thus, the main question was: how can we capture and compute such similarity to provide a score-based ranking? Being in accord with [8] that there is no universal way of measuring the distance and/or similarities between two formal queries, we based our approach on a linear combination of the following criteria[3]: (a) *Result Type Similarity* aiming at measuring the overlap of the types of results (URI or literal) between Q_o and any recommendations Q_{r_i}; (b) *Query Result Size Similarity* aiming at measuring the result size rate (normalized w.r.t datasets sizes) between Q_o and any recommended one Q_{r_i}; (c) *Query Root Distance* aiming at measuring the cost of each applied operation from the root node to the one containing the recommended Q_{r_i}. It takes into account the distance-based matching of the replaced entities and the structure (as a set of triple patterns) between Q_o and any recommended Q_{r_i}; and (d) *Query Specificity Distance* aiming at measuring the distance between Q_o and any recommended Q_{r_i} based on the sets of variables (total shared variables/total variables).

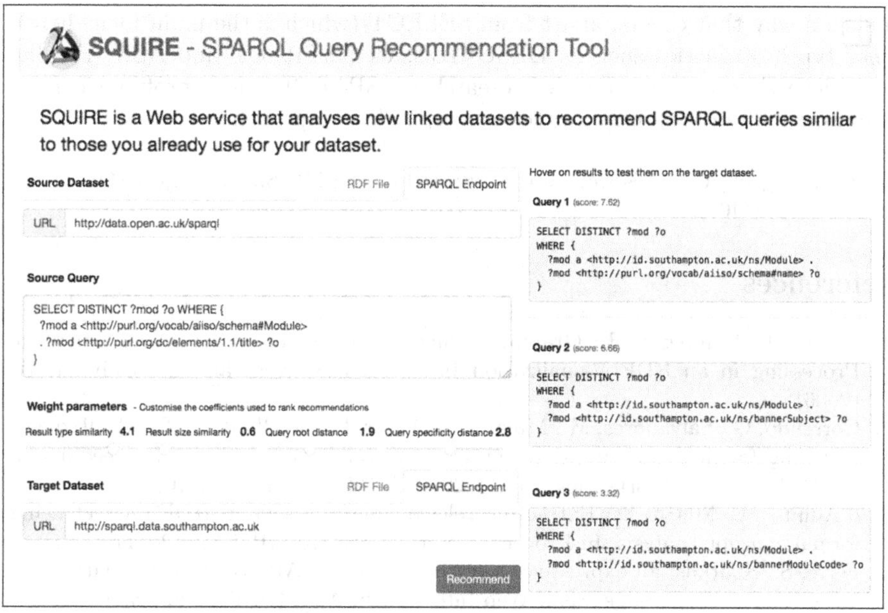

Fig. 2. SQRT prototype screenshot.

---

[3] The weights are options given to the user to set based on their preferences.

A screenshot of the implemented tool is shown in Fig. 2. Basically, SQUIRE allows the user to (1) refer to a source RDF dataset, either as an RDF file or as the URL of a SPARQL endpoint; (2) write down the query Q_o w.r.t D_s and (3) specify the target RDF dataset. Once the user clicks on the *Recommend* button, SQUIRE executes the method described above and returns a list of scored recommended queries, sorted high-to-low. Another distinctive characteristic of SQUIRE is that the recommended queries not only are expressed in terms of the target dataset, but also are guaranteed to be satisfiable (i.e. the result set is not empty) and can therefore be used to access and retrieve data from the target dataset.

4   Conclusion and Discussion

SQUIRE, as an approach and a tool, enables SPARQL query recommendations by reformulating a user query that is satisfiable w.r.t a *source* RDF dataset D_s, into others that are satisfiable w.r.t a *target* (and unmapped) RDF dataset D_t. One of the advantages of SQUIRE is that not only it helps learning the data model and content of a dataset, which usually requires a huge initial effort, but also enhances their use straightforwardly without the user's prior knowledge. Indeed, the problem is not fully solved. One of the aspects we have planned to investigate is the case where the reformulation is based on other types of operations (e.g. adding a triple pattern, or more generally replacing a graph pattern with another one, and so on). Moreover, we want to extend this work in such a way that covers, apart from SELECT (which is the main focus here), other types of queries such as DESCRIBE, CONSTRUCT and ASK. Finally, we believe that the outcomes of research on SPARQL query profiling can be combined with ours to improve the corresponding approaches.

Acknowledgments. This work was supported by the MK:Smart project (OU Reference HGCK B4466).

References

1. Borsje, J., Embregts, H.: Graphical Query Composition and Natural Language Processing in an RDF Visualization Interface. E.S. of E. and B., Univ., Rott. (2006)
2. Correndo, G., Salvadores, M., Millard, I., Glaser, H., Shadbolt, N.: SPARQL query rewriting for implementing data integration over linked data. In: Proceedings of the EDBT/ICDT Workshops, EDBT 2010. ACM, New York (2010)
3. d'Aquin, M., Motta, E.: Extracting relevant questions to an RDF dataset using formal concept analysis. In Proceedings of the 6th K-CAP, USA (2011)
4. Ferre, S., Sparklis: an expressive query builder for SPARQL endpoints with guidance in natural language. Sem. Web Inter. Usab. App. (2016, to appear)
5. Haag, F., Lohmann, S., Ertl, T.: SparqlFilterFlow: SPARQL query composition for everyone. In: Presutti, V., Blomqvist, E., Troncy, R., Sack, H., Papadakis, I., Tordai, A. (eds.) ESWC 2014. LNCS, vol. 8798, pp. 362–367. Springer, Heidelberg (2014). doi:10.1007/978-3-319-11955-7_49

6. Hogenboom, F., Milea, V., Frasincar, F., Kaymak, U: RDF-GL: a SPARQL-based graphical query language for RDF. In: Emergent Web Intelligence: Advanced Information Retrieval (2010)
7. Makris, K., Bikakis, N., Gioldasis, N., Tsinaraki, C., Christodoulakis, S.: Towards a mediator based on OWL and SPARQL. In: Lytras, M.D., et al. (eds.) WSKS 2009. LNCS, vol. 5736, pp. 326–335. Springer, Heidelberg (2009)
8. Picalausa, F., Vansummeren, S.: What are real SPARQL queries like? In: Proceedings of SWIM 2011, pp. 7:1-7:6. ACM, New York (2011)
9. Reddy, B.R.K., Kumar, P.S.: Efficient approximate SPARQL querying of web of linked data. In: URSW CEUR Workshop Proceeding, CEUR-WS.org (2010)
10. Seneviratne, O.: QueryMed: an intuitive SPARQL query builder for biomedical RDF data (2010)
11. Tzitzikas, Y., et al.: Integrating heterogeneous and distributed information about marine species through a top level ontology. In: Garoufallou, E., Greenberg, J. (eds.) MTSR 2013. CCIS, vol. 390, pp. 289–301. Springer, Heidelberg (2013). doi:10.1007/978-3-319-03437-9_29
12. Zenz, G., Zhou, X., Minack, E., Siberski, W., Nejdl, W.: From keywords to semantic queries-incremental query construction on the semantic web. Web Sem. **7**(3), 166–176 (2009)

Tabular Data Cleaning and Linked Data Generation with Grafterizer

Dina Sukhobok[1], Nikolay Nikolov[1], Antoine Pultier[1], Xianglin Ye[1],
Arne Berre[1], Rick Moynihan[2], Bill Roberts[2], Brian Elvesæter[1],
Nivethika Mahasivam[1], and Dumitru Roman[1(✉)]

[1] SINTEF, Forskningsveien 1a, 0373 Oslo, Norway
dina.suhobok@gmail.com, {nikolay.nikolov,antoine.pultier,arne.j.berre,
brian.elvesater,dumitru.roman}@sintef.no, lynnye1988mail@gmail.com,
nivemaham@gmail.com
[2] Swirrl IT LTD., Springbank Raod,
MacFarlane Gray House, Stirlingshire FK7 7WT, UK
{rick.m,bill}@swirrl.com

Abstract. Over the past several years the amount of published open data has increased significantly. The majority of this is tabular data, that requires powerful and flexible approaches for data cleaning and preparation in order to convert it into Linked Data. This paper introduces Grafterizer – a software framework developed to support data workers and data developers in the process of converting raw tabular data into linked data. Its main components include Grafter, a powerful software library and DSL for data cleaning and RDF-ization, and Grafterizer, a user interface for interactive specification of data transformations along with a back-end for management and execution of data transformations. The proposed demonstration will focus on Grafterizer's powerful features for data cleaning and RDF-ization in a scenario using data about the risk of failure of transport infrastructure components due to natural hazards.

Keywords: Open data · Linked data · Tabular data cleaning and preparation · Data transformation

1 Introduction

The potential gains from data analysis and knowledge discovery in the future is estimated to billions and even trillion dollars[1,2]. In order to get a broader view on a given problem and benefit from knowledge discovery from data, analysts need access to large amounts of data. This leads to an increase in demand for using data sources such as open data. Open data is commonly published in tabular formats, which are widely adopted and familiar to data workers. Nevertheless,

---

[1] http://www.irishexaminer.com/lifestyle/features/dell-chief-executive-says-data-is-the-next-trillion-dollar-opportunity-370608.html.

[2] http://www.idc.com/getdoc.jsp?containerId=prUS40560115.

© Springer International Publishing AG 2016
H. Sack et al. (Eds.): ESWC 2016 Satellite Events, LNCS 9989, pp. 134–139, 2016.
DOI: 10.1007/978-3-319-47602-5_27

only a small amount of published datasets are actually used for various reasons, primarily due to the lack of simple approaches to interconnect the data from various tables. Linked data can alleviate some of these problems by providing a set of standards for representing and connecting the data, therefore enabling data to be discovered and used by various applications [1].

To generate valid linked data we need data to be in a shape that it is easy to manipulate and convert to RDF ('RDF-ize'). Raw data in most cases contain a number of common data quality issues, such as missing values, invalid values, duplicate records, etc. Solving data quality issues is especially important when integrating heterogeneous data sources that should be addressed together with schema-related data transformations [2]. Data preparation provides a standard way of structuring data, which makes it easier to extract needed values [3]. Hence, data publishers first need to deal with data quality issues before data can be mapped to RDF. This process is usually considered as one of the most time- and cost-consuming – according to some sources it takes up to 80 % of the time [4]. We therefore need a unifying framework for data cleaning and RDF-ization, powerful enough to cope with common data cleaning problems, simple enough that non-programmers can use it, while at the same time flexible enough that data developers can easily work with it. A common framework for cleaning and RDF-ization can also simplify collaboration between users collaboratively working on the same data transformation (e.g., one doing the data cleaning, the another one the RDF-ization). In this demonstration we introduce Grafterizer as an example of such a framework.

2   The Grafterizer Framework

The main steps of transformation of raw tabular data into linked data supported by Grafterizer are depicted in Fig. 1.

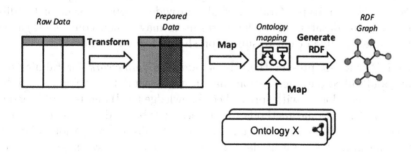

Fig. 1. Generating a semantic RDF graph from tabular data

Grafterizer is based on Grafter[3] – a library and DSL for producing linked data graphs from tabular data, which provides extensive support for data cleaning and powerful ETL data transformations, suitable for handling large datasets.

---

[3] http://grafter.org/.

The Grafter library is implemented in Clojure – a functional programming language that runs on the JVM. Grafter's properties give it a number of advantages: the Java virtual machine environment makes it possible to use numerous available libraries, and, Clojure, as a functional programming language, makes it natural to treat data structures as streams of immutable values.

The data transformation process is realized through a pipeline abstraction, i.e., each step of transformation is defined as a pipe. Thereby, a function performs simple data conversion on its input and the output of one pipe is the input of another. Grafter also provides a broad set of features for creating and managing linked data graphs out of these data.

Grafter's primary target are software developers, whereas Grafterizer makes the process of creating pipelines and ontology mappings for graphs more accessible. It provides a set of functions, that can be used to solve the most common data quality issues in a fast and effective way. A summary of the tabular transformation functions is given in Table 1.

Due to the large number of possible types of data quality issues, the operations on data are not limited to the functions listed above. Grafterizer makes it possible for users to define their own functions on data and involve them in the Grafterizer transformation pipeline, which makes transformations more flexible. Custom functions provide also a good way to encapsulate data modifications.

After data quality issues are solved, the dataset can be transformed to RDF. The RDF triple patterns that should appear in the resulting linked data are designed by the user, whereby the pattern for the subjects, predicates and objects of triples is specified through a mapping procedure. During the mapping process column headers are mapped to RDF nodes in order to produce a set of triples that corresponds to each data row. Grafterizer supports reuse of existing RDF ontologies by providing a searchable catalog of vocabularies and makes it possible to manage individual namespace prefixes. Each column in a dataset can be mapped as a URI node with namespace prefix assigned by user or literal node with a specified datatype. Grafterizer also provides support for error handling when casting to datatypes. Furthermore, users may assign condition(s) under which a node, a triple or entire sub-graph should be generated.

Related Work. At present there is no *unifying framework* for tabular data cleaning and linked data generation that targets both data developers and data workers (non-developers with spreadsheet knowledge level) and is flexible enough at the same time. Some methods of converting tabular data into linked data involve intermediate steps of importing raw tabular data to a relational database and transforming the resulting data into RDF format afterwards [5]. The set of recommendations[4] and advanced tools make a good base for using such methods. With the Grafterizer approach, this step is omitted – data are brought to semantic form directly from the raw tabular presentation[5]. OpenRefine[6] with the RDF

---

[4] http://www.w3.org/TR/rdb-direct-mapping/.
[5] http://www.w3.org/TR/csv2rdf/.
[6] http://openrefine.org/.

Table 1. Summary of tabular transformation functions supported by Grafterizer

Scope	Name	Short description
Rows	Add Row	Create a new record in a dataset
	Take/Drop Rows	Extract/delete selected row (sequence of rows)
	Shift Row	Change row's position inside a dataset
	Filter Rows	Filter rows for matches, regexes, empty values etc.
	Remove Duplicates	Remove similar rows based on certain column or set of columns
Entire dataset	Sort Dataset	Sorts dataset by given column names in given order
	Reshape dataset	Restructure a dataset
	Group and Aggregate	Group values by column or multiple columns and perform aggregation (get minimal, maximum or average value, count or sum values in every group) on the rest of columns
Columns	Add Column Derive Column	Add a column with a manually specified value, or as a result of some computations performed on other columns
	Take/Drop Columns	Take/drop selected column
	Shift Column	Arbitrarily change column's order
	Merge/Split Columns	Split or merge columns using custom separator
	Rename Columns	Change columns headers
	Map Columns	Apply function to all values in a column

Refine plugin[7] is relevant in the context of Grafterizer. OpenRefine functionalities are tightly-coupled to the service core, which hinders distribution and prevents its use "as-a-service". Furthermore, OpenRefine uses a memory-intensive multi-pass approach to data transformation functions, which is designed to operate with small to medium data volumes. Other platforms provide a broad and powerful functionality in tabular data cleaning (e.g. Trifacta Wrangler[8]), but do not support RDF data publication.

---

[7] http://refine.deri.ie/.
[8] https://www.trifacta.com/products/wrangler/.

3   Demonstration Outline

Grafterizer will be demonstrated in a real case for data cleaning and RDF-ization in the context of InfraRisk[9] — a project developing a framework to identify and track the impact of natural hazards on infrastructure networks (e.g. roads, rails).

During the demonstration, visitors will learn how to solve data quality issues and transform tabular data into a linked data graph. The scenario demonstrated will cover uploading raw data into Grafterizer's interactive user interface, constructing and testing a pipeline to transform that data (with the help of provided standard functions and embedded custom Clojure code), and generating a linked data graph out of an RDF mapping.

Within the demonstration several groups of data quality issues are addressed. For example, data rows that contain no useful information will be filtered out, values of rows that contain badly formatted data will be cleaned-up, and missing identifiers for each data row will be produced. An example of Grafterizer's transformation pipeline and preview results are shown in Fig. 2. For the RDF-ization process visitors will learn how to build an RDF mapping, manage namespace prefixes, link nodes to columns, specify conditions on triple generation, and cast cell values to literal data types.

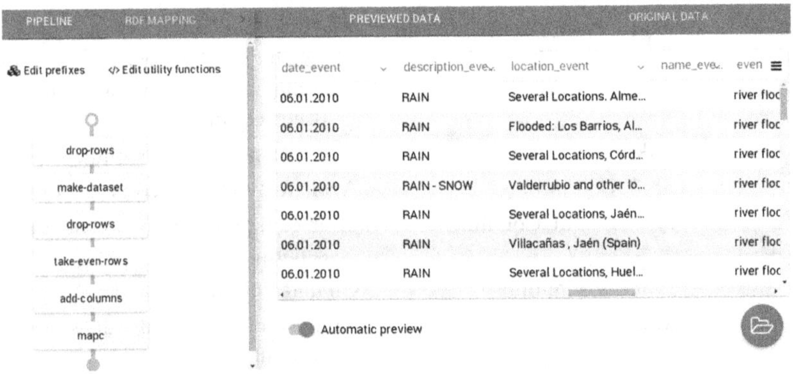

Fig. 2. Pipeline and transformation preview

Grafterizer is currently deployed within the DataGraft platform [6], available at https://datagraft.net/. A video showing Grafterizer in action can be found at https://youtu.be/zAruS4cEmvk.

Acknowledgements. This work was partly funded by the European Commission within the following research projects: DaPaaS (FP7 610988), SmartOpenData (FP7 603824), InfraRisk (FP7 603960), and proDataMarket (H2020 644497).

---

[9] http://www.infrarisk-fp7.eu/.

References

1. Bizer, C., Heath, T., Berners-Lee, T.: Linked data-the story so far. Emerg. Concepts, Semant. Serv. Interoperability Web Appl. 205–227 (2009)
2. Rahm, E., Do, H.H.: Data cleaning: problems and current approaches. IEEE Bull. Data Eng. **23**, 4 (2000)
3. Wickham, H.: Tidy Data. J. Stat. Softw. **59**(10), 1–23 (2011). Web. 1 Mar. 2016
4. Dasu, B.T., Johnson, T.: Exploratory Data Mining and Data Cleaning, 1st edn. Wiley, New York (2003)
5. Skjæveland, M.G., Lian, E.H., Horrocks, I.: Publishing the Norwegian petroleum directorate's factpages as semantic web data. In: Alani, H., Kagal, L., Fokoue, A., Groth, P., Biemann, C., Parreira, J.X., Aroyo, L., Noy, N., Welty, C., Janowicz, K. (eds.) ISWC 2013, Part II. LNCS, vol. 8219, pp. 162–177. Springer, Heidelberg (2013)
6. Roman, D., Nikolov, N., Putlier, A., Sukhobok, D., Elvester, B., Berre, A., Ye, X., Dimitrov, M., Simov, A., Zarev, M., Moynihan, R., Roberts, B., Berlocher, I., Kim, S., Lee, T., Smith, A., Heath, T.: DataGraft: one-stop-shop for open data management. Semant. Web J. (SWJ) (2016, to appear). http://www.semantic-web-journal.net/system/files/swj1428.pdf

SentiCircles: A Platform for Contextual and Conceptual Sentiment Analysis

Hassan Saif[1], Maxim Bashevoy[2], Steve Taylor[2], Miriam Fernandez[1(✉)], and Harith Alani[1]

[1] Knowledge Media Institute, Open University, Milton Keynes, UK
{h.saif,m.fernandez,h.alani}@open.ac.uk
[2] IT Innovation Centre, University of Southampton, Southampton, UK
{mvb,sjt}@it-innovation.soton.ac.uk

Abstract. Sentiment analysis over social streams offers governments and organisations a fast and effective way to monitor the publics' feelings towards policies, brands, business, etc. In this paper we present SentiCircles, a platform that captures feedback from social media conversations and applies contextual and conceptual sentiment analysis models to extract and summarise sentiment from these conversations. It provides a novel sentiment navigation design where contextual sentiment is captured and presented at term/entity level, enabling a better alignment of positive and negative sentiment to the nature of the public debate.

Keywords: Social media · Sentiment analysis

1 Introduction

With the emergence of social networking sites, fast and continuous streams of data are being generated by citizens about a variety of topics, including their opinions and arguments about policies, products, brands, etc. However, summarising and extracting sentiment from these large and continuous streams of data, as well as designing visualisation models to navigate such data, constitute difficult and important research problems.

Most current approaches for identifying the sentiment of posts can be categorised into one of two main groups: supervised approaches, such as [1,3], which use a wide range of features and labelled data for training sentiment classifiers, and lexicon-based approaches, such as [4,8], which make use of pre-built lexicons of words weighted with their sentiment orientations to determine the overall sentiment of a given text. Popularity of lexicon-based approaches is rapidly increasing since they require no training data, and hence they are more suited to a wider range of domains than supervised approaches [8]. Nevertheless, lexicon-based approaches have two main limitations. Firstly, the number of words in the lexicons is limited, which may constitute a problem when extracting sentiment from very dynamic environments such as the ones posed by social media, where

© Springer International Publishing AG 2016
H. Sack et al. (Eds.): ESWC 2016 Satellite Events, LNCS 9989, pp. 140–145, 2016.
DOI: 10.1007/978-3-319-47602-5_28

new terms constantly emerge. Secondly and more importantly, sentiment lexicons tend to assign a fixed sentiment orientation and score to words, irrespective of how these words are used in the text. For example, the word "`great`" conveys different sentiment when associated with the word "`problem`" than with the word "`smile`". The sentiment that a word expresses is not static, but depends on the semantics of the word in the particular context it is being used. In addition, relying on the context only for detecting terms' sentiment may also be insufficient. This is because the sentiment of a term may be conveyed via its conceptual semantics rather than by its context [2]. For example, the context of the word "Ebola" in "`Ebola continues spreading in Africa!`" does not indicate a clear sentiment for the word. However, "Ebola" is associated with the semantic type (concept) "`Virus/Disease`", which suggests that the sentiment of "Ebola" is likely to be negative.

As part of our previous work we have been investigating novel sentiment analysis models that account for contextual and conceptual semantics in order to enhance the accuracy of existing lexical-based sentiment classification methods [5–7]. These models have been integrated into the SentiCircles platform and a novel interface has been designed to navigate social media conversations based on the sentiment computed by these models. More specifically, the objectives of the SentiCircles platform are: (i) apply contextual and conceptual sentiment analysis models for identifying the sentiment expressed in social media discussions and, (ii) provide visualisation mechanisms that enable a fine-grained summarisation and exploration of the computed sentiment.

Demo: A fully working platform will be demoed at the conference, running over a total of 12,000 posts. These posts span conversations around several topics related with the UK National policy on renewable energy targets for 2030.[1] A tutorial of how to access and use the demo is available at: https://www. evernote.com/shard/s217/sh/8dba1de2-5353-4df2-89ce-2e4323a3eb36/36a950b b95b3deed

2   The SentiCircles Sentiment Analysis Approach

The sentiment analysis component behind the SentiCircles platform uses the SentiCircle approach [5]. This approach accounts for contextual and conceptual semantics of words when computing sentiment. It detects the context of a term from its co-occurrence patterns with other terms in tweets. In particular, the context for each term t in a tweet collection $\mathcal{T}$ is represented as a vector $c = (c_1, c_2, ..., c_n)$ of terms that occur with t in any tweet in $\mathcal{T}$. An example of this process is provided in Fig. 2. Given a tweet collection $\mathcal{T}$, the target term m_{great} is represented as a vector $c_{great} = (c_1, c_2, ..., c_n)$ of terms co-occurring with term m in any tweet in $\mathcal{T}$ (e.g., "pain", "loss", ..., "death"). The context vector c_{great} is transformed into 2d circle representation. The center of the circle represents the target term m_{great} and points within the circle denote the context terms of m_{great}.

---

[1] https://goo.gl/ykRD0A.

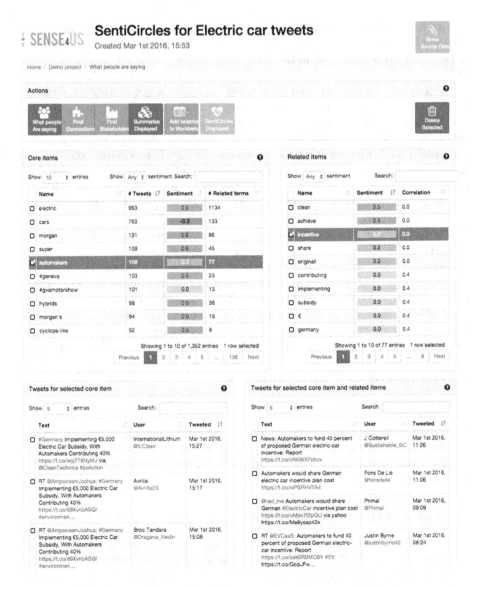

Fig. 1. SentiCircles visualisation for Twitter conversations around electric cars (Color figure online)

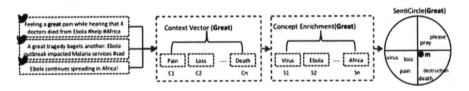

Fig. 2. Contextual and Conceptual sentiment of the word *great*

Conceptual semantics are incorporated into the approach by first extracting the entities from the posts (e.g., "Ebola", "Africa", and its conceptual types and subtypes (e.g., "Virus", "Continent") using AlchemyAPI[2] and incorporating this information $s = (s_1, s_2, ..., s_m)$ within the contextual vector $c_s = c + s = (c_1, c_2, ..., c_n, s_1, s_2, ..., s_m)$. The sentiment of t is then extracted by first transforming the term vector c_s into a 2d circle representation, and then extracting the geometric median of the points (context terms) within the circle. The position of the median within the circle represents the overall contextual sentiment of t. This simple technique has proven effective in calculating sentiment and entity as well as at tweet level (see [5–7] for more details)

3   The SentiCircles Sentiment Analysis Platform

In this section we present the designed visualisation for the SentiCircle Sentiment Analysis Platform. This visualisation has been designed to enable an easy navigation and exploration of the sentiment emerging from social media conversations. Figure 1 shows the sentiment emerging from a Twitter collection around electric cars. Each term emerging from the social media conversations is listed within the core items table (top left of the interface) alongside its corresponding contextual and conceptual semantics in the related items table (top right of the interface). These tables show the user to have a very quick overview of the issues people are discussing when talking about the topic in hand, along with their associated sentiment and relevant related information.

The core items table displays the terms and entities emerging from the social media conversations and for which sentiment has been computed, i.e., each core item is a SentiCircle (see Sect. 2), where the displayed term is at the center of the circle (e.g., electric, cars, automakers). Core items can be sorted and ranked according to: (i) the number of tweets in which they appear (i.e., how popular they are), (ii) their associated sentiment scores (from completely negative -1 to completely positive $+1$), and (iii) the number of related or contextual items associated to them. This information is represented in different columns of the core items table. Sorting capabilities are enabled on top of each column. For example, the sentiment column allows ranking the core items from more negative to more positive and vice-versa. Additionally, a range of colours is also provided to indicate positive (green), negative (red) and neutral (yellow) sentiment.

Contextual and conceptual semantics for each core item are presented in the related items table (top right of the figure) with their corresponding sentiment score, and the degree of correlation with respect to the core item. Examples of related items for the core item *automakers* include *clean, achieve, incentive,* etc. The degree of correlation represents how frequently do the co-occur in the post collection, which translates into the model on how strongly the related item influences the core item's sentiment. The top of each column enables to sort the rows based on their numeric value.

---

[2] www.alchemyapi.com.

The tables at the bottom part of the interface show examples of posts in which the selected core item appears (left bottom table), and examples of posts where both, the core item and the selected related item appear (right bottom table). The presented model and its visualisation enables the user to navigate social media conversations, observe what are the core emergent items around the topic at hand, what is the sentiment perceived towards those items, and what are the reasons (context) behind it.

4   Feedback and Future Work

The SentiCircles platform is part of the Sense4us project[3], a project focused on the development of tools that can support better policy making. The platform was showcased to 16 Members of Parliament (MPs) from the State Parliament of North Rhine Westphalia, the German Bundestag and the European Parliament. It was very well received as a tool that could enhance the collection of feedback, and speeding up the reaction to any concerns or challenges raised by citizens. MPs highlighted that positive and negative sentiment appeared better aligned to the nature of the public policy debate than in other tools, since SentiCircles shows them the key items under discussion within the conversations and allows them to investigate why sentiment is negative or positive by navigating between the core and related items. Transparency was raised as an important issue, since MPs need to be aware of the limitations, particularly in terms of data, model and users, when using social media analysis tools to inform policy making. We are currently working on preparing documentation, as well as various interface modifications to enhance the transparency of the results obtained by the tool.

5   Conclusions

This paper describes the SentiCircles platform; a web based tool for assessing and monitoring sentiment in public social media. The platforms applies contextual and conceptual sentiment analysis models to extract and summarise sentiment. Positive feedback has been received so far about the tool by several MPs when tested in a policy making context.

References

1. Barbosa, L., Feng, J.: Robust sentiment detection on twitter from biased and noisy data. In: Proceedings of COLING, Beijing, China (2010)
2. Cambria, E.: An introduction to concept-level sentiment analysis. In: Castro, F., Gelbukh, A., González, M. (eds.) MICAI 2013, Part II. LNCS, vol. 8266, pp. 478–483. Springer, Heidelberg (2013)
3. Kouloumpis, E., Wilson, T., Moore, J.: Twitter sentiment analysis: the good the bad and the omg! In: Proceedings of the ICWSM, Barcelona, Spain (2011)

---

[3] http://www.sense4us.eu/.

4. O'Connor, B., Balasubramanyan, R., Routledge, B.R., Smith, N.A.: From tweets to polls: linking text sentiment to public opinion time series. In: ICWSM, vol. 11, pp. 122–129 (2010)

5. Saif, H., Fernandez, M., He, Y., Alani, H.: SentiCircles for contextual and conceptual semantic sentiment analysis of Twitter. In: Presutti, V., d'Amato, C., Gandon, F., d'Aquin, M., Staab, S., Tordai, A. (eds.) ESWC 2014. LNCS, vol. 8465, pp. 83–98. Springer, Heidelberg (2014)

6. Saif, H., He, Y., Alani, H.: Semantic sentiment analysis of Twitter. In: Cudré-Mauroux, P., Heflin, J., Sirin, E., Tudorache, T., Euzenat, J., Hauswirth, M., Parreira, J.X., Hendler, J., Schreiber, G., Bernstein, A., Blomqvist, E. (eds.) ISWC 2012, Part I. LNCS, vol. 7649, pp. 508–524. Springer, Heidelberg (2012)

7. Saif, H., He, Y., Fernandez, M., Alani, H.: Contextual semantics for sentiment analysis of Twitter. Inf. Process. Manag. **52**(1), 5–19 (2016). doi:10.1016/j.ipm.2015.01.005. ISSN: 0306–4573

8. Thelwall, M., Buckley, K., Paltoglou, G.: Sentiment strength detection for the social web. J. Am. Soc. Inf. Sci. Technol. **63**(1), 163–173 (2012)

PepeSearch: Easy to Use and Easy to Install Semantic Data Search

Guillermo Vega-Gorgojo[1(✉)], Laura Slaughter[2], Martin Giese[1],
Simen Heggestøyl[1], Johan Wilhelm Klüwer[3], and Arild Waaler[1]

[1] Department of Informatics, University of Oslo, Oslo, Norway
{guiveg,martingi,simenheg,arild}@ifi.uio.no
[2] Oslo University Hospital, Oslo, Norway
laura.slaughter@gmail.com
[3] Det Norske Veritas (DNV), Høvik, Norway
Johan.Wilhelm.Kluewer@dnvgl.com

Abstract. Despite the increasing availability of RDF datasets, searching and browsing semantic data is still a daunting task for mainstream users. With PepeSearch, it is easy to query an arbitrary triple store without previous knowledge of RDF/SPARQL. PepeSearch offers a form-based interface with simple and intuitive elements such as drop-down menus or sliders that are automatically mapped from the ontological structures of the target dataset. In this demonstration we will show how to set up a PepeSearch instance, how to formulate queries and how to retrieve results.

1 Introduction

An increasing number of RDF datasets is available across all domains and, as a result, many non-programmers are expressing a need for exploring these datasets. The problem is that accessing semantic data requires proficiency in SPARQL, as well as familiarity with the specific vocabularies or ontologies employed by the dataset. Alternatives to searching directly with SPARQL are mainly visual query approaches, especially graph-based query editors, e.g. QueryVOWL [1], NITE-LIGHT [2]. While this type of interfaces can easily exploit the graph structure of RDF and SPARQL, mainstream users are not particularly comfortable with graph visualizations [3,4], making this approach questionable for this user group. Moreover, many common querying tasks do not require the expressivity of full graph-based querying.

We propose PepeSearch [5], a portable form-based search interface for querying semantic RDF datasets specifically aimed at helping mainstream users in their search tasks. Forms allow the user to exploit the ontology without manipulation of graph structures. Instead, the end-user employs drop-down menus, free-text entry fields, and sliders to specify classes, properties, strings, and data value ranges of their queries. This frees the user from having to invest a significant amount of time learning technical characteristics of the dataset, e.g., what an OWL class is or what ontologies are used to describe the data.

H. Sack et al. (Eds.): ESWC 2016 Satellite Events, LNCS 9989, pp. 146–150, 2016.
DOI: 10.1007/978-3-319-47602-5_29

Form-based interfaces tend to be designed for specific search tasks in a single domain. User experience and design work is therefore linked to a specific context. In contrast, PepeSearch exploits the self-describing nature of RDF and schema-level queries in SPARQL to develop a generic and portable solution that can run on any SPARQL endpoint. We allow the mainstream user to pose queries ranging from simply retrieving the members of a class, to queries joining multiple concepts and setting restrictions on datatype properties. So far, PepeSearch has been applied for use in two different contexts: government organizational data and healthcare. We will demonstrate PepeSearch at ESWC 2016: how to set up a PepeSearch instance, how to formulate queries and how to retrieve results.

2   Overview of PepeSearch

PepeSearch is an open source project under the Apache license developed at the University of Oslo[1]. It consists of the SPARQL analyzer[2], the PepeSearch component[3], and a text search engine – see Fig. 1. The provided GitHub repository also includes a screencast[4] and a live demo[5].

The analyzer is employed in a bootstrapping stage to gather information about the target data set. Through a series of generic SPARQL queries, the analyzer obtains the classes employed in the dataset, their datatype properties, and the connections to other classes through an object property or through a subclass relation. The result is a data schema in the JSON format.

The obtained data schema can then be used to configure a PepeSearch instance. The query builder component is in charge of preparing a suitable view for querying the dataset. For an arbitrary RDF class, a form block is created, in which datatype properties are mapped to widget elements. In order to support multi-class queries, a collapsible form block is included for each RDF class that is connected with an object property to the selected class – see Fig. 2(a) for an example. The results viewer element is in charge of sending the query to the SPARQL endpoint and presenting the results in a tabular representation – see Fig. 2(b). Browsing is supported through the instance viewer that obtains all the data about a particular individual with links to other connected instances – see Fig. 2(c).

The text search engine is an optional component that allows dynamic term suggestions during query specification. This is employed to provide autocomplete capabilities for the text fields of a class, e.g. to suggest names such as "Martin" or "Maria" after typing "mar" in a name textbox.

---

[1] http://www.uio.no/.
[2] https://github.com/simenheg/sparql-endpoint-analyzer.
[3] https://github.com/guiveg/pepesearch.
[4] http://folk.uio.no/simenheg/pepesearch.webm.
[5] http://sws.ifi.uio.no/project/semicolon/search/.

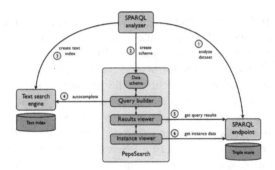

Fig. 1. Logical architecture of PepeSearch.

3   Hands on with PepeSearch

To illustrate the operation of PepeSearch, we will employ a sample dataset containing health records of fictitious patients. Anonymized patient data has been provided by our hospital project partner in the form of tables from a widely used hospital records application. It describes health care processes, with associated diagnoses and medical personnel in various roles, supported by a body of code lists. This data is mapped into RDF according to an ontology with three main parts: (i) excerpts from the Disease Ontology[6] to cover the medical conditions that appear in the data, (ii) the Information Artifact Ontology[7] for documents, and (iii) local extensions for measurements of vital signs and for a part/whole hierarchy of health care processes. Upper classes and relations are provided by the OBO Relations Ontology[8].

As an example, we show how to obtain a set of patients between 30–50 years of age that have suffered from an intestinal disease. Use of semantic technologies for cohort identification has been proposed [6], and is an important application area. We first run the SPARQL analyzer to generate the data schema out of the dataset structure with all the classes, properties and value types. PepeSearch can then be used to fulfill the aforementioned information need in this way:

1. PepeSearch presents a list of the top classes available in the dataset.
2. We select the concept "human being".
3. PepeSearch presents a form block for the "human being" class and a list of collapsibles corresponding to classes directly connected to "human being" in the dataset, e.g. "diagnosis" or "health care encounter".
4. We set the restrictions required for this search task: in the "human being" class we select "patient" as a more specific type; we use the age slider to set the appropriate range; and we select the "intestinal disease" after expanding the "disposition" collapsible. A snapshot of this query is shown in Fig. 2(a).

---

[6] http://disease-ontology.org/.
[7] https://github.com/information-artifact-ontology/IAO/.
[8] https://github.com/oborel/obo-relations.

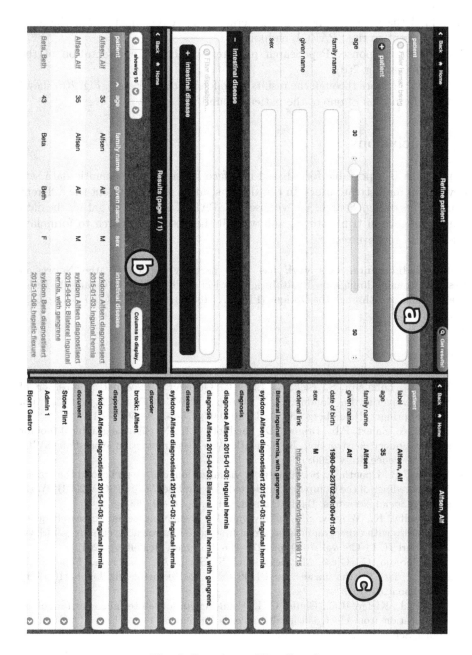

Fig. 2. Snapshots of PepeSearch.

5. We push the "Get results" button at the top right corner of the search interface.
6. Behind the scenes, PepeSearch generates a SPARQL query from the form that is sent to the SPARQL endpoint.
7. With the response, PepeSearch prepares a tabular representation of the results (see Fig. 2(b)).
8. We can navigate through the results by following the links, e.g. Fig. 2(c) shows the information of one of the patients found.

4   Conclusions

PepeSearch is a portable form-based interface for searching semantic data sets devised for mainstream users. In this demonstration we will present the different components of PepeSearch. We will use the SPARQL analyzer to gather the data schema of several triple stores, and we will then use PepeSearch to formulate queries and retrieve results.

Acknowledgements. This work has been partially funded by the Norwegian Research Council through the HealthInsight project (NFR 247784/O70), and the European Commission through the Optique (FP7 GA 318338), and BYTE (FP7 GA 619551) projects.

References

1. Haag, F., Lohmann, S., Siek, S., Ertl, T.: QueryVOWL: visual composition of SPARQL queries. In: Gandon, F., et al. (eds.) ESWC 2015. LNCS, vol. 9341, pp. 62–66. Springer, Heidelberg (2015). doi:10.1007/978-3-319-25639-9_12
2. Russell, A., Smart, P.R., Braines, D., Shadbolt, N.R.: Nitelight: a graphical tool for semantic query construction. In: Semantic Web User Interaction Workshop (SWUI 2008), Florence, Italy (2008)
3. Viégas, F.B., Donath, J.: Social network visualization: can we go beyond the graph? In: Proceedings of the Computer Supported Cooperative Work (CSCW 2004), Workshop on Social Networks, Banff, Canada, vol. 4, pp. 6–10 (2004)
4. Elbedweihy, K., Wrigley, S.N., Ciravegna, F.: Evaluating semantic search query approaches with expert and casual users. In: Cudré-Mauroux, P., et al. (eds.) ISWC 2012, Part II. LNCS, vol. 7650, pp. 274–286. Springer, Heidelberg (2012)
5. Vega-Gorgojo, G., Giese, M., Heggestøyl, S., Soylu, A., Waaler, A.: PepeSearch: semantic data for the masses. In: PLOS ONE (2016). http://dx.doi.org/10.1371/journal.pone.0151573
6. Pathak, J., Kiefer, R.C., Chute, C.G.: Using semantic web technologies for cohort identification from electronic health records for clinical research. AMIA Summits Transl. Sci. Proc. **2012**, 10–19 (2012)

AutoRDF - Using OWL as an Object Graph Mapping (OGM) Specification Language

Fabien Chevalier[(✉)]

AriadNEXT, 80 avenue des Buttes de Coëmes, 35700 Rennes, France
`fabien.chevalier@ariadnext.com`

Abstract. AutoRDF is an original open source framework that facilitates handling RDF data from a software engineering point of view. Built on top of the Redland software package, it bridges the gap between semantic web ontology and legacy object oriented languages, by providing transparent access to RDF resources from within standard C++ objects. Its use of widespread C++11, Boost and Redland makes it suitable not only for the desktop and server, but also for low computing power embedded devices. This framework is a result of the IDFRAud research project, where it is used to handle complex domain specific knowledge and make it available on smartphone-class devices.

1 Introduction

IDFRAud [4] project is an industrial research project led by French ID document verification leader company AriadNEXT. One of the objectives of IDFRAud is to propose an automatic solution for ID verification that can handle documents issued from a large set of countries. The solution will be able to execute specific controls according to the ID model (type, country, generation, etc.) thanks to a knowledge base. The core idea of IDFRAud project is to provide an automatic verification system for identity documents in order to replace existing manual verification processes. The different components of ID analysis and verification in IDFRAud are driven by a set of control rules. In order to guarantee an interpretable and adaptive behavior at each ID analysis step, the identity document descriptions are organized by a knowledge management module.

One of the requirements of our knowledge management module is its interoperability and portability. It is preferred to store the data in a standard way in order to be able to use other tools such as Sewelis [8] to navigate our data. Another strong requirements are that we must be able to easily extract a subset of the knowledge base to run on mobile platforms, where C/C++ language rules. After a thorough analysis of existing technologies, we decided to use RDF for its versatility/flexibility of knowledge modelling using graphs, coupled with its capability to bring formal structure to the knowledge using Web Ontology Language (OWL).

Providing an efficient, consistent, and descriptive enough model for ID documents proves to be a very challenging task. The biggest issues encountered is the

© Springer International Publishing AG 2016
H. Sack et al. (Eds.): ESWC 2016 Satellite Events, LNCS 9989, pp. 151–155, 2016.
DOI: 10.1007/978-3-319-47602-5_30

very high diversity of how ID documents look like, which makes it really difficult to design a data model that fits all cases. As such it is anticipated that the model will see serious evolutions with the number of supported documents. The second biggest issue is the fast evolving fraud patterns, which makes it necessary to add document characteristic attributes very easily in the system.

As such we needed something that could make code base maintenance easy by being able to follow a constantly evolving ontology. AutoRDF can be of some use to any kind of project that needs to manipulate RDF data where an ontology exists. The use of modern C++ make it portable to a wide variety of platforms, including all mobile phone platforms, as well as most of the embedded world systems.

2   Related Work

Code generators that are used in conjunction with UML design softwares such as BOUML [1] or IBM Rational Rose [3], are principally based on data modelling. Some other well known open source frameworks such as Hibernate OGM [2], or Neo4j OGM [5] provide easy read/write for Java objects to or from graph databases. They use a more code-centric approach, where annotations store the information the framework uses to know how object must be serialized. As such they are not model-centric, but more code-centric.

AutoRDF tries to promote a software design approach where data modelling is treated as a first class citizen. It shares also some ideas with Protégé code generator plugin [6], as it uses a similar code generating approach. However it goes further than all those tools, as it does not only build interface classes that would need to be implemented by the developer, but provides a ready to use object class hierarchy to read and write data to disk.

Compared to Redland [7], AutoRDF raises the abstraction level by understanding web ontologies, as well as providing a C++ object oriented design to RDF graphs API whereas Redland is a pure C library. It has some similarities with *owlcpp* [10], however this library has no C++ proxy code generation capability, and AutoRDF tries to keep the bar as low as possible for occasional users, by providing an easy to use C++11 API. It is quite similar in goal to the Automatic Mapping of OWL Ontologies described in [9], however transferred to C++/embedded world as an open source project.

3   Introducing AutoRDF

AutoRDF is an open source semantic web code generator for C/C++. It parses a Web Ontology Language file (OWL), builds an internal representation of the ontology, and generates a set of C++ classes that make it easy to read/write RDF data from within C++. It uses the well established Redland [7] library to perform all its input/output operations. AutoRDF uses a proxy approach, no data is copied into the C++ objects. C++ objects edit the underlying RDF graph in real time when the C++ object methods are called.

AutoRDF supports a subset of RDFS and OWL. Resources of type *rdfs:Resource* or *owl:Class* are identified as candidates for class generation, *owl:subClassOf* is used to generate inheritance relationship between generated C++ classes. *owl:oneOf* allows smart mapping to C++ enums. *owl:hasKey* generates a static object loading method, taking the key as parameter. Resources of type *owl:DatatypeProperty* and *owl:ObjectProperty* are used to generate appropriate getter/setter methods. *rdfs:domain* is used to target gettter/setter generation to the right C++ class, *rdfs:range* allows more specific C++ datatype selection. *owl:FunctionalProperty*, *owl cardinality* and *qualified cardinality* are used to choose if setters/getters should handle only single items, or if list of items should be supported. *owl:Restriction* with *owl:onDataRange* or *owl:onClass* allow to further specialize datatype of a property once applied to a given C++ object. Annotations of type *rdfs:comment, rdfs:label, rdfs:seeAlso, rdfs:isDefinedBy* are also used to generate documented C++ classes/methods. This subset is enough to generate C++ code that is most of the time as good as what a developer would have written by hand.

Figure 1 shows a simple UML diagram based on a simple geometry Ontology, and an corresponding OWL/RDF fragment is reproduced below.

```
@prefix geom: <http://example.org/geometry#> .

geom:topLeft a owl:ObjectProperty ;
    rdfs:domain geom:Rectangle .

geom:Point a owl:Class ;
    ...
    rdfs:subClassOf [ a owl:Restriction ;
        owl:onProperty geom:x ;
        owl:qualifiedCardinality"1"^^xsd:nonNegativeInteger ;
        owl:onDataRange xsd:double ] ;
    ... .

geom:Rectangle a owl:Class ;
    rdfs:subClassOf geom:Shape ;
    rdfs:subClassOf [ a owl:Restriction ;
        owl:onProperty geom:topLeft ;
        owl:qualifiedCardinality"1"^^xsd:nonNegativeInteger ;
        owl:onClass geom:Point ] .
```

AutoRDF provides a simple way to manipulate an RDF dataset from C++ code, as seen below:

```
geo::Rectangle r("http://example.org/myfancyrectangle");

// Set one of my rectangle coordinates - the long way
geo::Point tl;
tl.setX(1.0);
tl.setY(2.0);
tr.setTopLeft(tl);

// Set one of my rectangle coordinates - the short way
tr.setBottomRight(geo::Point().setX(11).setY(12));
```

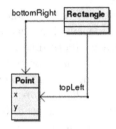

Fig. 1. Geometry ontology. This model is used as example to showcase AutoRDF code generation capabilities.

Those lines of code makes it very natural for a C++ developer to manipulate RDF data without even knowing its RDF.

```
@prefix rdf: <http://www.w3.org/1999/02/22-rdf-syntax-ns#> .
@prefix geo: <http://example.org/geometry#> .

<http://example.org/myfancyrectangle>
    geo:bottomRight [
        geo:x "11" ;
        geo:y "12" ;
        a geo:Point
    ] ;
    geo:topLeft [
        geo:x "1" ;
        geo:y "2" ;
        a geo:Point
    ] ;
    a geo:Rectangle .
```

4   Demonstration Content

The demonstration will run on a laptop and will feature Protege for ontology visualization, the geometry ontology and AutoRDF as code generator, and a simple load/save example. The demonstration will also be available with a more complex scenario, a draft of the IDFRAud project ontology.

This demonstration using demo geometry ontology is available as a screencast on the AutoRDF github page at https://github.com/ariadnext/AutoRDF.

5   Future Work

AutoRDF is still very young. It will be used in the coming years by the IDFRAud project, and as such we will add some more features, towards making working with RDF datasets easier:

- A data quality assessment API - Checking consistency of a given RDF dataset towards a reference ontology is of crucial importance to ensure correct applicative behaviour and avoid faults that are due to poor data quality.

– Some kind of C++ OWL API, in order to make creation of other tools on top of AutoRDF easier. For instance we plan to create a graphical document model editor for IDFRAud project, with user interface components generated automatically from the underlying ontology.

References

1. Bouml. http://www.bouml.fr/
2. Hibernate OGM-the power and simplicity of JPA for NoSQL datastores. http://hibernate.org/ogm/
3. IBM rational rose. http://www.ibm.com/software/products/fr/enterprise/
4. IDFRAud: an operational automatic framework for identity document fraud detection and profiling - joint research project with AriadNEXT, IRISA, ENSP and IRCGN funded by ANR grant ANR-14-CE28-0012. http://idfraud.fr/
5. Neo4j OGM-an object graph mapping library for Neo4j. http://neo4j.com/docs/ogm/java/stable/
6. Protégé code generator. http://protegewiki.stanford.edu/wiki/Protege-OWL-Code_Generator
7. Redland RDF libraries. http://librdf.org/
8. Ferré, S., Hermann, A.: Reconciling faceted search and query languages for the Semantic Web. Int. J. Metadata Semant. Ontol. **7**(1), 37–54 (2012)
9. Kalyanpur, A., Pastor, D.J., Battle, S., Padget, J.A.: Automatic mapping of owl ontologies into java. In: SEKE, vol. 4, pp. 98–103. Citeseer (2004)
10. Levin, M.K., Cowell, L.G.: owlcpp: a C++ library for working with OWL ontologies. J. Biomed. semant. **6**(1), 1 (2015)

JOPA: Efficient Ontology-Based Information System Design

Martin Ledvinka[(⊠)], Bogdan Kostov, and Petr Křemen

Czech Technical University in Prague, Prague, Czech Republic
{martin.ledvinka,bogdan.kostov,petr.kremen}@fel.cvut.cz

Abstract. Creating applications on top of linked data and ontologies brings many difficulties. The applications are either generic (and thus not appealing to end-users), or bound to ontology structure, change of which breaks the application. We present JOPA, a tool that formalizes the contract between the application and the ontology, combining advantages of both worlds. JOPA is a persistence framework for Java applications, providing formalized object-ontological mapping, transactions, access to multiple repository contexts, and producing linked data. The system is demonstrated on a real use-case of a reporting tool that we develop for the aviation industry.

1 Introduction

Keeping object-oriented applications aligned with underlying ontology commitments is a challenging task. Various ontology access libraries have been introduced over the last decade (these are briefly discussed in Sect. 2.1). Basically, developers either use too low-level API that is verbose and hard to use and maintain, or a high-level API that tries to map ontology structure to the object model which is necessarily lossy and dependent on the ontology structure.

We introduce the Java OWL Persistence API (JOPA), a tool that benefits from both approaches – it provides an object-ontological mapping, but also constructs to access all property values as well as inferred property values in an analogous manner. Furthermore, to allow easy maintenance of the application access to evolving ontologies, a formal contract between an object-oriented application and the ontology is set up. The formal contract consists of a set of integrity constraints describing the fixed part of the ontology relevant for the application, as introduced in [9]. An advantage of this explicit contract is that it allows rechecking ontology compliance with the application upon data update (a third-party change), i.e. *before* the application itself tries to access the data. Integrity constraint violation signalizes to the application designer the need for formal contract redesign (or even ontology redesign).

This demo shows the features of JOPA on a simplified application for aviation safety reporting. The full application is designed and implemented in cooperation with several stakeholders in the Czech aviation industry, including Air Navigation Services of the Czech Republic, or Prague Airport, for their future use.

© Springer International Publishing AG 2016
H. Sack et al. (Eds.): ESWC 2016 Satellite Events, LNCS 9989, pp. 156–160, 2016.
DOI: 10.1007/978-3-319-47602-5_31

2   Ontology Access Using JOPA

Let us briefly discuss first the JOPA framework itself and then delve into description of the example application and how it uses JOPA's features.

2.1   Application Access to Ontologies

There are two main approaches to application access to ontologies.

A *generic* one, where data in ontologies are manipulated without any assumptions about their nature. Such approach is represented for example by OWL API [7] or Sesame API [2]. This approach is suitable mostly for generic applications like ontology editors, because its use for domain-specific business logic requires a lot of boilerplate code.

A *domain-specific* approach to ontology access makes use of *object-ontological mapping* (OOM), which maps ontological constructs to concepts of the object-oriented paradigm. OOM enables the application to be written in object-oriented style, which is by far the most widespread programming paradigm nowadays. Frameworks exploiting OOM are for example Empire [5] or AliBaba [1].

More thorough discussion of both approaches can be found in [9] or [11].

2.2   JOPA Features

JOPA tries to take the best of both the *domain-specific* and *generic* approaches. It employs a formally defined object-ontological mapping, while providing a (limited) access to the more dynamic aspects of ontologies. Let us now briefly describe the main distinguishing features of JOPA. More detailed explanation of its architecture and features can be found in [9–11].

Formal OOM. In contrast to ad hoc mapping used by Empire or AliBaba, the object-ontological mapping in JOPA is based on a formally defined contract between the ontology and the object model. This contract is described by a set of OWL *integrity constraints* [13], which provide a closed-world view of a part of otherwise open-world assuming ontology. The OOM does not attempt to provide a complete mapping of OWL to Java, so for example only named classes and properties are supported.

Explicit Inferred Knowledge. JOPA provides explicit access to inferred knowledge in the object model. Inferred statements cannot be treated as asserted ones on the object level, because they cannot be directly changed. Therefore, JOPA enables the developer to explicitly mark attributes as inferred, which means they may contain inferred knowledge and are thus read-only.

Types and Properties. Besides mapping properties to attributes, JOPA also provides access to the more dynamic parts of the ontology. Namely, every instance can contain a set of ontological types (@Types field), to which the individual represented by this instance belongs. It can also contain a map of property values, which are not mapped by the object model. This gives, although limited, access to the ontological structure which is not directly compiled into the object model.

Separate Storage Access. By separating the actual storage access into the *Onto-Driver* layer, JOPA enables the application to easily switch between different storages. Such change thus comprises merely modifying a few lines in a configuration file. Similarly, Empire [5] uses pluggable storage access components.

JPA Features. JOPA was inspired by the JPA specification [8] for object-relational mapping in Java. As such, it supports transactional processing, caching, cascading. JOPA also supports executing SPARQL [6] and SPARQL Update [3] statements and mapping their results directly to entities. While the API of JOPA is inspired by JPA, it is not exactly the same. This is because it tries to take into account features specific to ontologies, like contexts and support for types and unmapped properties. Empire, on the other hand, goes even further and does actually implement a subset of the JPA specification.

Contexts. Some ontological storages support the notion of RDF *named graphs*, which enable data to be further structured. JOPA enables the application to exploit this feature both on object and attribute level.

2.3   Demo Application

The demo application showcases all of the features described in Sect. 2.2. The application is build for a use case in aviation safety. When a safety manager/aviation authority performs a safety audit, a checklist of several questions guides him/her through the audit agenda. The questions are linked to expected answers and whenever the actual answer does not match the expected one, it signalizes a possible safety issue.

The audit scenario is only a small part of a much larger field of aviation safety, which we are currently tackling in one of our projects[1]. The whole domain is described by a documentation ontology, which is based on the *unified foundational ontology* (UFO) [4].

For the purposes of our application, we create a set of integrity constraints [13], which restricts a part of the documentation ontology in order to make it suitable for an object-oriented application.

In the demo, a user can create audits, which are documented by reports. Every report contains a set of records, which are question-answer pairs. The records can be classified to express whether the answer was satisfactory or not.

From Ontology to Object Model. To give a glimpse of the design process, take for example the portion of the documentation ontology $\mathcal{O}_D$ shown in Table 1. A set $\mathcal{S}_{IC}$ of integrity constraints provides a closed-world view on $\mathcal{O}_D$ for the purpose of our application. When an integrity constraint is violated, the ontology becomes incompatible with the application.

---

[1] For JOPA, this is actually its second deployment. An early prototype was used in a tool called *StruFail* in the domain of structural failures of buildings.

Table 1. $\mathcal{O}_D$ represents an excerpt of the documentation ontology used in the demo application. $\mathcal{S}_{IC}$ depicts a set of OWL integrity constraints used as a contract between the application (its object model) and the ontology.

$$\mathcal{O}_D = \{Event \sqsubseteq Entity,$$
$$Report \sqsubseteq Entity,$$
$$Person \sqsubseteq Agent,$$
$$\top \sqsubseteq \forall hasAuthor \cdot Agent,$$
$$Report \sqsubseteq \exists documents \cdot Entity,$$
$$documents \equiv isDocumentedBy^- \}$$

$$\mathcal{S}_{IC} = \{Report \sqsubseteq \forall documents \cdot Event,$$
$$Report \sqsubseteq (= 1\, documents),$$
$$Report \sqsubseteq \forall hasAuthor \cdot Person,$$
$$Report \sqsubseteq (= 1\, hasAuthor),$$
$$Report \sqsubseteq \exists documents \cdot Entity,$$
$$Audit \sqsubseteq \forall isDocumentedBy \cdot Report\}$$

Based on $\mathcal{O}_D$ and $\mathcal{S}_{IC}$, transformation to the object-oriented paradigm yields a model shown in Fig. 1. The actual object model is generated from the integrity constraints by the *OWL2Java* tool, which is a part of JOPA.

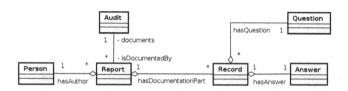

Fig. 1. Object model of the demo application. Due to space restrictions, $\mathcal{O}_D$ and $\mathcal{S}_{IC}$ capture only the Audit – Report – Person part of the model.

Demo Application Overview[2].

To list all audits and reports, a SPARQL query is used, whose results are directly mapped to the corresponding entities. While every report is related to an audit by an explicit assertion, the inverse relation is inferred, as it is not necessary to maintain both directions in the relationship. All operations on reports are cascaded to the records they contain, so for example when a report is persisted, all its records are persisted automatically as well. The same holds for the record-answer relationship. Questions are managed separately, because they can be reused by multiple reports.

Record classification is performed by adding the record individuals into OWL classes using the @Types field. In addition to the mapped attributes, every audit and record can also be enhanced with values of unmapped properties.

The demo application supports two storages - a Sesame storage and OWL files accessed by OWL API. The Sesame storage supports contexts, which is utilized by having the reports' authors stored in a dedicated context. The OWL API storage, on the other hand, is used by Pellet [12] to provide additional inferred knowledge. In our instance, it enables to show reports for each audit by exploiting the inverse *isDocumentedBy* property.

---

[2] The demo application can be found at http://onto.fel.cvut.cz/eswc2016, its source codes are available at https://github.com/kbss-cvut/jopa-examples.

3   Conclusions

We have discussed the difficulties of application access to ontologies and presented the JOPA framework as a possible solutions to these issues. We have demonstrated its viability as a persistence solution for ontology-based applications on a simplified demo application (a much more complex version of which is currently being evaluated by project partners in the Czech Republic), which nonetheless exploits most of the distinguishing features of JOPA.

Development of the aviation safety application has also shown us some shortcomings of JOPA, mainly its lack of support for OWL class subsumption (inheritance) and referential integrity. We plan to address these in our future work.

Acknowledgment. This work was supported by grant No. GA 16-09713S Efficient Exploration of Linked Data Cloud of the Grant Agency of the Czech Republic and by grant No. SGS16/229/OHK3/3T/13 Supporting ontological data quality in information systems of the Czech Technical University in Prague.

References

1. AliBaba. https://bitbucket.org/openrdf/alibaba/
2. Broekstra, J., Kampman, A., van Harmelen, F.: Sesame: a generic architecture for storing and querying RDF and RDF schema. In: Horrocks, I., Hendler, J. (eds.) ISWC 2002. LNCS, vol. 2342, p. 54. Springer, Heidelberg (2002)
3. Gearon, P., Passant, A., Polleres, A.: SPARQL 1.1 update. Technical report, W3C (2013)
4. Guizzardi, G.: Ontological foundations for structural conceptual models. Ph.D. thesis, University of Twente (2005)
5. Grove, M.: Empire: RDF & SPARQL meet JPA. semanticweb.com, April 2010. http://semanticweb.com/empire-rdf-sparql-meet-jpa_b15617
6. Harris, S., Seaborne, A.: SPARQL 1.1 query language. Technical report, W3C (2013)
7. Horridge, M., Bechhofer, S.: The OWL API: a Java API for OWL ontologies. Semant. Web Interoperability Usability Applicability **2**, 11 (2011)
8. JCP: JSR 317: Java^{TM} Persistence API, Version 2.0 (2009)
9. Křemen, P., Kouba, Z.: Ontology-driven information system design. IEEE Trans. Syst. Man Cybern. Part C **42**(3), 334–344 (2012). http://ieeexplore.ieee.org/xpl/freeabs_all.jsp?arnumber=6011704
10. Ledvinka, M., Křemen, P.: JOPA: developing ontology-based information systems. In: Proceedings of the 13th Annual Conference Znalosti (2014)
11. Ledvinka, M., Křemen, P.: JOPA: accessing ontologies in an object-oriented way. In: Proceedings of the 17th International Conference on Enterprise Information Systems (2015)
12. Sirin, E., Parsia, B., Grau, B.C., Kalyanpur, A., Katz, Y.: Pellet: a practical OWL-DL reasoner. Web Semant. Sci. Serv. Agents World Wide Web **5**(2), 51–53 (2007)
13. Tao, J., Sirin, E., Bao, J., McGuinness, D.L.: Integrity constraints in OWL. In: Fox, M., Poole, D. (eds.) AAAI. AAAI Press (2010). http://www.aaai.org/ocs/index.php/AAAI/AAAI10/paper/view/1931

A Visual Query System for Stream Data Access over Ontologies

Ahmet Soylu[1]([✉]), Martin Giese[2], Rudolf Schlatte[2], Ernesto Jimenez-Ruiz[3],
Özgür Özçep[4], and Sebastian Brandt[5]

[1] Norwegian University of Science and Technology, Gjøvik, Norway
ahmet.soylu@ntnu.no
[2] University of Oslo, Oslo, Norway
{martingi,rudi}@ifi.uio.no
[3] University of Oxford, Oxford, UK
ernesto.jimenez-ruiz@cs.ox.ac.uk
[4] University of Lübeck, Lübeck, Germany
oezcep@ifis.uni-luebeck.de
[5] Siemens AG, Munich, Germany
sebastian-philipp.brandt@siemens.com

Abstract. In this demo, we present an ontology-based visual query system, namely OptiqueVQS, extended for a stream query language called STARQL in the context of use cases provided by Siemens AG.

Keywords: Visual query formulation · Ontology · Streams · Sensors · OBDA

1 Motivation

Siemens runs several service centers for power plants, each responsible for remote monitoring and diagnostics of many thousands of gas/steam turbines and associated components such as generators and compressors. Diagnosis engineers working at the service centers are informed about any potential problem detected on site. They access a variety of raw and processed data with pre-defined queries in order to isolate the problem and to plan appropriate maintenance activities. For diagnosis situations not initially anticipated, new queries are required, and an IT expert familiar with both the power plant system and the data sources in question (e.g., up to 2.000 sensors in a part of appliance and static data sources) has to be involved to formulate these queries. Thus, unforeseen situations may lead to significant delays of up to several hours or even days.

By enabling diagnosis engineers to formulate complex queries on their own with respect to an expressive domain vocabulary, IT experts will not be required anymore for adding new queries, and manual preprocessing steps can be avoided. Yet they rarely have technical skills and knowledge on formal languages for querying streams. Ontology-based visual query formulation is promising as ontologies provide higher level abstractions closer to end users' understanding

© Springer International Publishing AG 2016
H. Sack et al. (Eds.): ESWC 2016 Satellite Events, LNCS 9989, pp. 161–166, 2016.
DOI: 10.1007/978-3-319-47602-5_32

(cf. [4,9]). Moreover, ontology-based data access (OBDA) technologies extend the reach of ontology-based querying from triple stores to relational databases (cf. [12]).

C-SPARQL [1], SPARQLstream [2], CQELS [5], and STARQL [6] are notable examples of semantic stream query languages. Although several visual tools exist for SPARQL (cf. [9]), the work is very limited for semantic stream query languages. An example is SPARQL/CQELS visual editor designed for Super Stream Collider framework [7]. However, it follows the jargon and structure of the underlying formal language closely and is not appropriate for end users.

In this demo, we present an ontology-based visual query system, namely OptiqueVQS [8], extended for a stream query language, called STARQL [6].

2 System Overview

OptiqueVQS is developed as a part of an OBDA system, Optique [4], which employs data virtualisation to enable in-place querying of legacy relational data sources over ontologies. This is realised through a set of mappings describing the relationships between the terms in the ontology and their representations in the data sources, and a query rewriting mechanisms for translating SPARQL to target language (e.g., SQL) [3,6]. In this demo, the focus is on visual query formulation and the underlying OBDA framework is out of scope. The following overviews the STARQL query language and OptiqueVQS with stream querying.

2.1 STARQL

STARQL [6] provides an expressive declarative interface to both historical and streaming data. In STARQL, querying historical and streaming data proceeds in an analogous way and in both cases the query may refer to static data. The relevant slices of the temporal data are specified with a window. In the case of historical data, this is a window with fixed endpoints. In the case of streaming data, it is a moving window and it contains a reference to the developing time NOW and a sliding parameter that determines the rate at which snapshots of the data are taken. The contents of the temporal data are grouped according to a sequencing strategy into a sequence of small graphs that represent different states. On top of the sequence, relevant patterns and aggregations are formulated in the HAVING-clause, using a highly expressive template language. In Fig. 1, an example STARQL query is given, which asks for a train with turbine named "Bearing Assembly", and queries for the journal bearing temperature readings in the generator. It uses a simple "echo" to display the results.

2.2 OptiqueVQS

OptiqueVQS[1] [8] is widget-based and supports three-shaped conjunctive queries. In Fig. 2, an example query is shown as a tree in the upper widget (W1), representing typed variables as nodes and object properties as arcs. Typed variables

---

[1] Demo video: https://youtu.be/TZTxujz5hCc.

```
PREFIX ns1 : <http://www.w3.org/1999/02/22-rdf-syntax-ns#>
PREFIX ns2 : <http://www.siemens.com/ontology/gasturbine/>
CREATE PULSE WITH FREQUENCY = "PT1s"^^xsd:duration

CREATE STREAM S_out AS
SELECT { ?_val0 ?Train_c1 ?Turbine_c2 ?Generator_c3 ?BearingHouse3_c4
    ?JournalBearing_c5 ?TemperatureSensor_c6 }
FROM STREAM measurement
    [NOW - "PT10s"^^xsd:duration, NOW]->"PT1s"^^xsd:duration
WHERE {
    ?Train_c1 ns1:type ns2:Train.
    ?Turbine_c2 ns1:type ns2:Turbine.
    ?Generator_c3 ns1:type ns2:Generator.
    ?BearingHouse3_c4 ns1:type ns2:BearingHouse3.
    ?JournalBearing_c5 ns1:type ns2:JournalBearing.
    ?TemperatureSensor_c6 ns1:type ns2:TemperatureSensor.
    ?Train_c1 ns2:hasTurbine ?Turbine_c2.
    ?Train_c1 ns2:hasGenerator ?Generator_c3.
    ?Generator_c3 ns2:hasBearingHouse3 ?BearingHouse3_c4.
    ?BearingHouse3_c4 ns2:hasJournalBearing ?JournalBearing_c5.
    ?JournalBearing_c5 ns2:isMonitoredBy ?TemperatureSensor_c6.
    ?Turbine_c2 ns2:hasName "Bearing Assembly"^^xsd:string.
}
SEQUENCE BY StdSeq AS seq
HAVING  EXISTS i IN seq
    ( GRAPH i { ?TemperatureSensor_c6 ns2:hasValue ?_val0 } )
```

Fig. 1. An example diagnostic task in STARQL.

can be added to the query by using the list in the bottom-left widget (W2). If a query node is selected, the faceted widget (W3) at the bottom-right shows controls for refining the corresponding typed variable, e.g. constraining a data property. Once a restriction is set on a data property or a data property is selected for output (i.e., using the eye icon), it is reflected in the label of the corresponding node in the query graph. The user can always jump to a specific part of the query by clicking on the corresponding variable-node in W1.

In W3, dynamic properties (i.e., whose extensions are time dependent) are colored in blue and as soon as one is selected OptiqueVQS switches to STARQL mode. A stream button appears on top of the W1 and lets the user configure parameters such as slide (i.e., frequency at which the window content is updated/moves forward) and window width interval. If the user clicks on the "Result Overview" button, a template selection widget (W4) appears for selecting a template for each stream attribute, which is by default "echo" (see Fig. 3). W4 is normally used for displaying example results in the SPARQL mode. The example query depicted in Fig. 2 and Fig. 3 represents the query example given in Fig. 1 with the exception that a "range" template is selected. The user can register the query in W4 by clicking on the "Register query" button.

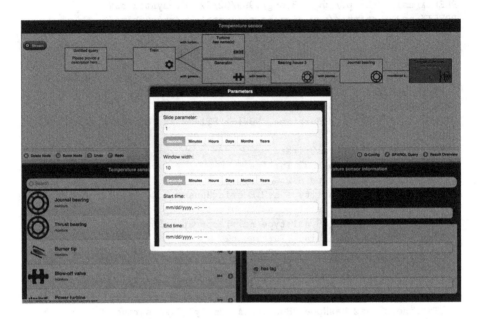

Fig. 2. OptiqueVQS with stream querying.

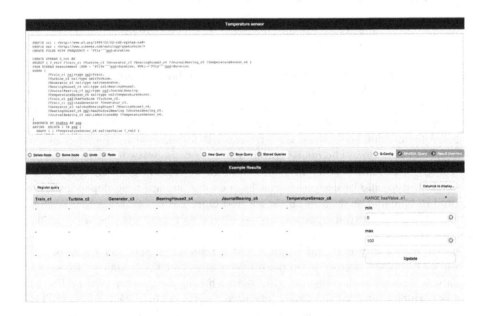

Fig. 3. OptiqueVQS with stream querying – template selection.

Several user experiments have been conducted over OptiqueVQS and the results suggest that OptiqueVQS is viable tool for end-user querying (e.g., [8, 10, 11]).

3 Demonstration Scenario

We will demonstrate OptiqueVQS over anonymised Siemens' relational stream data gathered from numerous gas and steam turbines and a set of representative diagnostic tasks. OptiqueVQS will run over the Optique platform and attendees will be able to formulate, register, and execute stream queries.

Acknowledgments. This research is funded by the "Optique" (FP7 - 318338).

References

1. Barbieri, D.F., Braga, D., Ceri, S., Della Valle, E., Grossniklaus, M.: C-SPARQL: SPARQL for continuous querying. In: Proceedings of the 18th International Conference on World Wide Web (WWW 2009), pp. 1061–1062. ACM (2009)
2. Calbimonte, J.-P., Corcho, O., Gray, A.J.G.: Enabling ontology-based access to streaming data sources. In: Patel-Schneider, P.F., Pan, Y., Hitzler, P., Mika, P., Zhang, L., Pan, J.Z., Horrocks, I., Glimm, B. (eds.) ISWC 2010, Part I. LNCS, vol. 6496, pp. 96–111. Springer, Heidelberg (2010). doi:10.1007/978-3-642-17746-0_7
3. Calvanese, D., Cogrel, B., Komla-Ebri, S., Kontchakov, R., Lanti, D., Rezk, M., Rodriguez-Muro, M., Xiao, G.: Ontop: answering SPARQL queries over relational databases. Semantic Web (in press)
4. Giese, M., Soylu, A., Vega-Gorgojo, G., Waaler, A., Haase, P., Jimenez-Ruiz, E., Lanti, D., Rezk, M., Xiao, G., Ozcep, O., Rosati, R.: Optique-zooming in on big data access. IEEE Comput. **48**(3), 60–67 (2015)
5. Le-Phuoc, D., Dao-Tran, M., Xavier Parreira, J., Hauswirth, M.: A native and adaptive approach for unified processing of linked streams and linked data. In: Aroyo, L., Welty, C., Alani, H., Taylor, J., Bernstein, A., Kagal, L., Noy, N., Blomqvist, E. (eds.) ISWC 2011, Part I. LNCS, vol. 7031, pp. 370–388. Springer, Heidelberg (2011). doi:10.1007/978-3-642-25073-6_24
6. Özçep, Ö.L., Möller, R., Neuenstadt, C.: A stream-temporal query language for ontology based data access. In: Lutz, C., Thielscher, M. (eds.) KI 2014. LNCS, vol. 8736, pp. 183–194. Springer, Heidelberg (2014). doi:10.1007/978-3-319-11206-0_18
7. Quoc, H.N.M., Serrano, M., Phuoc, D.L., Hauswirth, M.: Super stream collider: linked stream mashups for everyone. In: Proceedings of the Semantic Web Challenge at ISWC 2012 (2012)
8. Soylu, A., Giese, M., Jimenez-Ruiz, E., Vega-Gorgojo, G., Horrocks, I.: Experiencing OptiqueVQS - a multi-paradigm and ontology-based visual query system for end-users. Univers. Access Inf. Soc. **15**(1), 129–152 (2016)
9. Soylu, A., Giese, M., Kharlamov, E., Jimenez-Ruiz, E., Zheleznyakov, D., Horrocks, I.: Ontology-based end-user visual query formulation: why, what, who, how, and which? Univers. Access Inf. Soc (2016, in press). doi:10.1007/s10209-016-0465-0
10. Soylu, A., Giese, M., Schlatte, R., Jimenez-Ruiz, E., Ozcep, O., Brandt, S.: Domain experts surfing on stream sensor data over ontologies. In: Proceedings of the 1st International Workshop on Semantic Web Technologies for Mobile and Pervasive Environments (2016)

11. Soylu, A., Kharlamov, E., Zheleznyakov, D., Jimenez-Ruiz, E., Giese, M., Horrocks, I.: Ontology-based visual query formulation: an industry experience. In: Bebis, G., et al. (eds.) ISVC 2015. LNCS, vol. 9474, pp. 842–854. Springer, Heidelberg (2015). doi:10.1007/978-3-319-27857-5_75
12. Spanos, D.E., Stavrou, P., Mitrou, N.: Bringing relational databases into the semantic web: a survey. Semant. Web **3**(2), 169–209 (2012)

Extending RapidMiner with Data Search and Integration Capabilities

Anna Lisa Gentile[1]([✉]), Sabrina Kirstein[2], Heiko Paulheim[1],
and Christian Bizer[1]

[1] University of Mannheim, Mannheim, Germany
{annalisa,heiko,chris}@informatik.uni-mannheim.de
[2] RapidMiner, Dortmund, Germany
skirstein@rapidminer.com

Abstract. Analysts are increasingly confronted with the situation that data which they need for a data mining project exists somewhere on the Web or in an organization's intranet but they are unable to find it. The data mining tools that are currently available on the market offer a wide range of powerful data mining methods but hardly support analysts in searching for suitable data as well as in integrating data from multiple sources. This demo shows an extension to RapidMiner, a popular data mining framework, which enables analysts to search for relevant datasets and integrate discovered data with data that they already know. In particular, we support the iterative extension of data tables with additional attributes. We will demonstrate the usage of the extension with a large corpus of tabular data extracted from Wikipedia.

1 Introduction

The amount of data which is available within organizations and on the public Web has rapidly increased in recent years. Due to the large number of data sources, analysts face the challenge of collecting the data they need, which, although largely available, are difficult to find and integrate. The focus of many data mining projects is therefore increasingly shifting from the actual data analysis to the search for data which is suitable for a particular mining task, as well as their integration. The data mining tools currently available on the market offer analysts a wide variety of high-performance analytical methods, but they hardly support analysts in finding and integrating data. On the other hand, there are data management solutions such as Google Fusion Tables[1] or Microsoft Power Query for Excel[2] which provide for searching relevant data as well as for the manual integration of data from multiple sources. The major drawback for both solutions is that they are not full data mining frameworks and, while offering some data analysis facilities and dashboards, they are not originally intended as data mining tools and therefore lack advanced analytical methods.

---

[1] https://goo.gl/8uD4OB.

[2] https://goo.gl/Cj0Fnr.

© Springer International Publishing AG 2016
H. Sack et al. (Eds.): ESWC 2016 Satellite Events, LNCS 9989, pp. 167–171, 2016.
DOI: 10.1007/978-3-319-47602-5_33

In this demo, we show our efforts towards filling this gap by designing and implementing data search and data integration functionalities within Rapid-Miner[3], in order to assist the user in the process of finding relevant datasets for a given data mining task as well as in integrating newly discovered data with data that the user already knows. Our RapidMiner extension implements *SearchJoins*, i.e., a combined operation of searching and joining tables, similar to [1,3,5]. Starting with an initial *query table* and a *extension attribute* name (a column that the user wants to add to the table) our framework (i) retrieves *relevant tables* from previously indexed data sources and (ii) determines schema and instance correspondences between the *query table* and the retrieved *relevant tables*. The discovered tables and the correspondences between these tables and the query table are presented with a novel user interface within RapidMiner, which allows the user to iteratively correct and refine the results.

The main advantages from a user perspective are (i) the transparent search for relevant (table) data over multiple data sources, both on intranet and on the Web; (ii) the integration of retrieved *relevant tables* to the *query table* and (iii) the possibility to visually correct integration results and iterate the process.

In the following we first describe the demo from a user perspective as well as the dataset that we use for the demo (Sect. 2). We then provide a general description of our *SearchJoin* framework (Sect. 3).

2   SearchJoin: Demo and Dataset

The *SearchJoin* framework has two main steps: (i) Data Search, to find records that further describe an initial set of given entities; (ii) Data Integration, to interactively extend local tables with additional attributes based on the retrieved data records. During the demo, we will show how to perform a *SearchJoin* within RapidMiner. The user will be able to select a table she wants to enrich, e.g. a set of countries with country names and their capital and she wants to find out the population for each of them.

As table corpus for the *SearchJoin*, we will use a dataset originally consisting of 1.35 million Wikipedia tables, extracted in the WikiTables project [2]. After applying heuristics to detect useful tables (e.g. (i) it is a relational table, (ii) a subject column is detected, (iii) it contains at least 3 columns and 5 rows...) we retain and index 541 K tables. Table 1 shows some statistics of the dataset. The dataset is publicly available for download as json files[4].

The user will be free to perform arbitrary queries, i.e. choose any table from the corpus that she wants to enrich and specify the desired expansion attributes. We will also provide a ranked list of frequent classes and headers in the dataset to facilitate the formulation of the query.

---

[3] RapidMiner (https://rapidminer.com/) is recognised as a leader for Advanced Analytics Platforms (http://goo.gl/Qttrqb).

[4] All demo material, including a screencast of the demo, can be found at http://goo.gl/6MPHl0.

Table 1. Dataset statistics.

Number of tables	541, 611
Total number of cell values	56, 126, 735
...of which string	32, 931, 915
...of which numeric	18, 285, 894
...of which date	4, 908, 539
...of which link	387
Number of cells identified as subjects	9, 911, 733
Number of headers	158, 245

3   SearchJoin: A Table Extension Solution

3.1   SearchJoin: The Framework

The SearchJoin framework is specifically designed to retrieve tables that extend an initial user-provided *query table* and complement it with additional information when available. The search space can span to multiple and diverse sources of tabular data, coming from the intranet or from the Web. These are pre-processed offline, as a batch process, by the *TableIndexer* component. At the current status, the *TableIndexer* gets as input a *TableDataset*, a collection of tables $TD = t_1, t_2, ..., t_n$ and (i) detects if t_i is useful (we use a set of heuristics to discard non relational tables), (ii) pre-processes and semantically enriches all useful tables and (iii) stores the information in two Lucene[5] indexes, one for headers, one for values (to allow fast independent searches on headers or values).

In the pre-processing phase we reuse components from the T2K framework [4] to: (i) identify pseudo key attributes (the subject column); (ii) recognize table header structures; and (iii) identify data types. To identify the subject column we apply the heuristic proposed by [4] of choosing the column of type string with the highest number of unique values (in case of a tie, the left-most column is used). For detecting the headers, we currently assume that the header row is the first non-empty entity. For data type detection we use about 100 manually defined regular expressions to detect string, numeric value, date and link.

The *TableSearch* component incorporates searching and integration facilities. *TableSearch* receives as input a *query table*, where the user indicates which one is the subject column (the column in the table that contains entity names), and a set of extraction attributes (that she wants to complement her table with). It (i) retrieves a set of relevant tables from the connected indexes of tables (pre-compiled by the *TableIndexer*) and (ii) identifies correspondences between those and the query table, both at schema level (which columns in each retrieved table correspond to which columns in the query table) and at instance level (which rows in each retrieved table correspond to which rows the query table). As output

---

[5] https://lucene.apache.org.

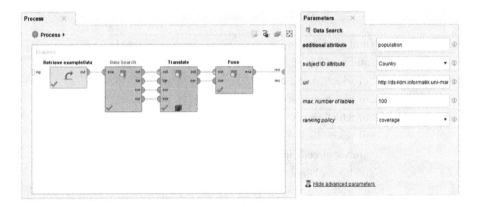

Fig. 1. RapidMiner operators *Data Search*, *Translate* and *Fuse* used to search for new attributes.

TableSearch returns all the identified correspondences and a confidence score for each matching.

3.2 SearchJoin: The RapidMiner Extension

TableSearch is provided as a RESTful service, for which a user interface is provided in RapidMiner as an extension, which is named the *Data Search Extension*. RapidMiner uses so called operators (building blocks for operations on data) to read data, transform data, learn models, apply models and write data and enables business users to build predictive analytics processes in a visual environment. Extensions can define new operators to provide additional functionality.

The new *Data Search Extension* adds three new operators to RapidMiner. The first operator, called *Data Search*, sends a request to *TableSearch* (cf. Sect. 3.1). The server response contains the target schema, names of relevant tables, and correspondences between those tables and the query table. The *Data Search* operator fetches each of those tables from the *TableSearch* Engine. The output of the operator is a collection of the relevant tables, as well as the target schema and the correspondences, at schema and at instance level.

The operator *Translate* receives the outputs from the *Data Search* operator and translates each of the relevant tables into the target schema. It delivers a collection of tables in form of the target schema, filled with the corresponding values from the relevant tables that were fetched from the *TableSearch* Engine. The operator *Fuse* uses a pre-defined order of fusion policies to decide for each row of the target schema, which value given from the relevant tables becomes the final value of this row. The result is the user *query table*, extended by an additional attribute. Figure 1 shows a RapidMiner process with the new operators *Data Search*, *Translate* and *Fuse* to add a new attribute.

4   Conclusions and Future Work

In this work, we present a data search and integration framework suitable for data coming from diverse data sources, being local storage, a company intranet or the Web. The framework is implemented as an Open Source RapidMiner extension. For demonstration purposes, we provide a 541 K dataset of tables, but the approach can be applied to any collection of tabular data.

In this prototype the integration of tables is based on strategies from [3] while the retrieval of relevant tables and extension attributes is mainly relying on keyword matching. As future work, we plan to develop a data search method that does not require the extension attributes to be known, but which makes it possible to search the attributes, e.g., based on their correlation with existing local attributes.

Acknowledgments. This paper describes the first public demo of the project DS4DM (Data Search for Data Mining) http://ds4dm.de, funded by the German Federal Ministry of Education and Research (BMBF).

References

1. Balakrishnan, S., Halevy, A., Harb, B., Lee, H., Madhavan, J., Rostamizadeh, A., Shen, W., Wilder, K., Wu, F., Yu, C.: Applying WebTables in practice. In: CIDR (2015)
2. Bhagavatula, C.S., Noraset, T., Downey, D.: Methods for exploring and mining tables on Wikipedia. In: Proceedings of the ACM SIGKDD Workshop on Interactive Data Exploration and Analytics - IDEA 2013, pp. 18–26 (2013)
3. Lehmberg, O., Ritze, D., Ristoski, P., Meusel, R., Paulheim, H., Bizer, C.: The Mannheim search join engine. Web Semant. Sci. Serv. Agents World Wide Web **35**(3), 159–166 (2015). Semantic Web Challenge 2014
4. Ritze, D., Lehmberg, O., Bizer, C.: Matching html tables to dbpedia. In: Proceedings of the 5th International Conference on Web Intelligence, Mining, Semantics, WIMS 2015, New York, NY, USA, pp. 10:1–10:6. ACM (2015)
5. Yakout, M., Ganjam, K., Chakrabarti, K., Chaudhuri, S.: Infogather: entity augmentation and attribute discovery by holistic matching with web tables. In: Proceedings of ACM SIGMOD 2012, New York, NY, USA, pp. 97–108. ACM (2012)

Smart City Artifacts Web Portal

Noorani Bakerally[(⊠)], Olivier Boissier, and Antoine Zimmermann

Univ Lyon, MINES Saint-Etienne, CNRS, Laboratoire Hubert Curien UMR 5516,
42023 Saint Etienne, France
{noorani.bakerally,olivier.boissier,antoine.zimmermann}@emse.fr

Abstract. In the smart city domain, many projects and works are generating essential information. Open and efficient sharing of this information can be beneficial for all parties ranging from researchers, engineers or even governments. To our knowledge, there is currently no full-fledged semantic platform which properly models this domain, publishes such information and allows data extraction using a standard query language. To complement this, we developed and deployed the Smart City Artifacts web portal. In this paper, we present our approach used within this platform and summaries some of its technical features and applications.

Keywords: Smart cities · Linked data · Semantic technologies · Semantic web

1 Introduction

Currently, as far as we know there is no semantic platform or central repository for keeping track of artifacts generated from projects or works in the field of smart cities. By artifact, we refer to any piece of work, for example a vocabulary, application or deliverable. Most well-known smart city projects, like km4City[1], could improve the way they publish their data. For instance, the entry points to their web portal are mostly human readable and can be enriched with a SPARQL endpoint to be machine queryable. Moreover the platform is not self-describing and one has to manually go through much details before discovering important things such as underlying model used to structure data, public datasets, RDF dumps or SPARQL endpoints. An exception is Ready4SmartCities [3]. They do provide a web platform which lists ontologies and datasets for smart cities[2] which both can be downloaded as RDF. However, the platform itself does not follow some of the Linked Data principles and best practices and does not make provenance information explicit for vocabularies. As a result, it may not be possible for a person or machine to explore their data, find specific resources and relate other data to it.

To complement existing smart city web portals, we propose the Smart City Artifacts (SCA) web portal[3] which gathers information about smart city projects

---

[1] http://www.disit.org/drupal/?q=node/6056 on 29/02/2016.

[2] http://www.ready4smartcities.eu/ on 29/02/2016.

[3] http://opensensingcity.emse.fr/scans/.

H. Sack et al. (Eds.): ESWC 2016 Satellite Events, LNCS 9989, pp. 172–177, 2016.
DOI: 10.1007/978-3-319-47602-5_34

and their artifacts while conforming to Linked Data principles and best practices. In the rest of this paper, we summarize some of the technical features and applications of the SCA web portal to demonstrate how we achieved the following tasks: (1) summarize the development of SCA ontology to provide a metamodel for smart city projects and artifacts in Sect. 2; (2) explain how we set up the SCA web portal to publish and answer queries related to smart city information in Sect. 3; (3) show how we conform to semantic standards, Linked Data principles and best practices in Sect. 4; (4) identify some use cases where the portal can be beneficial in Sect. 5.

2   SCA Ontology

The SCA web portal provides information about smart city projects and artifacts. We need a metamodel in the form of an ontology to structure all this information. No such ontology was found on ontology repositories but many such ontologies, like those shown in Table 1, were found which could be leveraged to describe some aspects of the metamodel.

To develop the SCA ontology, we chose to maximize vocabulary reuse. Most of the metadata we provide come from the domains of the ontologies listed in Table 1. When a relevant term could not be found in an existing vocabulary, we created the term within our ontology and ensured that the dereferenceable content provides the appropriate semantics. Part of the SCA ontology is shown in Fig. 1. All entities in the figure are linked to sca:Domain and muto:Tag via the dc:subject and muto:hasTag respectively. An instance of sca:Domain and/or muto:Tag is linked to a DBpedia resource via rdfs:seeAlso. More details about the whole ontology can be found on Github.[4]

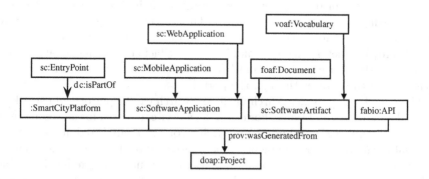

Fig. 1. Part of SCA ontology

---

[4] https://github.com/OpenSensingCity/Smart-City-Artifacts-Ontology.

Table 1. Reused vocabularies

Name	Prefix	Namespace IRI
Dublin core terms	dc	http://purl.org/dc/terms/
Description of a project	doap	http://usefulinc.com/ns/doap
The PROV ontology	prov	http://www.w3.org/ns/prov
Friend of a friend	foaf	http://xmlns.com/foaf/0.1/
Vocabulary of a friend	voaf	http://purl.org/vocommons/voaf
Schema.org	sc	http://schema.org/
Modular unified tagging Ontology	muto	http://purl.org/muto/core
FRBR-aligned bibliographic ontology	fabio	http://purl.org/spar/fabio/
DBpedia ontology	dbowl	http://dbpedia.org/ontology/
Ontology metadata vocabulary	omv	http://omv.ontoware.org/ontology#

3   SCA Web Portal

In this section, we describe the features of the SCA web portal and summarize its architecture.

3.1   User Features

All information on the SCA web portal comes from an RDF dataset structured according to the SCA ontology. The SCA web portal provides numerous entry points to visualize important resources from its dataset. For example, it provides an entry point to visualize a list of projects and then from this list, the user can choose a project to get more details about it. Viewing this project is like visualizing a specific resource in the dataset. The portal shows important links from that resource to other resources and literals. In the case of viewing a project, project's details like title and links to artifacts or documents are shown. From there, the user can choose another resource, like an artifact, and continue navigating in the dataset through the portal.

The portal also provides search facilities. Search can be performed using keywords, domain, tags and types. For example, a user can search for all resources of a particular type, e.g. vocabulary, or search resources related to domain or tags. External tools are used to augment services provided to the user. When viewing details about a particular vocabulary, the user can visualize it graphically (using Web VOWL[5]), detect ontology pitalls (using Ooops[6]) or validate namespaces (RDF Triple Checker[7]).

---

[5] http://vowl.visualdataweb.org/webvowl.html.

[6] http://oops.linkeddata.es.

[7] http://graphite.ecs.soton.ac.uk/checker/.

3.2   Architecture

Like most web application, the portal has 3 layers: a data, application and web Layer. At the data layer lies the dataset. It is structured as per the SCA ontology and enriched with links to resources from the DBpedia dataset. The portal communicates with the dataset through a SPARQL endpoint provided by Apache Fuseki[8], a SPARQL Server. At the Application layer lies the portal itself. It was developed using a Python microframework[9] and follows the Model-View-Controller pattern. The controller handles all requests, when required, it connects with the SPARQL server using a SPARQL endpoint interface[10] to fetch content and finally serves it in a particular format through content negotiation. The SPARQL endpoint is exposed on the web.[11]

At the Web layer resides Apache HTTP Server which expose the portal on the web. For all SPARQL queries over HTTP, Apache HTTP Server acts like a proxy server between the client and the SPARQL server. The portal also provides a web-based form[12] where users can directly make request on the SPARQL endpoint, get results on the same interface and download the result in different format.

4   Conformance to Semantic Standards and Best Practices

The conformance of a web platform to semantic standards and Linked Data principles and best practices can be evaluated at numerous levels. Lóscio et al. [4] list many principles and best practices which are currently in the standardization pipeline. Many principles and best practices we apply come from this list. At the vocabulary level, we have maximized vocabulary reuse to enhance semantic interoperability. At the dataset level, resources are linked to other resources from the LOD dataset, like DBpedia, for data enrichment and discoverability. It would have been ideal to link resources from our dataset to Ready4SmartCities's dataset.[13] However, this is not possible because none of the resources defined in their dataset come from their own namespace.

Moreover, all resources in our dataset which have an IRI within our namespace are dereferenceable with an almost equivalent representation of that resource both in human form (HTML) and machine readable form (rdf/xml,n3,nt,turtle) through content negotiation. Vocabulary of Interlinked Dataset (VoID) [1] is used to provide a self-description of the portal's dataset, facilitating automated data discovery. The VoID file can be obtained by requesting RDF data on the root IRI[14] of the portal.

---

[8] https://jena.apache.org/documentation/serving_data/.
[9] http://flask.pocoo.org/.
[10] https://rdflib.github.io/sparqlwrapper/.
[11] http://opensensingcity.emse.fr/scans/ds/query.
[12] http://opensensingcity.emse.fr/scans/query.
[13] http://smartcity.linkeddata.es/rdf/ontologyRDF.ttl on 29/02/2016.
[14] http://opensensingcity.emse.fr/scans.

5   Use Cases

In this section, we outline some scenarios in which the SCA web portal can be used to demonstrate the added value it can bring in.

Scenario 1: The usual scenario is navigating through the SCA web portal using a browser to obtain information. Users can go on the Projects Page[15] to view all projects. There, they can search using full-text search or through particular domain or tags. They can also, view and browse through all the details of artifacts, visualize vocabularies and apply external tools as mentioned above on vocabularies. Through this, anyone engaged in smart city projects can obtain an overview of the state of the arts in this field and find information or artifacts appropriate to their use.

Scenario 2: Consider a case where a user wants to create an HTML table of all smart city projects and the vocabularies they generated. Such content can be obtained using automated script which requests for RDF data on the *Projects Page*'s IRI. From there, after getting a list of projects' IRIs, the script exploits the provenance relationship to obtain all artifacts of type Vocabulary generated for each project.

Scenario 3: The European Union open data portal[16] provides data about many datasets which is queryable through a SPARQL Endpoint.[17] A user may want to search for all datasets from this portal and see if any of their dataset relates to a string literal containing a particular domain or tag found in SCA data. This is a possible SPARQL query[18] for such an information request.

Scenario 4: Consider an information request where someone from France wants to locate all smart city platforms in France having entry points which are related to bicycle and assume that the person only know the french word "Bicyclette" of bicycle. At first, it may seem trivial to incorporate multilingual and geographic operation. But, such an information request can be formalized as a SPARQL query[19] by exploiting the RDF links to resources from DBpedia.

6   Conclusion and Future Improvements

Smart city is becoming an important and hot topic of research in many countries. Much information about smart city projects and artifacts is scattered on the web. Through this work, we have shown that applying semantic standards, Linked Data principles and best practices has enabled us to efficiently centralize, enrich, publish and serve complex information requests using SPARQL queries and LOD dataset links. To ensure that the portal continues to benefit the international

---

[15] http://opensensingcity.emse.fr/scans/projects.

[16] https://open-data.europa.eu/fr/data.

[17] http://open-data.europa.eu/sparqlep.

[18] http://tinyurl.com/ztny6fh.

[19] http://tinyurl.com/gslgcfy.

community for smart cities, we intend to incorporate collaboration features such as providing a web form to submit new details about projects and artifacts. Also, to further enrich the dataset, we aim to densify the link set from SCA data to other LOD datasets. It is important to realize that when setting up a semantic platform, the Linked Data principles [2] are not the only 4 principles that developers have to follow. Instead, in a given context, these 4 principles generate a number of other principles, patterns and best practices which have to be considered to ensure the platform provides linked data and contributes to further realize the vision of the Semantic Web.

Acknowledgments. This work has been done within the OpenSensingCity project, supported by ANR (Agence Nationale de la Recherche), Project ID: ANR-14-CE24-0029.

References

1. Alexander, K., Cyganiak, R., Hausenblas, M., Zhao, Y.: Describing linked datasets with the VoID vocabulary, W3C interest group note 03. W3C IG note, World Wide Web Consortium (W3C), 3 March 2011
2. T Berners-Lee. http://www.w3.org/designissues/linkeddata.html. Accessed 11 June 2012
3. Garcıa-Castro, R., Gómez-Pérez, A., Corcho, O.: Ready4smartcities: ICT roadmap and data interoperability for energy systems in smart cities. In: 11th Extended Semantic Web Conference (ESWC 2014) (2014)
4. Lóscio, B.F., Burle, C., Calegari, N.: Data on the web best practices. Working draft, W3C, January 2016. https://www.w3.org/TR/dwbp/

Workflow Supporting Toolset for Diagram-Based Collaborative Ontology Development Implemented in the Open Budget Domain

Dmitry Mouromtsev[1], Dmitry Pavlov[1,2(✉)], Yury Emelyanov[1,2],
Alexey Morozov[2], Daniil Razdyakonov[2], and Olga Parkhimovich[2]

[1] ITMO University, St. Petersburg, Russia
{mouromtsev@mail.ifmo.ru, yuvemelyanov@corp.ifmo.ru
[2] Vismart Ltd., St. Petersburg, Russia
{dmitry.pavlov,alexey.morozov,daniil.razdyakonov}@vismart.biz,
olya.parkhimovich@gmail.com

Abstract. We present a live demo of a use case and a technical solution that addresses the problem of organizing the collaborative ontology development with deliverables including the diagrams and various views of the data model. The use case describes the real life situation, in which the geographically distributed team was challenged with a task of producing the open budget ontology and consequently was to select the tool set to support such development. The technical solution is based on the combination of 3 basic tools: Protégé - to provide a collaborative environment for ontology creation and modification, Ontodia.org - to visualize and publish results in a form of diagrams and GitHub - to host the repository of the project, whilst Ontodia is integrated with the last. The preliminary version of the produced ontology can be accessed at: https://github.com/k0shk/pfontology. Ontodia with GitHub integration capabilities is fully operational and can be tested here: http://www.ontodia.org.

Keywords: Semantic data visualization · Collaborative ontology development · Ontology development tool set · Ontology production tools

1 Introduction

The main objective of our project was to transform large amounts of spreadsheet data in the field of budgeting available on the web to RDF linked data in order to make it accessible for cross analysis and federated queries [3]. We realized that in order to succeed one needs to have a conceptual model to apply to the transformation. One of the most critical tasks was to produce an ontology to represent the knowledge accumulated in open budget domain. The extra complications to overcome was that our team is distributed between Russia and Great Britain and not all experts are familiar with Semantic Web standards. We assume that many ontology development teams are facing the same challenges

© Springer International Publishing AG 2016
H. Sack et al. (Eds.): ESWC 2016 Satellite Events, LNCS 9989, pp. 178–182, 2016.
DOI: 10.1007/978-3-319-47602-5_35

and the solution that we introduce can be beneficial and possesses a sufficient scalability.

As for ontology engineering methodology, the team has chosen to work according to METHODOLOGY: it supports re-usability and suggests a clear life cycle with evolving prototype model, thus ensuring effective process for distributed teams.

1.1 Workflow Scenario

At the initiation stage our team adopted the following workflow scenario:

1. Collecting the appropriate data sets of open budget data.
2. Analysis of collected data and production of generalized concept maps.
3. Construction of domain model fragments in sketches and drawings.
4. Development of the ontology in authoring tool.
5. Community review and approval.

On Stages 1, 2 and 3 selection of tools is quite trivial and stays beyond the scope of the demo. For Stages 4 and 5 we prescribed the tool category with regard to the deliverables that we committed to ship (see Table 1). On the next step for each tool category we specified the requirements based on deliverables and team limitations (see Table 2).

Table 1. Stage goals and tool categories

Stage	Goal	Deliverables	Tool category
4	To assemble the ontology in ontology authoring tool	OWL file, slides with fragments of ontology	Ontology authoring tool
			Ontology visualization tool
			Online repository hosting service
5	To receive a review from the community	Documentation for the ontology and slides	Ontology visualization tool
			Online repository hosting service

1.2 Requirements Justification

Free-to-use. Our project is non-profit and community driven one and therefore we cannot afford any pay-for products at the present stage. *Publicity and visibility.* We are strongly interested in contributions, for which the openness is crucial.

Support for collaborative work. We as a team exist in different time zones and work remotely from each other and hence have a strong need for collaborative work solutions.

Table 2. Tool categories and requirements

Tool category	Requirement
Ontology authoring tool	(1) Free to use; (2) Stable and reliable; (3) Supporting all basic OWL versions; (4) Rich import/export capabilities; (5) Well known to most user or easy to learn
Ontology visualization tool	(1) Free to use; (2) Simple and web-based for non-coding domain experts; (3) Enables access control and publishing of content; (4) Stable and reliable; (5) Rich import/export capabilities; (6) Supports collaborative work; (7) Integrates with repository hosting service
Online repository hosting service	(1) Free to use; (2) Stable and reliable; (3) Enables access control and publishing of content

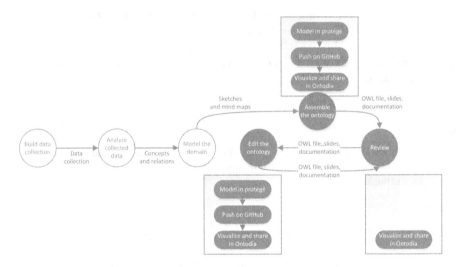

Fig. 1. Workflow and the choice of corresponding tools

Diagrams as the main artifact. Since our team members vary greatly in their expertise and specialization the only common language they can use is diagramming of the domain knowledge.

2   Technical Solution

2.1   Tool Selection

Ontology authoring tool. Initially the most preferable candidate was WebProtégé due to its support for collaborative work and online accessibility of its projects and files. The testing revealed that WebProtégé constituted too much effort to migrate to from offline solution since it has some peculiarities related to sign in and assignment of URI's to objects and we rolled back to limited usage of the offline Protégé. Due to Protégé being not precisely suited for generation of diagrams from ontologies and publishing on the web we still required a separate tool to visualize and publish ontologies for domain experts and community. The presence of Protégé in the stack also forced us to couple it with online repository hosting service for us to be able to exchange data and version control the content.

Ontology visualization tool. The tool selection is quite diverse for this task and we made use of the thorough and most detailed survey of available visualization tools [1] and found no particular tool to fulfill our vision. We made a commitment to use what we already had at our disposal: one of the team members was the contributor to ontology visualization tool development project for ITMO University. The tool was already demonstrated at ISWC2015 under the name "Ontodia" and collected good feedback [2]. Since the tool met most of the requirements except for integration with GitHub and we entirely maintain its code, we decided to extend its functionality with integration means. The description of the integration solution is provided in the next subsection.

Online repository hosting service. Provided our requirements regarding visibility and publicity we employed GitHub to serve as our file publishing medium. The need for it arose from the decision to have Protégé as our only option for modelling environment. The presence of GitHub also resulted in additional requirement for Ontodia to be integrated with GitHub.

2.2 Integration of Ontodia with GitHub

The introduced integration solution is rather obvious. It was tested not only with GitHub but with WebProtégé as well. It was discovered that Github generates the permanent URL of the file page with the following structure: https://github.com/repoowner/reponame/blob/branchname/folder/subfolder/filename. In order to get access to the file itself one needs to have the link of the following structure: https://raw.githubusercontent.com/repoowner/reponame/repobranch/folder/subfolder/filename. The desired transformation was made with a simple regular expression operation.

Ontodia has a feature of a file upload already implemented, therefore once the link leading to the file is provided Ontodia can use it as a new data source for building diagrams upon it. From the user perspective we created a new type of data source, which we tagged "GitHub source file" and that can be configured by providing a link to one or several files.

The other important feature for integration with any type of online data source is synchronization. We implemented two means of syncing with GitHub: (i) forced sync - it is when the user knows that some changes were made to the source file and wants to have the updated version in Ontodia, so he initiates the file update and (ii) regular sync - Ontodia once in 30 min syncs with the latest version of the file.

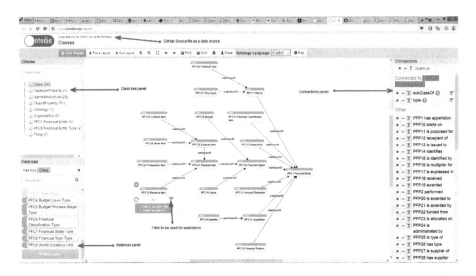

Fig. 2. Visualization of a part of the open budget ontology in Ontodia from Github OWL file

As a result the user can connect Ontodia to GitHub ontology, visualize all of it or its certain part, share it with his colleagues via their email addresses, publish it with a permanent link on the web. Ontodia can be used for preparing presentation slides by utilizing its bitmap and vector file export feature. The user invited to view the published ontology may explore the data with the use of filter button located underneath each node - See Fig. 2.

Figure 1 illustrates the proposed and well-tested workflow solution for collaborative ontology development with support for visibility and iterative approach.

The proposed toolset covers the full cycle of ontology production: (a) making changes to the ontology in Protégé; (b) pushing new file version on GitHub; (c) obtaining and publishing automatically updated diagrams in Ontodia.

Acknowledgements. This work was partially financially supported by Government of Russian Federation, Grant 074-U01.

References

1. Dudáš, M., Zamazal, O., Svátek, V.: Roadmapping and navigating in the ontology visualization landscape. In: Janowicz, K., Schlobach, S., Lambrix, P., Hyvönen, E. (eds.) EKAW 2014. LNCS, vol. 8876, pp. 137–152. Springer, Heidelberg (2014)
2. Mouromtsev, D., Pavlov, D., Emelyanov, Y., Morozov, A., Razdyakonov, D., Galkin, M.: The simple, web-based tool for visualization and sharing of semantic data and ontologies (2015)
3. Vlasov, V., Parkhimovich, O.: Development of the open budget format. In: 2014 16th Conference of Open Innovations Association (FRUCT 2016), pp. 129–136. IEEE (2014)

Qanary – The Fast Track to Creating a Question Answering System with Linked Data Technology

Kuldeep Singh[1]([envelope]), Andreas Both[2], Dennis Diefenbach[3], Saedeeh Shekarpour[4], Didier Cherix[6], and Christoph Lange[1,5]

[1] Fraunhofer IAIS, Sankt Augustin, Germany
kuldeep.singh@iais.fraunhofer.de
[2] Mercateo AG, Köthen, Germany
andreas.both@mercateo.com
[3] Laboratoire Hubert Curien, Saint-Etienne, France
dennis.diefenbach@univ-st-etienne.fr
[4] Knoesis Center, Fairborn, USA
saeedeh@knoesis.org
[5] University of Bonn, Bonn, Germany
langec@cs.uni-bonn.de
[6] FLAVIA IT-Management GmbH, Kassel, Germany
didier.cherix@gmail.com

Abstract. Question answering (QA) systems focus on making sense out of data via an easy-to-use interface. However, these systems are very complex and integrate a lot of technology tightly. Previously presented QA systems are mostly singular and monolithic implementations. Hence, their reusability is limited. In contrast, we follow the research agenda of establishing an ecosystem for components of QA systems, which will enable the QA community to elevate the reusability of such components and to intensify their research activities.

In this paper, we present a reference implementation of the Qanary methodology for creating QA systems. Qanary relies on linked data vocabularies and provides a fast track to integrating QA components into a light-weight, message-driven, component-oriented architecture.

Keywords: Software reusability · Question answering · Semantic search · Ontology · Annotation model

1 Motivation

The Web of Data is every day. Researchers have developed a variety of monolithic Question Answering (QA) systems (e.g., [2,3]) to make sense out of the enormous amount of available web data. Although the field of QA is large and many state-of-the-art QA systems exist, researchers are facing difficulties to reuse them because of their focus on implementation details and for lack of a generic approach for designing QA systems. For example, PowerAqua [3] links information available across distributed semantic resources to answer queries whereas

H. Sack et al. (Eds.): ESWC 2016 Satellite Events, LNCS 9989, pp. 183–188, 2016.
DOI: 10.1007/978-3-319-47602-5_36

TBSL [8] presents an approach that parse the question to produce SPARQL template that depictis the internal structure of the question. However, TBSL provides better results regarding linguistic analysis of questions, whereas Power-Aqua is limited w.r.t. linguistic coverage of questions. Combining the capabilities of both systems will provide better functionalities. However, these systems are monolithic and they cannot easily be combined, which reduces their applicability to new domains and the options for synergy effects.

In other research areas, such as service-oriented architectures or cloud computing, the vision of building an ecosystem of components within a dedicated field has already proven its significance for the rapid advancement of research. Therefore, establishing a methodology – on a conceptual and implementation level – is considered crucial for managing the challenges of question answering. The Qanary approach [1] provides such a methodology. Driven by linked data technology and particularly by vocabularies, it integrates the knowledge of QA components into an overall component-based QA system. However, the conceptual layer for QA systems leaves the implementation of the QA system open.

We have implemented the Qanary using a message-driven and light-weight architecture that provides a fast track for integrating QA components, and uses standard RDF technology. We present a reference implementation of a framework for QA systems, manifesting the abstract/conceptual layer of Qanary. The framework covers the main features of component-based systems, i.e., interoperability, exchangeability and reusability, flexible granularity, as well as isolation of components (cf. [1]). Therefore, a sophisticated framework level is achieved while hiding implementation details of the integrated components and establishing the qa vocabulary [7] as representation of the knowledge about the user's question and the search query derived from it.

Following our long-term research agenda, this framework provides a significant step towards a best-of-breed approach for integrating the most suitable QA components for the planned domain of application. As components integrated by this framework, we initiate hereby an ecosystem for QA components and promote the reusability of existing technology. Hence, efficiency for establishing new QA systems is increased while the effort for providing reusable components is reduced.

The next section covers our approach to create a QA system following the Qanary methodology. We also briefly introduce the qa vocabulary. Section 3 presents a methodology for vocabulary-driven integration of QA components. Section 4 concludes.

2   Approach

2.1   Requirements for Open Question Answering Systems

We have identified four key requirements, namely, *interoperability, exchangeability and reusability, flexible granularity,* and *isolation* for open QA systems [1]. The QA components are heterogeneous in their implementation, therefore, we have identified that a consistent standard interaction level, i.e., a (self-describing) abstraction of the implementation is needed to promote interoperability. This abstraction will further promote reusability to enhance efficiency of the user to build a new QA system. Hence, exchangeability and reusability are important requirements. Isolation is another identified requirement where each component should run independently of other components, i.e., it is enabled to be loosely coupled with QA systems. Flexible granularity of the components is required so they can be integrated at any step of the QA process, i.e., in contrast to other QA frameworks the granularity is not pre-defined and therefore open for future (special or general) components. To the best of our knowledge, no existing QA system or framework meets these requirements. Therefore, in our concrete implementation of the Qanary methodology, we aim at meeting these requirements.

2.2 The qa Vocabulary

In [7], we presented a vocabulary for question answering (abbreviated as qa), for representing the knowledge about a question within a QA system. Following the Qanary methodology, the qa vocabulary[1] is used to represent transitional results during the QA process, i.e., each component increases the knowledge about the given question by creating or enriching instances of the concepts `qa:Question`, `qa:Dataset` or `qa:Answer`. The qa vocabulary provides the main concepts needed to express the information for annotating a question with knowledge that was computed during the QA process. Each time a component is executed, properties (or information) such as provenance of annotation, score of annotation, relation between annotations, etc., are annotated to the message to make it available for subsequent components in the QA process. Hence, after every step of the QA process, the knowledge base (short: KB) is enriched with additional information about the question.

2.3 Integration by Vocabulary Alignment

We consider the fact that Qanary should not overrule existing (domain-specific) vocabularies. Therefore, it is intended to align existing vocabularies to the qa vocabulary, s.t., the computed data is available in a normalized representation and can easily be reused by other component just by knowing the concepts of qa. This can be done by using axioms or rules. The OWL subclass/sub property or class/property equivalence might be used to implement alignment axioms or rules. A reasoner or a rule engine can be used to map information from the Qanary KB to the input representation understood by a QA component (if the latter is RDF-based). A reasoner further translates the RDF output of a QA component to the extended vocabulary for uniformity, then adds it to the KB. An alternative option is to use SPARQL CONSTRUCT or INSERT queries to translate the data computed by a component to a representation that is aligned with the qa vocabulary.

---

[1] cf., https://github.com/WDAqua/QAOntology.

3 Methodology for Vocabulary-Driven Integration of Question Answering Components

To illustrate the power of the Qanary methodology, we took three independent components – DBpedia Spotlight [4], PATTY [5], and SINA [6] – arranged in the same order in the pipeline to build an exemplary QA system (cf. [1]). Here, we describe our approach of integrating them using Qanary with minimal programming effort. Our aim here was not to develop an actual QA system or to answer some specific questions by depicting a QA process, but to support and evaluate our claim that it is possible to reuse existing QA components by creating a new abstraction level for interoperability.

Qanary enriches a process-independent KB in each step. Unlike in a traditional QA pipeline, the output of the first component, DBpedia Spotlight, is not directly passed to the second component, PATTY, but is fed into a KB via the abstract level defined by the qa vocabulary and by aligning existing vocabularies to it. The second component needs particular input, and it fetches required input directly from the KB and pushes its output back to KB. The third component does the same. Each component can access all the messages generated by the previous components stored in a triple store through SPARQL SELECT queries and can update that information using SPARQL UPDATE queries. We follow a three-step process to implement an exemplary QA system:

1. Information gathering: In general, every component has a particular need for information as input. To ensure free access to the required information, every QA component is enabled to execute SPARQL queries and can thus retrieve any knowledge about the question. As the qa vocabulary provides a normalized representation of the data, each component only has to know qa to access the data. For example, DBpedia Spotlight needs a text query as its input. It might fetch it from the question URI (`<URIQuestion> a qa:Question`), following linked data principles. Additional RDF information about the question can be retrieved by executing a SPARQL query, e.g., to fetch named entities already annotated within a textual question. To access existing components, we have implemented light-weight wrappers that send information to the particular component to wrap around and perform its action. The sample code[2] is shown below:

```
// Execute a SPARQL query to retrieve the question URI
String sparqlQuery = "PREFIX qa: <http://www.wdaqua.eu/qa#>
                      SELECT ?questionURI FROM " + namedGraph + "
                      WHERE {?questionURI a qa:Question}";
QueryExecution qExe = QueryExecutionFactory.sparqlService(endpoint,
    QueryFactory.create(sparqlQuery));
ResultSet result = qExe.execSelect();
URL uriQuestion = result.next().getResource("questionURI");

// Retrieve the question using an HTTP request
RESTClient myRestClient = new RESTClient();
String question = rstclnt.getResults(uriQuestion.toString());
```

---

[2] using Apache Jena: https://jena.apache.org/.

```
// Send the question to the DBpedia Spotlight (local, port 8099)
String serviceUrl = "http://localhost:8099/" +
    URLEncoder.encode(question, "UTF-8");
String serviceResult = myRestClient.getResults(serviceUrl);
```

2. Information retrieval: Each component performs actions on extracted information and produces some results. In the next step, the wrapper retrieves the computed information from the component. Before pushing it to the KB, it is stored in a temporary location and the defined bindings to the qa vocabulary are applied.

3. Store results in triple store: After binding is applied on the retrieved information, the information is pushed to the KB, i.e. a triple store. Hence, following Qanary all QA components are independent from each other and reusable. For example, if a new state-of-the-art named entity disambiguation (NED) method evolves, or new input types come into the picture, researchers just need to replace the NED (in our case study this is DBpedia Spotlight), following above mentioned three steps and the new component can easily be integrated in the QA system. Additionally, it becomes reusable for any other QA system following the Qanary methodology.

A possible extension of the described QA system might incorporate support for spoken questions. Hence, a component C1 is required that translates an audio stream to a textual question, which is required by DBpedia Spotlight. The qa vocabulary is extensible and already covers the requirements for audio streams. Now the individual vocabulary of C1 needs to be aligned to qa. To integrate C1 into the QA system, a light-weight wrapper has to be implemented that fetches the required information and passes it to C1. The above mentioned three-step process will be followed and C1 can be integrated easily and efficiently into the QA system.

For details of the implementation of our exemplary QA system, please refer to our case study at https://github.com/WDAqua/Pipeline.

4   Conclusion

Qanary establishes a methodology independent from the process actually implemented by concrete QA systems. Hence, it is open for extension and ready for any new idea of how to solve QA tasks. Additionally our approach is built on top of formal logic to support reasoning and querying in a well-defined way and is independent from the actual implementation (the case study has to be considered as just one possible implementation). When a new requirement evolves, or a new component needs to be included in the pipeline, this can be accomplished via a "fast track" with minimal programming effort. Following the Qanary methodology, we meet all the requirements for a vital ecosystem of QA system components that are actually reusable. Hence, Qanary constitutes the first logical step towards actual open QA systems.

Acknowledgements. Parts of this work received funding from the European Union's Horizon 2020 research and innovation programme under the Marie Skłodowska-Curie grant agreement No. 642795, project: Answering Questions using Web Data (WDAqua).

References

1. Both, A., Diefenbach, D., Singh, K., Shekarpour, S., Cherix, D., Lange, C.: Qanary – a methodology for vocabulary-driven open question answering systems. In: Sack, H., Blomqvist, E., d'Aquin, M., Ghidini, C., Ponzetto, S.P., Lange, C. (eds.) ESWC 2016. LNCS, vol. 9678, pp. 625–641. Springer, Heidelberg (2016). doi:10.1007/978-3-319-34129-3_38
2. Damljanovic, D., Agatonovic, M., Cunningham, H.: Freya: an interactive way of querying linked data using natural language. In: García-Castro, R., Fensel, D., Antoniou, G. (eds.) ESWC 2011. LNCS, vol. 7117, pp. 125–138. Springer, Heidelberg (2012)
3. Lopez, V., Fernández, M., Motta, E., Stieler, N.: PowerAqua: supporting users in querying and exploring the semantic web. Semant. Web **3**(3), 249–265 (2011)
4. Mendes, P.N., Jakob, M., García-Silva, A., Bizer, C., DBpedia spotlight: shedding light on the web of documents. In: I-SEMANTICS (2011)
5. Nakashole, N., Weikum, G., Suchanek, F.M.: PATTY: a taxonomy of relational patterns with semantic types. In: EMNLP-CoNLL (2012)
6. Shekarpour, S., Marx, E., Ngomo, A.-C.N., Auer, S.: SINA: semantic interpretation of user queries for question answering on interlinked data. Web Semant. Sci. Serv. Agents WWW **30**, 39–51 (2015)
7. Singh, K., Both, A., Diefenbach, D., Shekarpour, S.: Towards a message-driven vocabulary for promoting the interoperability of question answering systems. In: 10th IEEE International Conference on Semantic Computing (ICSC) (2016)
8. Unger, C., Bühmann, L., Lehmann, J., Ngomo, A.-C.N., Gerber, D., Cimiano P.: Template-based question answering over RDF data. In: WWW (2012)

Computing Linked Data On-Demand Using the VOLT Proxy

Blake Regalia$^{(\boxtimes)}$ and Krzysztof Janowicz

STKO Lab, University of California, Santa Barbara, USA
blake@geog.ucsb.edu, janowicz@ucsb.edu

Abstract. The Linked Data paradigm has changed how data on the Web is published, retrieved, and interlinked, thereby enabling modern question answering systems and contributing to the spread of open data. With the increasing size, interlinkage, and complexity of the Linked Data cloud, the focus is now shifting towards strategies and technologies to ensure that Linked Data can also succeed as an infrastructure. This raises questions about the sustainability of query endpoints, the reproducibility of scientific experiments conducted using Linked Data, the lack of established quality metrics, as well as the need for improved ontology alignment and query federation techniques. One core issue that needs to be addressed is the trade-off between storing data and computing them on-demand. Data that is derived from already stored data, changes frequently in space and time, or is the output of some workflow, should be computed. However, such functionality is not readily available on the Linked Data cloud today. To address this issue, we have developed a transparent SPARQL proxy that enables the on-demand computation of Linked Data together with the provenance information required to understand how the data were derived. Here, we demonstrate how the proxy works under the hood by applying it to the computation of cardinal directions between geographic features in DBpedia.

1 Motivation

Despite the rapid growth of Linked Data and tool chains for publishing, storing, interlinking, and querying Linked Data, large-scale, real-world success stories are emerging slowly. There are many reasons for this, e.g., the difficulty of maintaining a scale-able SPARQL endpoint, and several approaches have been introduced within the last few years to make the Linked Data cloud more sustainable [2].

We believe that one challenge that has not been addressed sufficiently is the accuracy, completeness, and up-to-dateness of the data as such. From a user's perspective, it is often unclear why certain data are present while other data are not, how much uncertainty is inherent in the queried (RDF) statements, and how stable these statements are. More technically speaking, parts of this challenge are rooted in the open question of which Linked Data triples should be stored and which should be computed. Intuitively, data that is derived from already stored data, that changes frequently in space and time, or is the result

© Springer International Publishing AG 2016
H. Sack et al. (Eds.): ESWC 2016 Satellite Events, LNCS 9989, pp. 189–193, 2016.
DOI: 10.1007/978-3-319-47602-5_37

of a sequence of workflow steps, should be computed. To give a simple example, a property such as population density can be computed based on the area and population. In contrast, if the density is stored, it needs to be kept in sync with the changing population (and area) especially if multiple population values are given, e.g., from different years or on the municipality and the metropolitan level.

A similar example can be given for the completeness and accuracy cases. DBpedia stores 133,941 cardinal direction triples such as dbr:San_Francisco dbp:northeast dbr:Berkeley,_California. Why those triples exist and all the billions of cardinal direction triples between other pairs of places do not, remains unclear to the end user. Clearly, due to combinatorial explosion, it is impossible to store all possible triples so they should instead be computed on-demand. Furthermore, a simple experiment reveals that more than half of all cardinal direction relations currently stored in DBpedia are invalid or incorrect [1].

To address these and related challenges, we have developed a transparent proxy that can sit in front of arbitrary SPARQL 1.1 endpoints to augment them with computational capabilities by deriving triples and provenance records on-demand.

In this demonstration, we show how to: define a *computable property* procedure using the VOLT language, compile it to RDF, invoke computation on the procedure using real data from DBpedia, and inspect the results along with their provenance metadata. The full video is available at: https://youtu.be/EO2cD6Qy-Hc.

2 Demonstrating VOLT

In this section, we demonstrate how the transparent VOLT proxy works and how it interacts with arbitrary SPARQL endpoints by applying it to the cardinal directions use case introduced before.

2.1 Interface and Encapsulation

The VOLT proxy behaves transparently, making it appear to the user as though they are directly querying the encapsulated SPARQL endpoint. However, a single query issued by the end-user may prompt several interactions between the proxy and the actual endpoint. As depicted in Fig. 1, results from the original input query are directly returned to the client.

2.2 The VOLT Language and Compiler

The VOLT proxy requires RDF statements in the triplestore's *model graph* to follow a strict ontology because those triples ultimately get translated into machine code. Since the number of triples and nested blank nodes grows quickly with the more sophisticated procedures, directly encoding procedural logic into RDF can

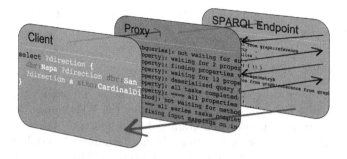

Fig. 1. A depiction of the interaction that occurs between the client, VOLT proxy and encapsulated SPARQL endpoint.

be a tedious and error-prone task for humans. Similar to the reasons that programming languages were invented to provide a layer of abstraction between the developer and assembly code, we have created a custom abstraction language designed specifically to streamline the process of coding procedures and their embedded SPARQL queries. Aside from significantly reducing the amount of code a developer has to write as well as improving its readability, the compiler also performs optimizations such as data locality, parallelization, refactoring, strength reduction, and so forth.

2.3   Creating a Procedure

For this demonstration, we start by defining a *computable property* using the VOLT programming language. A computable property represents a potential relation between two individuals. It defines all the criteria required by each the subject and object in order for a particular directed relation to hold between them. An *input triple* refers to a triple that functions as an input to a computable property. It has an explicit RDF term for each the subject, predicate, and object (i.e., it is not a *pattern*). We start the computable property definition by declaring its name as an IRI, which acts as the predicate of an input triple anytime the procedure gets invoked by a SPARQL query. In this demonstration, the computable property is named stko:south, which will test if the object of an input triple is south of its subject.

The PostGIS[1] plugin we use for this demonstration simply translates Extensible Value Testing (EVT) function calls on RDF literals from the SPARQL query into SQL strings and pipes them into a PostgreSQL[2] child process on the host machine, enabling a user to access the full domain of spatial functions provided by PostGIS. For computing cardinal directions, we employ the ST_Azimuth[3] function to compute the clockwise angle from A to B in radians relative to true north. With the help of some expressions, we can test if the radians returned by

---

[1] http://postgis.net/.
[2] http://www.postgresql.org/.
[3] http://postgis.net/docs/ST_Azimuth.html.

the `postgis:azimuth` call falls within the range that our model will consider to be `stko:south` [1].

2.4 Embedded SPARQL Queries

In order to compute the azimuth between two places, we need to obtain their coordinates. Using DBpedia data, we can extract the Well-known Text (WKT) from the `geo:geometry` property contained by each of the places we want to compare. Instructing our procedure to fetch this data from the triplestore means we will need to execute a SPARQL query at runtime. The VOLT language has a few syntactic variations when it comes to embedding SPARQL queries into a procedure's code. In the video demonstration, we show the implicit syntax which is the simplest way to obtain a value from one of the subject's or object's own triples. In this case, we select the triple(s) having `geo:geometry` as the predicate. Since a subject or object may contain more than one triple that share the same predicate, VOLT will fork the execution of a procedure anytime a SPARQL query returns more than one result. This ensures that every combination of selections is treated to a procedure's computation. Also, if any subject or object in a computable property does not yield a result from the SPARQL query, for example a place does not have the `geo:geometry` property, execution on that input triple is cancelled.

2.5 Testing Relations Existentially

Computable properties can be used to explicitly test if a certain relation holds between two named individuals. For our example, we will be asking if San Diego is south of Yosemite National Park. The computable property we wrote, `stko:south`, will be evaluated using these two places as the object and subject, respectively. Since VOLT intends to act as transparently as possible, our modus operandi is to issue SPARQL queries as though we are querying for triples that may or may not exist in the triplestore. Therefore, to invoke computation on the computable property method we created, a client simply issues an ASK query containing a single triple whose predicate is the name of the property we want to test, `stko:south`, and puts the URIs for the places of Yosemite National Park and San Diego as the subject and object of the input triple. The benefit of this approach is that a user does not need to be aware that they may be invoking procedural computation in their query — although, they are able to discover all procedures that the proxy can offer. If a user intends to discover which relations exist between two entities, the proxy would have to test all possible procedures against that subject-object pair to determine which ones should be materialized. Such a request could cause unwanted delays during query execution. To address this issue, the proxy will only test procedures when an explicit pattern is stated in the query. This pattern requires a triple having a variable in the predicate position, where that variable also appears in a triple that asserts its `rdf:type`. The object of the predicate variable's `rdf:type` represents the class of procedures to invoke testing on. For example, `dbr:San_Diego ?rel dbr:Napa. ?rel a` `stko:CardinalDirection` tells the proxy to test all procedures that are declared to be a type of `stko:CardinalDirection` and materialize the ones that evaluate positive.

If the input triple already exists in either the source graph (e.g., the relation is explicit) or the output graph (e.g., the relation was computed earlier), then the proxy does not evaluate the associated procedure since the solution is already present in the triplestore. On the other hand, if the computable property is determined to be viable for a given subject and object, i.e., if its procedure evaluates to `true`, then the input triple is materialized and stored to a temporary graph in the triplestore. By the time the proxy has finished evaluating all procedures invoked by the client's query, the proxy terminates its session by issuing the original SPARQL query on the union of the source graph and output graph(s), combining all source data with any triples that were derived by computation.

3 Summary and Outlook

In this demo paper he have showcased the transparent VOLT proxy by applying it to the computation of cardinal directions between geographic features in DBpedia. We have argued why it is important and often necessary to compute data on-demand instead of storing it and have highlighted some of the implementation details underlying VOLT. Future work will focus on improving the performance of the VOLT proxy (see [1] for first results), a library of commonly used computable properties for the geo-sciences and beyond, and on an ontological framework for caching.

References

1. Regalia, B., Janowicz, K., Gao, S.: VOLT: a provenance-producing, transparent SPARQL proxy for the on-demand computation of linked data and its application to spatiotemporally dependent data. In: Sack, H., Blomqvist, E., d'Aquin, M., Ghidini, C., Ponzetto, S.P., Lange, C. (eds.) ESWC 2016. LNCS, vol. 9678, pp. 523–538. Springer, Heidelberg (2016). doi:10.1007/978-3-319-34129-3_32
2. Rietveld, L., Verborgh, R., Beek, W., Vander Sande, M., Schlobach, S.: Linked data-as-a-service: the semantic web redeployed. In: Gandon, F., Sabou, M., Sack, H., d'Amato, C., Cudré-Mauroux, P., Zimmermann, A. (eds.) ESWC 2015. LNCS, vol. 9088, pp. 471–487. Springer, Heidelberg (2015). doi:10.1007/978-3-319-18818-8_29

The Song Remains the Same: Lossless Conversion and Streaming of MIDI to RDF and Back

Albert Meroño-Peñuela[1,2(✉)] and Rinke Hoekstra[1,3]

[1] Department of Computer Science,
Vrije Universiteit Amsterdam, Amsterdam, Netherlands
{albert.merono,rinke.hoekstra}@vu.nl
[2] Data Archiving and Networked Services, KNAW, The Hague, Netherlands
[3] Faculty of Law, University of Amsterdam, Amsterdam, Netherlands

Abstract. In this demo, we explore the potential of RDF as a representation format for digital music. Digital music is broadly used today in many professional music production environments. For decades, MIDI (Musical Instrument Digital Interface) has been the standard for digital music exchange between musicians and devices, albeit not in a Web friendly way. We show the potential of expressing digital music as Linked Data, using our `midi2rdf` suite of tools to convert and stream digital music in MIDI format to RDF. The conversion allows for lossless round tripping: we can reconstruct a MIDI file identical to the original using its RDF representation. The streaming uses an existing, novel generative audio matching algorithm that we use to broadcast, with very low latency, RDF triples of MIDI events coming from arbitrary *analog* instruments.

Keywords: MIDI · RDF · Music streams · Linked Data

1 Introduction

The Semantic Web is all about data diversity. The use of Linked Data principles [4, a.o.] boosted the growth of the Web of Data in a wide variety of domains. At the heart of Linked Data lies RDF, the Resource Description Framework (RDF)[1], which is a data model for expressing metadata *about any* resource. The Linked Open Data Cloud [6] and LOD Laundromat [1] consist of thousands of millions of such *metadata* statements. With a focus on what these RDF statements are *about*, we can distinguish between two types of Linked Data datasets. First of all, datasets that directly describe a domain – the statements are about real world facts. Examples are Bio2RDF, GeoNames and DBPedia.[2] The second

---

[1] See http://www.w3.org/RDF.
[2] See http://bio2rdf.org, http://www.geonames.org and http://dbpedia.org, respectively.

© Springer International Publishing AG 2016
H. Sack et al. (Eds.): ESWC 2016 Satellite Events, LNCS 9989, pp. 194–199, 2016.
DOI: 10.1007/978-3-319-47602-5_38

type capture *metadata* only – the statements are about data artifacts that themselves describe data. Examples are L3S DBLP and Semantic Web Dogfood[3] on bibliographic data, and – in our case – Linked Data about *music*.

In the case of the latter type of Linked Data, the actual *digital objects* represented by RDF statements are not (yet) machine interpretable from a semantic perspective (texts, images, video, audio), or require the use of legacy tooling to do so. Thus, although the resources that *describe* digital objects are promoted to first-class citizens of the Web – the objects themselves remain, to a large extent, non-interoperable.

One of the most prominent types of digital objects are *music* pieces. Music metadata has received a lot of attention by the Linked Data community. For example, DBTune.org links several music-related data sources on the Semantic Web [7], among them MusicBrainz [8], MySpace, and BBC Music. Other works have looked into publishing music recording metadata, and results from audio analysis algorithms [2]. Although the publication of music metadata about artists, songs, albums, musical events, and experimental results is an obvious contribution to the variety of Linked Data, music itself is currently only composed, published, and exchanged offline or using monolithic systems.

Currently, *musicians* compose and interpret music in a data-silo setup, with no use of scalable, global, machine readable, or resource-linkable formats. Consequently, existing musical data is not shared nor reused when musicians create, mix, combine, and publish music. Reuse only happens at an abstract, hardly reproducible level – listening, reading scores, transcribing, etc. A large-scale analysis of musical compositions and artifacts, and the relationships between them, their properties, fundamental nature, and intended meaning, is, with current musical representations, impossible. The fact that Linked Music Data is limited to metadata proper means that musicians cannot exploit Web technologies to address these issues.

Many works have studied symbolic representations for music suitable for the Web. The Notation Interchange File Format (NIFF), the Music Encoding Initiative (MEI), and MusicXML [3] are standards aimed at exchanging digital sheet music in a machine-readable way. Beyond music scores, MIDI (Musical Instrument Digital Interface) allows machines to exchange, manipulate and interpret musical events, and is to date the only symbolic music interchange format with wide adoption. The W3C Audio Working Group is working on a Web MIDI API[4] for "enabling web applications to enumerate and select MIDI input and output devices on the client system, and send and receive MIDI messages". This relates to the actual music in the same way that semantic web services relate to Linked Data. Some works presented at the International Workshop on Semantic Music and Media[5] (SMAM) specifically address the task of broadcasting chord and other recognized information from analog musical events in RDF on the Web.

---

[3] See http://dblp.l3s.de/d2r/ and http://data.semanticweb.org.

[4] https://www.w3.org/TR/webmidi/.

[5] See http://semanticmedia.org.uk/smam2013/.

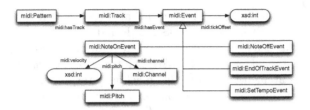

Fig. 1. Data model of MIDI concepts and their relations: patterns, tracks, events, and their attributes.

In this paper, we describe the first steps towards representing music in MIDI format as Linked Data. We investigate whether RDF is a suitable format to represent the *content* of digital music, and we explore the potential of such a representation. The novelty of the demonstration lies in three aspects. First, we study the essential concepts of MIDI to create a conversion workflow of MIDI music to RDF. Second, we invert this transformation, finding that it is possible to recompose the original MIDI from its RDF representation, thus providing a lossless round-trip. Third, we reuse a novel generative audio matching algorithm [5] to create RDF streams of music from any analog instrument, using MIDI as a discretization proxy.

The rest of the paper is organized as follows. In Sect. 2 we describe the basic concepts of MIDI, and we propose two methods: one to transform MIDI files to RDF and back; and another to stream live music from analog instruments encoded in RDF, using a novel generative audio matching algorithm. We also detail the key technology used in their implementations. In Sect. 3 we illustrate the contents of the demonstration, focusing on the lessons that can be learned from it, and we briefly discuss future work.

2   midi2rdf

MIDI is a standard that allows communication between a wide variety of electronic musical instruments, computers, and other devices. As a universal synthesizer interface, it abstracts musical events from the hardware, facilitating music exchange in a hardware independent manner.

Figure 1 depicts the fundamental MIDI concepts of *patterns*, *tracks* and *events*. A pattern acts as a high level container of a musical work (e.g. a song). It contains global MIDI metadata, such as resolution and format, and a list of tracks. Tracks are the logical divisions of a pattern, and typically represent musical instruments that play simultaneously. Finally, tracks contain events, which are sequential occurrences of instrumental actions, such as "a note starts playing" (`mid:NoteOnEvent`) or "a note stops playing" (`mid:NoteOffEvent`). MIDI defines 28 different types of events that control various aspects of the musical play. All MIDI events have associated a *tick* offset, which indicates the relative distance between events in discrete units. Each event type has its own attributes.

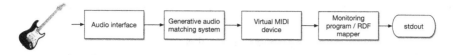

Fig. 2. Workflow for converting instrument analog signals to MIDI-like RDF streams.

For example, a `mid:NoteOnEvent` has a pitch (the note to be played), a velocity (its duration), and a channel (the tone or "instrument").

We use the data model of Fig. 1 to implement a conversion workflow transforming the internal components of a MIDI to an RDF model. `midi2rdf`[6] is an open-source suite of programs for encoding and decoding digital music between the MIDI and RDF formats. `midi2rdf` consists of several tools addressing two tasks: *conversion*, and *streaming*.

Conversion. The conversion programs are `midi2rdf` and `rdf2midi`. Both are Python scripts written on top of the `python-midi`[7] (a comprehensive abstraction over MIDI that facilitates reading and writing contents from and to MIDI files) and `rdflib`[8] libraries. We use `midi2rdf` to iterate over all tracks and contained events in the pattern of any MIDI file, and transform them to an RDF graph as shown in Fig. 1. The inverse process is carried out by `rdf2midi`, which transforms any RDF graph compliant with the model of Fig. 1 back to MIDI. Auxiliary tools for displaying contents of MIDI files, and for playing RDF files with MIDI contents in just one command, are also supplied.

Streaming. To stress the importance of a symbolic and Web-friendly representation of music during a *performance*, we propose the workflow depicted in Fig. 2. We implement this workflow in the program `stream-midi-rdf`, also included in `midi2rdf`. The basic idea is to create a stream of RDF data according with the model of Fig. 1, using a discretization of input analog instruments through MIDI[9]. A key step here is the generative audio matching system: an algorithm that can translate analog, continuous input audio to digital, discrete MIDI events. For this, we use the algorithm provided in Guitar MIDI 2[10] [5]. We route the output of this algorithm to an IAC (Inter-Application Communication Driver) virtual MIDI device, dumping raw MIDI event data. To read them, we use `pygame`[11], a set of Python modules designed for writing video games with good support of MIDI interfaces. Finally, we use again `rdflib` to send streams of RDF triples according to Fig. 1 to the standard output.

---

[6] Source code available at https://github.com/albertmeronyo/midi2rdf.
[7] See https://github.com/vishnubob/python-midi.
[8] https://github.com/RDFLib/rdflib.
[9] Instrument and audio interfaces can be replaced by any analog input (microphone).
[10] See also http://www.jamorigin.com/products/midi-guitar/.
[11] See http://pygame.org/.

3   Demonstration

The demonstration of the `midi2rdf` suite will consist of two parts: a validation on the lossless round-trip conversion between MIDI and RDF; and a live performance, with real instruments, showing the RDF streaming workflow of Fig. 2. In the first part, visitors of the demonstration will evaluate whether the round-trip conversion is lossless in a variety of MIDI files. They will learn the basic features of symbolic music representation, as well as the atomic concepts of MIDI, and how this explains that both representations are equivalent. We will show why the translation of these concepts to RDF is challenging, especially regarding issues like preserving the order of MIDI events. Furthermore, visitors will be challenged to think over a Web of Linked Music Data beyond musical metadata. We are interested in the adsvantages that Linked Data provides over current approaches like MusicXML, in e.g. uniquely (and globally) representing notes using URIs – thus enabling a more standard way of music comparison and combination –, the content of their dereferencing, and the ability to query across (streamed) music – something that is not possible with current standards.

In the second part, visitors will see the state of the art ins generative audio matching in action, witnessing how a live performance with a real instrument is turned into an RDF stream of triples describing MIDI events of the music they hear. The audience will learn that, despite the precision and performance of these methods, detecting correct notes at real-time latency is a problem with an impact on the usefulness of such RDF streams. We hope to engage in discussions about the pros and cons of discretizing music this way.

We plan to improve this work in several ways. First, we plan on issuing PROV triples indicating which `prov:Agent` performed which `prov:Activity` on what `prov:Entity` to better document performances. Second, we will extend `midi2rdf` to better support less common types of MIDI events. Finally, we plan to use the proposed tools to deploy a Linked Dataset of MIDI music, link it to related resources in the LOD cloud, compose new pieces by combining existing ones with SPARQL, and explore the semantics of music.

References

1. Beek, W., Rietveld, L., Bazoobandi, H.R., Wielemaker, J., Schlobach, S.: LOD laundromat: a uniform way of publishing other people's dirty data. In: Mika, P., et al. (eds.) ISWC 2014, Part I. LNCS, vol. 8796, pp. 213–228. Springer, Heidelberg (2014). doi:10.1007/978-3-319-11964-9_14
2. Cannam, C., Sandler, M., Jewell, M., Rhodes, C., d'Inverno, M.: Linked data and you: bringing music research software into the semantic web. J. New Music Res. **39**(4), 313–325 (2010)
3. Good, M.: MusicXML: An Internet-Friendly Format for Sheet Music (2001). http://citeseerx.ist.psu.edu/viewdoc/summary?doi=10.1.1.118.5431
4. Heath, T., Bizer, C.: Linked Data: Evolving the Web into a Global Data Space, 1st edn. Morgan and Claypool, New York (2011)

5. Kristensen, O.: Generative audio matching game system, 3 March 2011. https://www.google.com/patents/WO2010142297A3?cl=en, wO Patent App. PCT/DK2010/050,132
6. Schmachtenberg, M., Christian Bizer, A.J., Cyganiak, R.: Linking Open Data cloud diagram 2014 (2014). http://lod-cloud.net/
7. Raimond, Y., Sandler, M.: A web of musical information. In: Ninth International Conference on Music Information Retrieval (ISMIR2008) (2008)
8. Swartz, A.: Musicbrainz: a semantic web service. IEEE Intell. Syst. **17**, 76–77 (2002)

Workshop on Emotions, Modality, Sentiment Analysis and the Semantic Web

An Empirical, Quantitative Analysis
of the Differences Between Sarcasm and Irony

Jennifer Ling and Roman Klinger[(✉)]

Institut für Maschinelle Sprachverarbeitung, Universität Stuttgart,
Pfaffenwaldring 5b, 70569 Stuttgart, Germany
{jennifer.ling,roman.klinger}@ims.uni-stuttgart.de

Abstract. A variety of classification approaches for the detection of ironic or sarcastic messages has been proposed in the last decade to improve sentiment classification. However, despite the availability of psychologically and linguistically motivated theories regarding the difference between irony and sarcasm, these typically do not carry over to a use in predictive models; one reason might be that these concepts are often considered very similar. In this paper, we contribute an empirical analysis of Tweets and how authors label them as irony or sarcasm. We use this distantly labeled corpus to estimate a model to distinguish between39 both classes of figurative language with the aim to, ultimately, improve the semantically correct interpretation of opinionated statements. Our model separates irony from sarcasm with 79 % accuracy on a balanced set. This result suggests that the task is harder than separating irony or sarcasm from regular texts with 89 % and 90 % accuracy, respectively. A feature analysis shows that ironic Tweets have on average a lower number of sentences than sarcastic Tweets. Sarcastic Tweets contain more positive words than ironic Tweets. Sarcastic Tweets are more often messages to a specific recipient than ironic Tweets. The analysis of bag-of-words features suggests that the comparably high classification performance to distinguish irony from sarcasm is supported by specific, reoccurring topics.

1 Introduction

Irony and sarcasm are rhetoric devices present in everyday life. With the advent of social media platforms, the interest increases to differentiate ironic and sarcastic textual contributions from regular ones. This is particularly necessary to correctly link opinions to semantic concepts. For instance the review fragment [1]

> *"i would recomend this book to friends who have insomnia or those who i absolutely despise."*

contains the positive phrase "recomend this book". However, the irony markers "have insomnia" and "who i absolutely despise" suggest a non-literal meaning. The detection of these semantic differences between ironic, sarcastic and regular language is a prerequisite for a subsequent aggregation of harvested information.

© Springer International Publishing AG 2016
H. Sack et al. (Eds.): ESWC 2016 Satellite Events, LNCS 9989, pp. 203–216, 2016.
DOI: 10.1007/978-3-319-47602-5_39

Many previous approaches to detect irony and sarcasm assume that both concepts are sufficiently similar to not make an explicit differentiation when distinguishing them from regular utterances. In this paper, we challenge this assumption: We do not focus on the difference between figurative (*e.g.*, ironic and sarcastic, amongst other) and non-figurative utterances but on an empirical evaluation and analysis of the *difference between sarcasm and irony*. We develop a data-driven model to distinguish sarcasm and irony based on authors' labels of their own Tweets. We assume that such an approach can potentially help to improve polarity detection in sentiment analysis settings [2], as sarcasm is often considered a more drastic form of irony which is used to attack something or someone [3]. We provide a data-driven interpretation of the difference between sarcasm and irony based on the author's point of view. Similarly to previous work in sentiment classification on Twitter [4], we use information attached to each Tweet to distantly label it. Such approach might lead to noisy labels, but this is not an issue in our setting: We are interested in what authors consider to be ironic and sarcastic, though these users might not be aware of formal definitions of these concepts.

Therefore, our main contributions are:

(1) We retrieve and publish a corpus of 99000 English Tweets, 33000 of which contain the hashtag #irony or #ironic and 33000 contain #sarcasm or #sarcastic. The remaining ones contain none of those.
(2) We compare typical features used for sentiment analysis and figurative language detection in a classification setting to differentiate between irony and sarcasm.
(3) We perform a feature analysis with the aim to discover the latent structure of ironic and sarcastic Tweets.

The remainder of the paper is organized as follows: We briefly review previous work in Sect. 2. In Sect. 3, we explain our corpus collection and introduce the features and the classifiers. In Sect. 4, we discuss our findings and conclude in Sect. 5 with a summary and present the key results of our experiments.

2   Background and Related Work

Irony and sarcasm are important devices in communication that are used to convey an attitude or evaluation towards the content of a message [5]. Between the age of six and eight years, children gain the ability to recognize ironic utterances [6–8]. The principle of inferability [9] states that figurative language is used if the speaker is confident that the addressee will interpret the utterance and infer the communicative intention of the speaker/author correctly. Irony is ubiquitous, with up to 8 % of utterances exchanged being ironic [10].

Utsumi proposed that an ironic utterance can only occur in an *ironic environment* [11]. The theories of the *echoic account* [12], the *pretense theory* [13] or the *allusional pretense theory* [14] have challenged the understanding that an ironic utterance typically conveys the opposite of its literal propositional content. However, in spite of the fact that the attributive nature of irony is widely accepted (see [15]), no formal definition of irony is available as of today which is instantiated across operational systems and empirical evaluations.

Commonly, the difference between irony and sarcasm is not made explicit (for instance in the context of an ironic environment, which holds for sarcasm analogously [16]). [17] state that irony and sarcasm are often used interchangeably, other publications mention a high similarity between sarcasm, satire, and irony [18]. Sarcasm is often considered a specific case of irony [19], often more negative and "biting" [20,21]. In other approaches, it is defined as a synonym to verbal irony [9,22].

Systems for automatic recognition of irony and sarcasm typically focus on one of both classes (irony *or* sarcasm) or do not explicitly encode the difference (combine irony *and* sarcasm) [23–25].

There is only little work on automated systems which aim at learning the difference between sarcasm and irony, neither of data-driven analyzes. One approach we are aware of assumes in a corpus-based analysis that the hashtag #irony refers to situational irony and #sarcasm to the intended opposite of the literal meaning [2]. In a corpus of 257 Tweets with the hashtag #irony, only 2 refer to verbal irony. About 25 % of this corpus involve clear situational irony. However, this work focused on the impact of irony detection on opinion mining. Similarly, one goal of the SemEval 2015 Task 11 [26] was to evaluate sentiment analysis systems separately on sarcastic and ironic subcorpora. It has been shown that figurative language-specific methods improve the result [27].

Wang [28] focused on the analysis of irony vs. sarcasm – to our knowledge the first work with this goal. Her approach applies sentiment analysis and performs a manual, qualitative sub-corpus analysis. She finds that irony is used in two senses: One which is equivalent to the use of sarcasm and intends to attack something or someone, and one which is to describe an event; therefore refers to situational irony. However, in contrast to our work, she does not perform an automatic classification of irony and sarcasm, nor a detailed feature analysis. Our work, in contrast to her work, focuses more on quantitative aspects than qualitative aspects.

Very recently, Sulis et al. [29] analyzed the differences between the hashtags #irony, #sarcasm, and #not. Their approach is similar to our experiments. However, the feature sets described in their work and in our work complement each other.

3   Methods

3.1   Corpus

Our assumption is that users on Twitter annotate their messages ("Tweets") to be ironic or sarcastic using the respective hashtags. These hashtags are therefore the *irony markers*. It cannot necessarily be expected that additional irony markers exist. Therefore, our distant supervision assumption is different from the findings and conclusions in which the labels should approximate a ground truth. On the contrary, our research goal is to understand the specific use of the difference of the authors' labels and not to generalize to Tweets without such labels.

The basis for our analysis of irony and sarcasm is a corpus of 99 000 messages crawled between July and September 2015. The corpus consists of 33 000 Tweets (30 000 training and 3 000 test data) from each of the categories *irony*, *sarcasm*, and *regular*, selected with respective hashtags #irony/#ironic, #sarcasm/#sarcastic. Frequent other hashtags in these subcorpora are #drugs, #gopdebate, #late, #news and #peace which we use as search terms to compile the corpus of regular Tweets together with #education, #humor and #politics, which were used in previous work (*cf.* [23,24]).

Each Tweet is preprocessed as follows: We store the post date, the author name, the Tweet ID, the text and, if available, the geolocation (available in 1241 cases) and the provided location (5860 cases). Retweets (marked with "RT"), Tweets shorter than five tokens and near-duplicates (based on token-based Jaccard similarity with a threshold of 0.8) are discarded [30]. Tweets are assigned to the ironic subcorpus if they contain #irony/#ironic but not #sarcasm/#sarcastic and vice versa. If both classes of figurative language are present, the Tweet is discarded (0.24 % of the Tweets with at least one label also contain the other). The text is tokenized with a domain-specific approach [31] (with small adaptations to keep UTF8 emoticons, acronyms and URLs each in one token) and part-of-speech-tagged with the GATE Twitter PoS Tagger [32]. The corpus is available at http://www.romanklinger.de/ironysarcasm.

3.2 Features

In line with previous work with the aim to distinguish between figurative language and regular language [24,33,34], we employ a set of features to measure characteristics of each text, which we list and explain briefly in the following.

Bag-of-Words/Unigrams, Bigrams. All words (unigrams) and the 50 000 most frequent bigrams are used as features individually. We refer to this feature as *BoW*.

Figurative Language. Each of the following features is instantiated in four versions, the count ($c \in \mathbb{N}$), Boolean (true iff $c > 0$), the count normalized by the number of tokens $|\mathbf{t}|$ in the Tweet ($r = \frac{c}{|\mathbf{t}|}$), and a series of Boolean features representing r by stacked binning in steps of 0.2.

The feature *Emoticon* counts the number of emoticons corresponding to facial expressions in the Tweet [18,24]. We use all UTF8 encoded emoticons and symbols from [35]. For instance, the Tweet "the joke's on me 🫠🬀😄😊" has an *Emoticon* count of 2, the Boolean version is true, the normalized value is $\frac{2}{8}$, and stacked binning Boolean features are true for >0 and for >0.2. Analogously, we use features for *Emoticon Positive* and *Emoticon Negative* and *Symbol* for non-facial symbols [36]. The feature *Emoticon-Sequence* is analogously defined as the length of the longest sequence of consecutive emoticons (the Boolean version is true if count is greater 1).

The feature *Capitalization* counts the number of capitalized words per Tweet [23]. The feature *Interjection* measures the count of occurrences of phrases like "aha", "brrr" and "oops" [37,38]. The feature *Additional Hashtags* measures the occurrence of hashtags in addition to the label hashtags.

The following features are implemented Boolean-only: *User* holds if a Tweet contains a string starting with "@", which typically refers to an addressee [18,36]. The feature *URL* captures the occurrences of a URL. *Ellipsis and Punctuation* holds if ". . ." is directly followed by a punctuation symbol [23,33]. In addition, we check if *Punctuation Marks* [23,36,37], a *Laughing Acronym* (`lol`, `lawl`, `luls`, `rofl`, `roflmao`, `lmao`, `lmfao`), a *Grin* (`*grin*`, `*gg*`, `*g*`), or *Laughing Onomatopoeia* (`[mu-|ba-]haha`, `hihi`, `hehe`) occur [33].

Sentiment. The expression of a sentiment or opinion is a central method to formulate emotions and attitudes. We calculate a dictionary-based sentiment score to classify Tweets into positive or negative [39]. The feature *Sentiment Score Positive* and *Sentiment Score Negative* are summed-up scores of positive words and negative words, respectively [28]. The *Sentiment Score Gap* is the absolute difference between the scores of the most positive and the most negative word [40]. Boolean features are *Valence Shift*, which holds if polarity switches within a four word window. *Pos/Neg Hyperbole* holds if a sequence of three positive or three negative words occurs, *Pos/Neg Quotes* if up to two consecutive adjectives, adverbs or nouns in quotation marks have a positive or negative polarity, *Pos/Neg &Punctuation* if a span of up to four words with at least one positive/negative word occurs but no negative/positive word ends with at least two exclamation marks or a sequence of a question mark and an exclamation mark. *Pos/Neg &Ellipsis* is defined analogously [33].

Syntactic. In the four versions mentioned above, we measure the number of *Stopwords, Nouns, Verbs, Adverbs, Adjectives* [37], *Pronomina* ("I/we", "my/our", "me/us", "mine/ours"). The *Tweet Length* as an integer is measured in addition and, as Boolean feature only, the occurrence of *Negations* ("n't", "not") [36]. *Sentence Length Gap* measures the difference between the shortest and the longest sentence. The Boolean version of this feature holds if the sentences of the Tweet have different lengths. The Boolean feature *Repeated Word* holds if one word is occurring more than once in a window of four [37].

4 Results

We aim at answering the following questions:

(1) Is it possible to predict if a Tweet has been labeled to be ironic or sarcastic by the author (without having access to the actual label)?
(2) Which features have an impact on making this prediction?
(3) Can we get qualitative insight with this approach what users consider to be sarcasm or irony?

We apply statistical models to distinguish between these classes to answer these questions. Further, we analyze how our features are distributed in the different subcorpora and qualitatively analyze words with high (pointwise) mutual information.

4.1 Classification of Irony, Sarcasm and Regular Tweets

In the experiments for classification of Tweets, subsets of features described in Sect. 3.2 and labels based on hashtags (the feature extraction does not have access to them) are used as input for different classification methods, namely support vector machines (SVM, in the implementation of liblinear [41]) [42], decision trees [43] (in the Weka J48 implementation [44]) and maximum entropy classifiers (MaxEnt) [45]. Consistent with previous experiences in the domain [33] and throughout our meta-parameter optimization and feature selection with 10-fold cross validation on the training set, the MaxEnt model outperforms the other approaches (drop of performance in decision trees of up to 14 % points, SVM up to 2 % points). Therefore, we only report on the MaxEnt model in the following.

Table 1 shows results for four different classification tasks: Throughout all feature settings (unigram, unigram+bigram, all, all−BoW (=only domain-specific features)) the differences between the classification tasks are comparable. Distinguishing sarcasm and regular Tweets leads to the highest performance with an accuracy of 0.90, closely followed by irony vs. regular Tweets with 0.89 accuracy. Separating figurative (without making the distinction between ironic and sarcastic) from regular text is slightly harder than the separate tasks with up to 0.88 accuracy. Distinguishing irony from sarcasm leads to a lower performance when compared to the other three tasks: The accuracy is ≈10 % points lower with 0.79 accuracy with all features. The performance in this task with domain-specific features only (All−BoW) of 0.64 reveals that the actual words in the text have a high impact on the classification performance. However, this result might be surprisingly high, given that both classes are often considered synonymous or at least similar. To achieve a better understanding of the structure of the task and the importance of the different features, we analyze their contribution in more detail in the following.

Table 1. Accuracy of Maximum Entropy Classifier on independent test set with all features described in Sect. 3.2 in comparison to unigram and bigram bag-of-words models. The data sets are balanced.

Feature set	Irony vs. Sarcasm	Irony vs. Regular	Sarcasm vs. Regular	Figurative vs. Regular
Baseline	0.50	0.50	0.50	0.50
All–BoW	0.64	0.81	0.83	0.82
BoW Unigram	0.76	0.87	0.89	0.87
BoW Uni+Bigram	0.78	0.88	0.90	0.88
All	0.79	0.89	0.90	0.88

4.2 Feature Analysis

We discuss the features with highest impact on one of the classification tasks in the following. A complete list of counts and values is depicted in Table 2.

The most important features to distinguish figurative from regular Tweets are *URL, Additional Hashtags, Verbs, Stopwords, Adverbs, User* and the Boolean *Sentiment Score Gap, Sentence Length Gap,* and the *Tweet Length* (these features are marked with FR in Table 2). The most important features to distinguish irony from sarcasm are *Nouns, Sentiment Score Positive,* and *Stopwords, Sentence Length Gap, Tweet Length, Verbs, Sentiment Score Negative, Interjection, Sentiment Score Gap,* (all marked in table with IS).

Emoticons are used more frequently in regular Tweets, and more frequent in their positive versions. Sarcastic Tweets use more of them than ironic Tweets. *Laughing Onomatopoeia* are used more often for figurative Tweets. Further, sarcastic Tweets are more often positive, and ironic tweets are more often negative. This suggests that both contexts are commonly used to express the opposite of the literal meaning, as sarcasm can be considered a more negative version of irony (*cf.* Sect. 2). Ironic Tweets tend to be longer than sarcastic and regular Tweets (*Tweet Length*). One reason might be that irony often refers to descriptions of situations, which are more complex to be described than verbal irony [2,28]. *User* is used more often in figurative Tweets (which is in line with the interpretation of [28]).

4.3 Bag-Of-Words Analysis

The bag-of-words model is a very strong baseline (*cf.* Sect. 4.1), and these features have a high impact in comparison to our domain-specific feature set. In the following, we discuss a selection of words in a ranked top-50 list of pointwise mutual information for the different classes in the whole corpus of 99000 Tweets. The words can be divided into two interesting subclasses: Those which express a sentiment and those which are corpus or topic-specific (presumably because of a limited time-frame of crawling).

Table 2. Counts of features and corpus statistics for irony, sarcasm and non-figurative Tweets. All numbers refer to the number of occurrences of the Boolean versions except where denoted with ∗. Top features for Figurative/Regular discrimination are marked with FR, those for Irony/Sarcasm discrimination with IS.

	Feature	Irony	Sarcasm	Regular	
Stat.	# Tweets	33000	33000	33000	
	Average # Sentences	1.82	2.06	1.43	
Figurative-specific	Emoticon	1155	1453	1925	
	Emoticon Positive	670	754	1101	
	Emoticon Negative	316	547	504	
	Symbol	331	363	928	
	Emoticon-Sequence	663	796	1681	
	Capitalization	11798	12453	13324	
	Interjection	411	1152	97	IS
	Additional Hashtags	14224	14024	25939	FR
	User	11866	13138	5619	FR
	URL	8660	7704	26523	FR
	Ellipsis and Punctuation	129	165	25	
	Punctuation Marks	273	546	119	
	Laughing Acronym	622	568	1227	
	Grin	0	2	0	
	Laughing Onomatopoeia	179	194	91	
Sentiment	Sentiment Score Positive	8423	13156	6128	IS
	Sentiment Score Negative	9096	6481	4443	IS
	Sentiment Score Gap	18396	20473	10940	FR IS
	Valence Shift	1729	1905	719	
	Pos Hyperbole	7	22	5	
	Neg Hyperbole	17	11	10	
	Pos Quotes	110	93	10	
	Neg Quotes	157	61	11	
	Pos & Punctuation	85	336	66	
	Neg & Punctuation	107	186	43	
	Pos & Ellipsis	413	655	233	
	Neg & Ellipsis	427	412	225	
Syntactic	Negations	5044	5495	1905	
	Sentence Length Gap	14531	18222	9901	FR IS
	Repeated Word	7706	7948	6549	
	Stopwords∗	3.27	2.69	1.74	FR IS
	Nouns∗	3.78	3.03	3.91	IS
	Verbs∗	2.77	2.46	1.54	FR IS
	Adverbs∗	0.69	0.77	0.32	FR
	Adjectives∗	1.05	1.08	0.87	
	Pronomina∗	0.25	0.26	0.13	
	Tweet Length∗	17.05	15.71	15.39	FR IS

High-PMI words which can be considered negatively connotated from the irony corpus are: *complains, claiming, criticizing, illegally, betrayed, burnt, aborted, accusing,* and *however.* Positive are *#karma* and *giggle.*

In the sarcasm subcorpus, words with a high PMI which express an opinion are *Yay, Duh, thrilled, goody, #kidding, #yay, C'mon, Soooo, AWESOME, Mondays, exciting, Yup, nicer, #not, shocked, woah,* and *shocker.* It is less straightforward to detect a dominating polarity in these words, but an investigation of Tweets reveal that these are used for instance in patterns like "[Statement about something negative] [Positive Word] [#sarcasm]". This suggests, similarly to the feature analysis of sentiment characteristics in Sect. 4.2, that irony and sarcasm are commonly used to express the opposite, but irony more often to express something positive with a negative literal meaning and sarcasm to express something negative with a positive literal meaning.

Examples for topic specific words with high PMI for sarcasm are #creativity (because of a popular news article about a relationship between sarcasm and creativity), #Royals (related to a sports team), @businessinsider (answering to different posts on a news aggregator site), playoff, Research, @Etsy, @LenKaspar. An interesting special case is the word "coworker", which does not seem to relate to a specific event nor can be considered to be of specific opinion. However, the respective Tweets are mostly negative.

Examples for specific topics in the irony subcorpus are *Independence* and *Labor* (both mentioning reasons for not celebrating the respective holiday), *Dismaland, #refugees, Syria, hypocrisy, extremists.* These topics are more clearly discussed in a negative context than those marked with #sarcasm.

5 Discussion and Future Work

In this paper, we have shown that a classification model can distinguish ironic and sarcastic Tweets with substantial accuracy. Without taking into account word-specific features, the performance is limited (0.64 accuracy). When words are taken into account in addition, the accuracy increases (0.79). Compared to distinguishing ironic and sarcastic Tweets from regular Tweets (0.88 accuracy), this result is comparably low. This suggests that this task is, as expected, more challenging than recognizing figurative Tweets without making this difference.

To get a better understanding of the specific task structure, we performed an analysis of the importance of features. In summary, the key results of our experiments are as follows:

Key Result 1: Lengths of Tweets. Regular Tweets have fewer sentences than ironic Tweets. However, most sentences occur in sarcastic Tweets. On the contrary to this observation, ironic Tweets contain more tokens than sarcastic Tweets, followed by regular posts (sentence counts: Sarcasm: 2.06 > Irony: 1.82 > Regular: 1.43; token counts: Regular: 15.39 < Sarcasm: 15.71 < Irony: 17.05).

One interpretation is that the irony hashtag is commonly used to describe and explain ironic situations; for instance as in *"Mum signed me up to so many job sites as a hint to get a job, all they do is spam my email &now I can't find an email from my job. #Irony"* [46]. Whereas irony demands for many words to illustrate the situation's circumstances and convey the irony, sarcasm (as an instance of verbal irony) can be expressed using the interaction of different phrases or sentences to introduce the sarcastic meaning incrementally. An example is *"You can smile, you know. Whoa, really?? I had no fucking idea!! Please, tell me more of what I can do. I'm so interested #sarcasm"* [47].

Key Result 2: Polarity. Sarcastic Tweets use more positive words than ironic Tweets (13146 positive sarcastic Tweets vs. 8423 positive ironic Tweets). On the other side, irony is used more often with negative words (9096 vs. 6481). This is in line with the common understanding of irony to express the opposite of the literal meaning and sarcasm as a generally negative version of it. An example is *"Another wonderful day #blessed #sarcasm I love it when people are nice... #sarcasm"* [48]. This characteristic of formulating the opposite is also supported by the frequencies of negations: Whereas 5044 ironic and 5495 sarcastic Tweets contain negations, they are only used in 1905 regular Tweets (for instance *"Nah, I'm sure they aren't mad for being ignored. #Sarcasm #Never-Forget #Venezuela"* [49]).

Key Result 3: Hashtags, Usernames, URLs. Sarcastic Tweets are comparably often messages to somebody and therefore mention usernames more often than ironic Tweets (13138 sarcastic vs. 11866 ironic Tweets that contain usernames). Usernames are less frequently mentioned in regular Tweets (with 5619). The following example supports one possible interpretation of sarcastic Tweets being more often targeted to a specific entity: *"Thanks @sonicdrivein for giving me so many onion rings on my meal. #sarcasm #sonic #hungry"* [50].

Regular Tweets refer more often to other things: Hashtags are most frequent in them (25939 vs. 14223 in irony and 14024 in sarcasm) as well as are URLs (26523 vs. 8660 and 7704).

These results support existing theories of irony and sarcasm with task-specific features and help to understand the actual use of these devices by authors. However, our study also revealed that word-based features are of high impact for the automated classification task. This suggests that topic-specific background knowledge is helpful to detect these devices. Only a subset of these words support interpretations of [28] for irony and sarcasm and for figurative vs. regular language [36], namely that irony and sarcasm are commonly used to express the opposite of the literal meaning, in line with the common understanding of these devices.

On the contrary, many words refer to specific topics under discussion. This suggests that the high performance in the classification is at least partially a result of overfitting to the data. Another interpretation is that these words build an environment for irony or sarcasm such it can actually be understood. To investigate this further, it is important to focus on a further data-driven analysis of topics in the different subcorpora and aim at discovering latent patterns in them. This will support the analysis of concept drift: training a classifier and testing it on Tweets from a different (and distant) time frame will reveal differences between features which generalize over specific events.

References

1. Amazon Review: "worst book i have ever read" (2010). http://www.amazon.com/review/R86RAMEBZSB11. Accessed 29 Feb 2016
2. Maynard, D., Greenwood, M.: Who cares about sarcastic tweets? investigating the impact of sarcasm on sentiment analysis. In: Calzolari, N., Choukri, K., Declerck, T., Loftsson, H., Maegaard, B., Mariani, J., Moreno, A., Odijk, J., Piperidis, S. (eds.) Proceedings of the Ninth International Conference on Language Resources and Evaluation (LREC'14), Reykjavik, Iceland, European Language Resources Association (ELRA). pp. 4238–4243. ACL Anthology Identifier: L14–1527, May 2014
3. Kreuz, R.J., Glucksberg, S.: How to be sarcastic: the echoic reminder theory of verbal irony. J. Exp. Psychol. Gen. **118**(4), 374 (1989)
4. Go, A., Bhayani, R., Huang, L.: Twitter sentiment classification using distant supervision. Technical report, Stanford University (2009). http://www.stanford.edu/alecmgo/papers/TwitterDistantSupervision09.pdf
5. Abrams, M.H.: A Glossary of Literary Terms, 7th edn. Heinle & Heinle, Thomson Learning, Boston (1999)
6. Nakassis, C., Snedeker, J.: Beyond sarcasm: intonation and context as relational cues in children's recognition of irony. In: Proceedings of the Twenty-sixth Boston University Conference on Language Development. Cascadilla Press (2002)
7. Creusere, M.A.: A developmental test of theoretical perspective on the understanding of verbal irony: children's recognition of allusion and pragmatic insincerity. In: Raymond W Gibbs, J., Colston, H.L. (eds.) Irony in Language and Thought: A Cognitive Science Reader, 1st edn, pp. 409–424. Lawrence Erlbaum Associates, New York (2007)
8. Glenwright, M.H., Pexman, P.M.: Children's perceptions of the social functions of verbal irony. In: Raymond W Gibbs, J., Colston, H.L. (eds.) Irony in Language and Thought: A Cognitive Science Reader, 1st edn, pp. 447–464. Lawrence Erlbaum Associates, New York (2007)
9. Kreuz, R.J.: The use of verbal irony: cues and constraints. In: Mio, J.S., Katz, A.N. (eds.) Metaphor: Implications and Applications. Lawrence Erlbaum Associates, Mahwah (1996)
10. Gibbs, R.W.J.: Irony in talk among friends. In: Raymond W Gibbs, J., Colston, H.L. (eds.) Irony in Language and Thought: A Cognitive Science Reader, 1st edn. Lawrence Erlbaum Associates, New York (2007)
11. Utsumi, A.: A unified theory of irony and its computational formalization. In: Proceedings of the 16th Conference on Computational Linguistics - vol. 2. COLING 1996, pp. 962–967. Association for Computational Linguistics, Stroudsburg (1996)
12. Wilson, D., Sperber, D.: On verbal irony. Lingua **87**, 53–76 (1992)
13. Clark, H.H., Gerrig, R.J.: On the pretense theory of irony. J. Exp. Psychol. Gen. **113**(1), 121–126 (1984)
14. Kumon-Nakamura, S., Glucksberg, S., Brown, M.: How about another piece of pie: the allusional pretense theory of discourse irony. J. Exp. Psychol. Gen. **124**(1), 3–21 (1995)

15. Wilson, D., Sperber, D.: Explaining Irony. Cambridge University Press, Cambridge (2012)
16. Utsumi, A.: Verbal irony as implicit display of ironic environment: distinguishing ironic utterances from nonirony. J. Pragmatics **32**(12), 1777–1806 (2000)
17. Clift, R.: Irony in conversation. Lang. Soc. **28**(4), 523–553 (1999)
18. Ptáček, T., Habernal, I., Hong, J.: Sarcasm detection on czech and english twitter. In: Proceedings of COLING 2014, the 25th International Conference on Computational Linguistics: Technical Papers, Dublin, Ireland, Dublin City University and Association for Computational Linguistics, pp. 213–223, August 2014
19. Filatova, E.: Irony and sarcasm: Corpus generation and analysis using crowdsourcing. In: Chair, N.C.C., Choukri, K., Declerck, T., Doan, M.U., Maegaard, B., Mariani, J., Moreno, A., Odijk, J., Piperidis, S. (eds.) Proceedings of the Eight International Conference on Language Resources and Evaluation (LREC 2012), Istanbul, Turkey, European Language Resources Association (ELRA), May 2012
20. Reyes, A., Rosso, P., Veale, T.: A multidimensional approach for detecting irony in Twitter. Lang. Resour. Eval. **47**(1), 239–268 (2013)
21. Rajadesingan, A., Zafarani, R., Liu, H.: Sarcasm detection on twitter: a behavioral modeling approach. In: Proceedings of the Eighth ACM International Conference on Web Search and Data Mining, WSDM 2015, 97–106. ACM, New York (2015)
22. Tepperman, J., Traum, D., Narayanan, S.S.: "Yeah right": sarcasm recognition for spoken dialogue systems. In: Proceedings of InterSpeech, Pittsburgh, PA, pp. 1838–1841, September 2006
23. Tsur, O., Davidov, D., Rappoport, A.: ICWSM - a great catchy name: semi-supervised recognition of sarcastic sentences in online product reviews. In: International AAAI Conference on Web and Social Media (ICWSM), Washington D.C., USA (2010)
24. Barbieri, F., Saggion, H.: Modelling irony in twitter. In: Proceedings of the Student Research Workshop at the 14th Conference of the European Chapter of the Association for Computational Linguistics, Gothenburg, Sweden, pp. 56–64. Association for Computational Linguistics, April 2014
25. Riloff, E., Qadir, A., Surve, P., De Silva, L., Gilbert, N., Huang, R.: Sarcasm as contrast between a positive sentiment and negative situation. In: Proceedings of the 2013 Conference on Empirical Methods in Natural Language Processing, pp. 704–714. Association for Computational Linguistics, Seattle, October 2013
26. Ghosh, A., Li, G., Veale, T., Rosso, P., Shutova, E., Barnden, J., Reyes, A.: SemEval-2015 task 11: sentiment analysis of figurative language in twitter. In: Proceedings of the 9th International Workshop on Semantic Evaluation (SemEval 2015), pp. 470–478. Association for Computational Linguistics, Denver, June 2015
27. Van Hee, C., Lefever, E., Hoste, V.: LT3: sentiment analysis of figurative tweets: piece of cake #NotReally. In: Proceedings of the 9th International Workshop on Semantic Evaluation (SemEval 2015), pp. 684–688. Association for Computational Linguistics, Denver, June 2015
28. Wang, P.Y.A.: #Irony or #Sarcasm - a quantitative and qualitative study based on twitter. In: 27th Pacific Asia Conference on Language, Information, and Computation (PACLIC), Taiwan, Taipei, pp. 349–356 (2013)
29. Sulis, E., Farías, D.I.H., Rosso, P., Patti, V., Ruffo, G.: Figurative messages and affect in Twitter: differences between #irony, #sarcasm and #not. Knowledge-Based Systems, May 2016. (in Press)
30. Jaccard, P.: Étude comparative de la distribution florale dans une portion des Alpes et des Jura. Bull. Soc. Vaudoise des Sci. Nat. **37**, 547–579 (1901)

31. Potts, C.: Twitter-aware tokenizer (2011). http://sentiment.christopherpotts.net/code-data/happyfuntokenizing.py. Accessed 03 Mar 2016

32. Derczynski, L., Ritter, A., Clark, S., Bontcheva, K.: Twitter part-of-speech tagging for all: Overcoming sparse and noisy data. In: Proceedings of the International Conference Recent Advances in Natural Language Processing RANLP 2013, Hissar, Bulgaria, pp. 198–206. Incoma Ltd., Shoumen, September 2013

33. Buschmeier, K., Cimiano, P., Klinger, R.: An impact analysis of features in a classification approach to irony detection in product reviews. In: Proceedings of the 5th Workshop on Computational Approaches to Subjectivity, Sentiment and Social Media Analysis, pp. 42–49. Association for Computational Linguistics, Baltimore, June 2014

34. Davidov, D., Tsur, O., Rappoport, A.: Semi-Supervised Recognition of Sarcasm in Twitter and Amazon. In: Proceedings of the Fourteenth Conference on Computational Natural Language Learning, pp. 107–116. Association for Computational Linguistics, Uppsala, July 2010

35. Whitlock, T.: Emoji unicode tables (2015). http://apps.timwhitlock.info/emoji/tables/unicode. Accessed 26 Feb 2016

36. González-Ibáñez, R., Muresan, S., Wacholder, N.: Identifying sarcasm in twitter: A closer look. In: Proceedings of the 49th Annual Meeting of the Association for Computational Linguistics: Human Language Technologies, pp. 581–586. Association for Computational Linguistics, Portland, June 2011

37. Kreuz, R., Caucci, G.: Lexical influences on the perception of sarcasm. In: Proceedings of the Workshop on Computational Approaches to Figurative Language, pp. 1–4. Association for Computational Linguistics, Rochester, April 2007

38. Holen, V.: Dictionary of interjections (2016). http://www.vidarholen.net/contents/interjections/. Accessed 03 Mar 2016

39. Nielsen, F.: Afinn, March 2011. http://www2.imm.dtu.dk/pubdb/p.php?6010. Accessed 18 Mar 2016

40. Barbieri, F., Saggion, H.: Modelling irony in twitter: feature analysis and evaluation. In: Calzolari, N., Choukri, K., Declerck, T., Loftsson, H., Maegaard, B., Mariani, J., Moreno, A., Odijk, J., Piperidis, S. (eds.) Proceedings of the Ninth International Conference on Language Resources and Evaluation (LREC 2014), pp. 4258–4264. European Language Resources Association (ELRA), Reykjavik, May 2014

41. Fan, R.E., Chang, K.W., Hsieh, C.J., Wang, X.R., Lin, C.J.: Liblinear: a library for large linear classification. J. Mach. Learn. Res. 9, 1871–1874 (2008)

42. Cristianini, N., Shawe-Taylor, J.: An Introduction to Support Vector Machines: And Other Kernel-based Learning Methods. Cambridge University Press, Cambridge (2000)

43. Utgoff, P.E.: Incremental induction of decision trees. Mach. Learn. 4(2), 161–186 (1989)

44. Witten, I.H., Frank, E., Hall, M.H.: Data Mining: Practical Machine Learning Tools and Techniques. Morgan Kaufmann, San Francisco (2010)

45. Nigam, K., Lafferty, J., McCallum, A.: Using maximum entropy for text classification. In: IJCAI Workshop on Machine Learning for Information Filtering, Stockholm, Sweden (1999)

46. @Katie_Blair_: Mom signed me up... Twitter (2015). https://twitter.com/Katie_Blair_status/624580420668104704

47. @Bellastar12597: You can smile... Twitter (2015). https://twitter.com/Bellastar12597/status/630851771599069184

48. @MrHoffman9: Another wonderful day... Twitter (2015). https://twitter.com/Mr Hoffman9/status/635539576426262529
49. @CidHialeah: Another wonderful day... Twitter (2015). https://twitter.com/Cid Hialeah/status/623120702221103104
50. @Deadstitch: Another wonderful day... Twitter (2015). https://twitter.com/Dea dstitch/status/622931551064453120

Polarity Classification for Target Phrases in Tweets: A Word2Vec Approach

Andi Rexha[1(✉)], Mark Kröll[1], Mauro Dragoni[2], and Roman Kern[1]

[1] Know-Center GmbH, Graz, Austria
{arexha,mkroell,rkern}@know-center.at
[2] FBK-IRST, Trento, Italy
dragoni@fbk.eu

Abstract. Twitter is one of the most popular micro-blogging services on the web. The service allows sharing, interaction and collaboration via short, informal and often unstructured messages called tweets. Polarity classification of tweets refers to the task of assigning a positive or a negative sentiment to an entire tweet. Quite similar is predicting the polarity of a specific target phrase, for instance *@Microsoft* or *#Linux*, which is contained in the tweet.

In this paper we present a Word2Vec approach to automatically predict the polarity of a target phrase in a tweet. In our classification setting, we thus do not have any polarity information but use only semantic information provided by a Word2Vec model trained on Twitter messages. To evaluate our feature representation approach, we apply well-established classification algorithms such as the Support Vector Machine and Naive Bayes. For the evaluation we used the *Semeval 2016 Task #4* dataset. Our approach achieves F1-measures of up to ∼90 % for the positive class and ∼54 % for the negative class without using polarity information about single words.

1 Introduction

With the growing popularity of online social media services, different types and means of communication are available nowadays. There is an observable trend towards microblogging and shorter text messages (snippets) which often are unstructured and informal. One of the most popular microblogging platforms is Twitter which allows for spreading short text messages (140 characters) called tweets. The language used in these messages often is very informal, with creative spelling and punctuation, misspellings, slang, URLs and abbreviations. In short, a challenge as well as opportunity for every researcher in the NLP area.

Automatically predicting the polarity of tweets represents an ongoing endeavor and relates to the task of assigning a positive or a negative sentiment to an entire tweet. Quite similar is predicting the polarity of a specific target phrase which is contained in the tweet. Consider following example where the references to *@Microsoft* and to *#Linux*) are called target phrases:

*New features **@Microsoft** suck. Check them back! **#Linux** solutions are awesome*

© Springer International Publishing AG 2016
H. Sack et al. (Eds.): ESWC 2016 Satellite Events, LNCS 9989, pp. 217–223, 2016.
DOI: 10.1007/978-3-319-47602-5_40

The overall polarity of this tweet might turn out neutral, since the first part "New features **@Microsoft** suck" expresses a negative sentiment while the last part of the tweet "**#Linux** solutions are awesome." expresses a positive one. So, the averaging of sentiment assignments might lead to loss of information, i.e. the individual attitude towards products or the like.

In this paper we present an algorithm which automatically predicts the polarity of target phrases in a tweet. In the previous example, our algorithm will return a positive rating about the target *@Microsoft* and a negative one about the target *#Linux*. To do that, we explore using but semantic information (cf. [8]) given by a Word2Vec[1] model trained on Twitter messages, i.e. without using polarity information about single words. Word2Vec models provide a feature space representation of words which reflects their relation to other words in the training corpus. To evaluate our algorithm, we use the test and golden standard dataset of the *Semeval 2016 Task #4*[2] challenge about Twitter sentiment mining.

The paper is structured as follows. In Sect. 2 we present and discuss related work. In Sect. 3 we describe two different feature representations of tweets using the Word2Vec model. Section 4 evaluates our algorithm on the Semeval dataset. The paper concludes and presents future work in Sect. 5.

2   Related Work

The task of Sentiment Analysis, also known as opinion mining (cf. [7,9]), is to classify textual content according to expressed emotions and opinions. Sentiment classification has been a challenging topic in Natural Language Processing (cf. [14]). It is commonly defined as a binary classification task to assign a sentence either positive or negative polarity (cf. [10]). Turneys work was among the first ones to tackle automatic sentiment classification [13]. He employed an information-theoretic measure, i.e. mutual information, between a text phrase and the words excellent and poor as a decision metric.

Micro-blogging data such as tweets differs from regular text as it is extremely noisy, informal and does not allow for long messages (which might not be a disadvantage (cf. [3]). As a consequence, analyzing sentiment in Twitter data poses a lot of opportunities. Traditional feature representations such as part-of-speech information or the usage of lexicon features such as SentiWordNet have to be re-evaluated in the light of Twitter data. In case of part-of-speech information, Gimpel et al. [4] annotated tweets and developed a tagset and features to train an adequate tagger. Kouloumpis et al. [6] investigated the usefulness of existing lexical resources and other features including part-of-speech information in the analysis task.

Go et al. [5], for instance, used emoticons as additional features, for example, ":)" and ":-)" for the positive class, ":(" and ":-(" for the negative class.

---

[1] Word2Vec models provide a representation of words in a feature space reflecting their relation to other words in the corpus.

[2] http://alt.qcri.org/semeval2016/task4/data/uploads/semeval2016_task4_report.pdf.

They then applied machine learning techniques such as support vector machines to classify the tweets into a positive and a negative class. Agarwal et al. [1] introduced POS-specific prior polarity features along with using a tree kernel for tweet classification. Barbosa and Feng [2] present a robust approach to Twitter sentiment analysis. The robustness is based on an abstract representation of tweets as well as the usage of noisy/biased labels from three websites to train their model.

Last but not least, recent years have seen a lot of participation in the annual SemEval tasks on Twitter Sentiment Analysis (cf. [11,12,15]). This event provides optimal conditions to implement novel ideas and is a good starting point to catch up on the latest trends in this area.

3   A Word2Vec Approach

In this section we describe our algorithm's feature engineering which encompasses three steps (1) pre-processing, (2) feature representation (two approaches), and (3) post-processing.

3.1   Pre-processing

As preprocessing step, we add additional information to the words, i.e. part-of-speech, by using the Tweet NLP library[3]. Furthermore, we extract the Word2Vec vector[4] representation for each word of the tweet by using a Twitter model trained on ~400 million tweets.

3.2   Feature Representation

In this paper, we experimented with two approaches of representing a tweet using a Word2Vec trained model on Twitter messages. In the first approach, we do not consider the position of target phrases and use Word2Vec information for every word in a tweet (see Fig. 1).

In the second approach, we consider only the neighborhood of a target phrase in our polarity classification task (see Fig. 2) by using a window of size n.

The tweets are preprocessed as described in Sect. 3.1, target phrases of the tweets are located according to their annotation in the dataset and window of size "n" is determined. Per tweet, there is only one target phrase. In case that a target phrase occurs several times in a tweet, only the first occurrence is taken into account resulting in exactly one window per tweet. For each word in the window, the Word2Vec information is extraced and an average vector is formed.

---

[3] http://www.cs.cmu.edu/~ark/TweetNLP/.
[4] http://www.fredericgodin.com/software/.

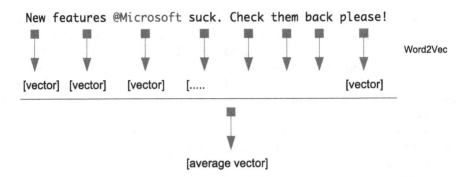

Fig. 1. In this approach, we extract Word2Vec information for all words in a tweet and form an average vector (post-processing step). We do not take into account the position of the target phrase

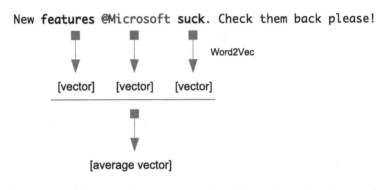

Fig. 2. In the second approach, we extract Word2Vec information for words in the neighborhood of a target phrase (covered by a window of size 1 in this case) and form an average vector (post-processing step)

3.3 Post-processing

In the postprocessing step we generate one average vector per tweet - either from every word (approach 1) or only from words within the window (approach 2).

As last step we introduce a binary feature which is set to 1 if in the tweet exists any negation word (don't, not, ...). Out of our experience, this feature provides valuable information to the subsequent learning step.

4   Results

We used the *Semeval 2016 Task 4* dataset to evaluate our two feature extraction approaches. The training set is composed by 3858 entries and the evaluation set by 10551 entries. Both datasets are skewed, i.e. the training set contains 17 % of negative and 83 % of positive and the evaluation set of 22 % of negative and 78 % of positive examples. In our experiments, we applied four different well-established classification models, i.e. Naive Bayes, Support Vector Machines,

Logistic Regression and Random Trees, to our feature representations. For each feature representation version, we present precision, recall and F1-measures for the positive and negative classes: Tables 1 and 2 contain performance values of the positive and negative classes for the full text approach. Tables 3 and 4 contain performance values for the second approach with a window size n of 3. Other window sizes did not lead to better results. We remark that we experimented with small window sizes rather than large ones to focus on the proximity aspect.

Table 1. Precision, Recall and F1-Measure results for the negative class when using Word2Vec information of all words in a tweet to generate the average vector

	Precision	Recall	F1-Measure
Naive Bayes	0.396	**0.733**	0.514
Support vector machine	**0.724**	0.347	0.469
Logistic regression	0.606	0.481	**0.536**
Random tree	0.321	0.266	0.291

Table 2. Precision, Recall and F1-Measure results for the positive class when using Word2Vec information of all words in a tweet to generate the average vector

	Precision	Recall	F1-Measure
Naive Bayes	**0.900**	0.681	0.775
Support vector machine	0.838	**0.962**	**0.896**
Logistic regression	0.860	0.911	0.885
Random tree	0.801	0.840	0.820

Table 3. Precision, Recall and F1-Measure results for the negative class when using Word2Vec information of a target phrases' neighboring words (a window size of 3) in a tweet to generate the average vector

	Precision	Recall	F1-Measure
Naive Bayes	0.303	**0.732**	**0.429**
Support vector machine	**0.442**	0.273	0.338
Logistic regression	0.396	0.391	0.394
Random tree	0.260	0.281	0.271

Using the Word2Vec information for all words in a tweet yielded high F1-measures (up to ∼90 %) for the positve class and average F1-measures (up to ∼54 %) for the negative class. The performance difference might be due to the difference in available training examples for both classes. Table 1 reveals that the Support Vector Machine is capable of identifying instances of the negative class with a precision of ∼72 %.

Table 4. Precision, Recall and F1-Measure results for the positive class when using Word2Vec information of a target phrases' neighboring words (a window size of 3) in a tweet to generate the average vector

	Precision	Recall	F1-Measure
Naive bayes	**0.872**	0.520	0.652
Support vector machine	0.813	**0.902**	**0.855**
Logistic regression	0.827	0.830	0.829
Random tree	0.791	0.773	0.781

Tables 3 and 4 show worse performance values for both classes when compared to Tables 1 and 2, i.e. using the Word2Vec information of all the words in a tweet. However, in particular for the positive class, the proximity of a target phrase often contains sufficient semantic information for the prediction task as taking the entire tweet into account. Both approaches do quite well with the positive class. Both approaches yield considerably lower F1-measures for the negative class than for the positive one - probably due to the skewness in the dataset.

5   Conclusion

In this paper we introduced an algorithm which automatically predicted the polarity of a target phrase in a tweet. Our algorithm uses only semantic information provided by a Word2Vec model trained on Twitter messages. Evaluating our algorithm on the *Semeval 2016 Task #4* dataset shows that F1-measures of up to ∼90 % for the positive class and ∼54 % for the negative class are achievable without using polarity information about single words.

In future work we intend to exploit information provided by dependency trees of tweets for the polarity classification task. We hypothesize that going beyond textual proximity, i.e. taking into account remoter structures, might improve the performance of the classification algorithm.

Acknowledgments. This work is funded by the KIRAS program of the Austrian Research Promotion Agency (FFG) (project number 840824). The Know-Center is funded within the Austrian COMET Program under the auspices of the Austrian Ministry of Transport, Innovation and Technology, the Austrian Ministry of Economics and Labour and by the State of Styria. COMET is managed by the Austrian Research Promotion Agency FFG.

References

1. Agarwal, A., Biadsy, F., Mckeown, K.R.: Contextual phrase-level polarity analysis using lexical affect scoring and syntactic n-grams. In: Proceedings of the 12th Conference of the European Chapter of the Association for Computational Linguistics, EACL 2009 (2009)

2. Barbosa, L., Feng, J.: Robust sentiment detection on twitter from biased and noisy data. In: Proceedings of the 23rd International Conference on Computational Linguistics: Posters, COLING 2010 (2010)
3. Bermingham, A., Smeaton, A.F.: Classifying sentiment in microblogs: is brevity an advantage? In: Proceedings of the 19th ACM International Conference on Information and Knowledge Management, CIKM 2010. ACM (2010)
4. Gimpel, K., Schneider, N., O'Connor, B., Das, D., Mills, D., Eisenstein, J., Heilman, M., Yogatama, D., Flanigan, J., Smith, N.A.: Part-of-speech tagging for twitter: annotation, features, and experiments. In: Proceedings of the 49th Annual Meeting of the Association for Computational Linguistics: Human Language Technologies: Short Papers - vol. 2, HLT 2011 (2011)
5. Go, A., Bhayani, R., Huang, L.: Twitter sentiment classification using distant supervision. Technicak report, Stanford (2009)
6. Kouloumpis, E., Wilson, T., Moore, J.D.: Twitter sentiment analysis: the good the bad and the omg! In: Proceedings of the Fifth International Conference on Weblogs and Social Media, Barcelona, Spain (2011)
7. Liu, B., Zhang, L.: Mining text data. In: Aggarwal, C.C., Zhai, C. (eds.) A Survey of Opinion Mining and Sentiment Analysis, pp. 415–463. Springer, New York (2012)
8. Mikolov, T., Chen, K., Corrado, G., Dean, J.: Efficient estimation of word representations in vector space. arXiv preprint arXiv:1301.3781 (2013)
9. Pang, B., Lee, L.: Opinion mining and sentiment analysis. Found. Trends Inf. Retrieval **2**(1–2), 1–135 (2008)
10. Pang, B., Lee, L., Vaithyanathan, S.: Thumbs up? sentiment classification using machine learning techniques. In: Proceedings of the Conference on Empirical Methods in Natural Language Processing, pp. 79–86, EMNLP 2002. Association for Computational Linguistics, Stroudsburg (2002). http://dx.doi.org/10.3115/1118693.1118704
11. Rosenthal, S., Nakov, P., Kiritchenko, S., Mohammad, S., Ritter, A., Stoyanov, V.: Semeval-2015 task 10: sentiment analysis in twitter. In: Proceedings of the 9th International Workshop on Semantic Evaluation (SemEval 2015). Association for Computational Linguistics (2015)
12. Rosenthal, S., Ritter, A., Nakov, P., Stoyanov, V.: Semeval-2014 task 9: sentiment analysis in twitter. In: Proceedings of the 8th International Workshop on Semantic Evaluation (SemEval 2014). Association for Computational Linguistics and Dublin City University (2014)
13. Turney, P.D.: Thumbs up or thumbs down? Semantic orientation applied to unsupervised classification of reviews. In: Proceedings of the 40th Annual Meeting on Association for Computational Linguistics. Association for Computational Linguistics (2002)
14. Wiebe, J., Wilson, T., Cardie, C.: Annotating expressions of opinions and emotions in language. Lang. Resour. Eval. **1**(2), 165–210 (2005)
15. Wilson, T., Kozareva, Z., Nakov, P., Alan, R., Rosenthal, S., Stoyonov, V.: Semeval-2013 task 2: sentiment analysis in twitter. In: Proceedings of the 7th International Workshop on Semantic Evaluation. Association for Computation Linguistics (2013)

5th Workshop on Knowledge Discovery and Data Mining Meets Linked Open Data (Know@LOD)

PageRank on Wikipedia: Towards General Importance Scores for Entities

Andreas Thalhammer[✉] and Achim Rettinger

AIFB, Karlsruhe Institute of Technology, Karlsruhe, Germany
{andreas.thalhammer,achim.rettinger}@kit.edu

Abstract. Link analysis methods are used to estimate importance in graph-structured data. In that realm, the PageRank algorithm has been used to analyze directed graphs, in particular the link structure of the Web. Recent developments in information retrieval focus on entities and their relations (i.e., knowledge graph panels). Many entities are documented in the popular knowledge base Wikipedia. The cross-references within Wikipedia exhibit a directed graph structure that is suitable for computing PageRank scores as importance indicators for entities. In this work, we present different PageRank-based analyses on the link graph of Wikipedia and according experiments. We focus on the question whether some links—based on their context/position in the article text—can be deemed more important than others. In our variants, we change the probabilistic impact of links in accordance to their context/position on the page and measure the effects on the output of the PageRank algorithm. We compare the resulting rankings and those of existing systems with page-view-based rankings and provide statistics on the pairwise computed Spearman and Kendall rank correlations.

Keywords: Wikipedia · DBpedia · PageRank · Link analysis · Page views · Rank correlation

1 Introduction

Entities are omnipresent in the landscape of modern information extraction and retrieval. Application areas range from natural language processing over recommender systems to question answering. For many of these application areas it is essential to build on objective importance scores of entities. One of the most successful amongst different methods is the PageRank algorithm [4]. It has been proven to provide objective relevance scores for hyperlinked documents, for example in Wikipedia [6,8,11]. Wikipedia serves as a rich source for entities and their descriptions. Its content is currently used by major Web search engine providers as a source for short textual summaries that are presented in knowledge graph panels. In addition, the link structure of Wikipedia has been shown to exhibit the potential to compute meaningful PageRank scores: connected with semantic background information (such as DBpedia [1]) the Page-Rank scores computed on the Wikipedia link graph enable rankings of entities

© Springer International Publishing AG 2016
H. Sack et al. (Eds.): ESWC 2016 Satellite Events, LNCS 9989, pp. 227–240, 2016.
DOI: 10.1007/978-3-319-47602-5_41

Listing 1.1. Example: SPARQL query on DBpedia for retrieving top-10 scientists ordered by PageRank (can be executed at http://dbpedia.org/sparql).

```
PREFIX v:<http://purl.org/voc/vrank#>

SELECT ?e ?r
FROM <http://dbpedia.org>
FROM <http://people.aifb.kit.edu/ath/#DBpedia_PageRank>
WHERE {
?e rdf:type dbo:Scientist;
v:hasRank/v:rankValue ?r.
} ORDER BY DESC(?r) LIMIT 10
```

of specific types, for example for scientists (see Listing 1.1). Although the provided PageRank scores [11] exhibit reasonable output in many cases, they are not always easily explicable. For example, as of DBpedia version 2015-04, "Carl Linnaeus" (512) has a much higher PageRank score than "Charles Darwin" (206) and "Albert Einstein" (184) together in the result of the query in Listing 1.1. The reason is easily identified by examining the articles that link to the article of "Carl Linnaeus":[1] Most articles use the template Taxobox[2] that defines the field binomial_authority. It becomes evident that the page of "Carl Linnaeus" is linked very often because Linnaeus classified species and gave them a binomial name (see [9]). In general, entities that help to structure the geographic and biological domains have distinctively higher PageRank scores than most entities from other domains. While, given the high inter-linkage of these domains, this is expected to some degree, according to our computations articles such as "Bakhsh" (1914), "Powiat" (1408), "Chordate" (1527), and "Lepidoptera" (1778) are occurring in the top-50 list of all things in Wikipedia (see Table 5, column "DBP 2015-04"). These observations led us to the question whether these rankings can be improved. Unfortunately, this is not a straight forward task as a gold standard is missing and rankings are often subjective.

In this work we investigate on different link extraction[3] methods that aim to address the root causes of the observed effects. We focus on the question whether some links—based on their context/position in the article text—can be deemed more important than others. In our variants, we change the probabilistic impact of links in accordance to their context/position on the page and measure the effects on the output of the PageRank algorithm. We compare these variants and the rankings of existing systems with page-view-based rankings and provide statistics on the pairwise computed Spearman and Kendall rank correlations.

---

[1] Articles that link to "Carl Linnaeus" – https://en.wikipedia.org/wiki/Special:WhatLinksHere/Carl_Linnaeus.

[2] Template:Taxobox – https://en.wikipedia.org/wiki/Template:Taxobox.

[3] With "link extraction" we refer to the process of parsing the wikitext of a Wikipedia article and to correctly identify and filter hyperlinks to other Wikipedia articles.

2   Background

In this section we provide additional background on the used PageRank variants, link extraction from Wikipedia, and redirects in Wikipedia.

2.1   PageRank Variants

The PageRank algorithm follows the idea of a user that browses Web sites by following links in a random fashion (random surfer). For computing PageRank, we use the original PageRank formula [4] and a weighted version [2] that accounts for the position of a link within an article.

- Original PageRank [4] – On the set of Wikipedia articles W, we use individual directed links $link(w_1, w_2)$ with $w_1, w_2 \in W$, in particular the set of pages that link to a page $l(w) = \{w_1 | link(w_1, w)\}$ and the count of out-going links $c(w) = |\{w_1 | link(w, w_1)\}|$. The PageRank of a page $w_0 \in W$ is computed as follows:

$$pr(w_0) = (1 - d) + d * \sum_{w_n \in l(w_0)} \frac{pr(w_n)}{c(w_n)} \tag{1}$$

- Weighted Links Rank (WLRank) [2] – In order to account for the relative position of a link within an article, we adapt Formula (1) and introduce link weights. The idea is that the random surfer is likely not to follow every link on the page with the same probability but may prefer those that are at the top of a page. The WLRank of a page $w_0 \in W$ is computed as follows:

$$wlr(w_0) = (1 - d) + d * \sum_{w_n \in l(w_0)} \frac{pr(w_n) * lw(link(w_n, w_0))}{\sum_{w_m} lw(link(w_n, w_m))} \tag{2}$$

The link weight function lw is defined as follows:

$$lw(link(w_1, w_2)) = 1 - \frac{first\_occurrence(link(w_1, w_2), w_1)}{|tokens(w_1)|} \tag{3}$$

In order to form a correct probability model, the individual link weight is normalized in accordance to the link weights of all outgoing links of a page in Formula 2. If we set the link weight of every incoming link to the same value (e.g., 1) we obtain the original PageRank formula (see Formula 1). The used helper functions of Formula 3 can be described as follows:

- $first\_occurrence(link(w_1, w_2), w_1)$ – the token number of the first occurrence of a $link(w_1, w_2)$ at the respective Wikipedia page w_1. The token numbering starts at 1 (i.e., the first word/link in the wikitext).
- $tokens(w_1)$ – the total number of tokens of the Wikipedia page w_1. Tokenization is performed as follows: we split the article text in accordance to white spaces but do not split up links (e.g., [[brown bear|bears]] is treated as one token).

Both Formulas (1) and (2) are iteratively applied until the scores converge. The variable d is called "damping factor": in the random surfer model, it accounts for the possibility of accessing a page via the browser's address bar instead of accessing it via a link from another page.

For reasons of presentation, we use the non-normalized version of PageRank in both cases. In contrast to the normalized version, the sum of all computed PageRank scores is the number of articles (instead of 1) and, as such, does not reflect a statistical probability distribution. However, normalization of the PageRank scores does not influence the final ranking (i.e., the resulting ordering relation between the Wikipedia articles does not change).

2.2 Wikipedia Link Extraction

In order to create a Wikipedia link graph we need to clarify which types of links are considered. The input for the rankings of [11] is a link graph that is constructed by the DBpedia Extraction Framework[4] (DEF). The DBpedia extraction is based on Wikipedia database backup dumps[5] that contain the non-rendered wikitexts of the Wikipedia articles and templates. From these sources, DEF builds a link graph by extracting links of the form [[article|anchor text]]. We distinguish between two types of links with respect to templates:[6]

1. Links that are defined in the Wikipedia text but do not occur within a template, for example "[[brown bear|bears]]" outside {{and}}.
2. Links that and provided as (a part of) a parameter to the template, for example "[[brown bear|bears]]" inside {{and}}.

DEF considers only these two types of links and not any additional ones that result from the rendering of an article. It also has to be noted that DEF does not consider links from category pages. This mostly affects links to parent categories as the other links (i.e., links to all articles of that category) are presented only in the rendered version of the category page (i.e., they do not occur in the wikitext). As an effect, the accumulated PageRank of a category page would be transferred almost 1:1 to its parent category. This would lead to a top-100 ranking of things with mostly category pages only. In addition, DEF does not consider links in references (denoted via <ref> tags).

In this work, we describe how we performed more general link extraction from Wikipedia. Unfortunately, in this respect, DEF exhibited certain inflexibilities as it processes Wikipedia articles line by line. This made it difficult to regard links in the context of an article as a whole (e.g., in order to determine the relative position of a link). In consequence, we reverse-engineered the link extraction parts of DEF and created the SiteLinkExtractor[7] tool. The tool enables to

---

[4] DBpedia Extraction Framework – https://github.com/dbpedia/extraction-frame work/wiki.
[5] Wikipedia dumps – http://dumps.wikimedia.org/.
[6] Template inclusions are marked by double curly brackets, i.e. {{ and }}.
[7] SiteLinkExtractor – https://github.com/TBritsch/SiteLinkExtractor.

$$A \xrightarrow{\;PL\;} B \xrightarrow{\;PL^R\;} C$$

$$A \xrightarrow{\hspace{3cm}PL\hspace{3cm}} C$$

Fig. 1. Transitive resolution of a redirect in Wikipedia. A and C are full articles and B is called a "redirect page", PL are page links, and PL^R are page links marked as a redirect (e.g., #REDIRECT [[United Kingdom]]). The two page links from A to B and from B to C are replaced by a direct link from A to C.

execute multiple extraction methods in a single pass over all articles and can also be extended by additional extraction approaches.

2.3   Redirected vs. Unredirected Wikipedia Links

DBpedia offers two types of page link datasets:[8] one in which the redirects are resolved and one in which they are contained. In principle, also redirect chains of more than one hop are possible but, in Wikipedia, the MediaWiki software is configured not to follow such redirect chains (that are called "double redirect" in Wikipedia)[9] automatically and various bots are in place to remove them. Therefore, we assume that only single-hop redirects are in place. However, as performed by DBpedia, also single-hop redirects can be resolved (see Fig. 1). Alternatively, for various applications (especially in NLP) it can make sense to keep redirect pages as they also have a high number of inlinks in various cases (e.g., "Countries of the world")[10]. In that case, with reference to Fig. 1 and assuming that redirect pages only link to the redirect target, B passes most of its own PageRank score on to C (note that the damping factor is in place). Thus, we assume that the PageRank score of pages of type C is not heavily influenced by resolving/not resolving redirects.

3   Link Graphs

We implemented five Wikipedia link extraction methods that enable to create different input graphs for the PageRank algorithm. In general we follow the example of DEF and consider type 1 and 2 links for extraction (which form a subset of those that occur in a rendered version of an article). The following extraction methods were implemented:

---

[8] DBpedia PageLinks – http://wiki.dbpedia.org/Downloads2015-04.
[9] Wikipedia: Double redirects – https://en.wikipedia.org/wiki/Wikipedia:Double_redirects.
[10] Inlinks of "Countries of the world" – https://en.wikipedia.org/wiki/Special:What LinksHere/Countries_of_the_world.

All Links (ALL) This extractor produces all type 1 and 2 links. This is the reverse-engineered DEF method. It serves as a reference.

Article Text Links (ATL) This measure omits links that occur in text that is provided to Wikipedia templates (includes type 1 links, omits type 2 links). The relation to ALL is as follows: $ATL \subseteq ALL$.

Article Text Links with Relative Position (ATL-RP) This measure extracts all links from the Wikipedia text (type 1 links) and produces a score for the relative position of each link (see Formula 3). In fact, the link graph ATL-RP is the same as ATL but uses edge weights based on each link's position.

Abstract Links (ABL) This measure extracts only the links from Wikipedia abstracts. We chose the definition of DBpedia which defines an abstract as the first complete sentences that accumulate to less than 500 characters.[11] This link set is a subset of all type 1 links (in particular: $ABL \subseteq ATL$).

Template Links (TEL) This measure is complementary to ATL and extracts only links from templates (omits type 1 links, includes type 2 links). The relation to ALL and ATL is as follows: $TEL = ALL \setminus ATL$.

Redirects are not resolved in any of the above methods. We executed the introduced extraction mechanisms on a dump of the English Wikipedia of February 5, 2015. This date is in line with the input of DEF with respect to DBpedia version 2015-04.[12] Table 1 provides an overview of the number of extracted links per link graph.

Table 1. Number of links per link graph. Duplicate links were removed in all graphs (except in ATL-RP where multiple occurrences have different positions).

ALL	ATL	ATL-RP	ABL	TEL
159 398 815	142 305 605	143 056 545	32 887 815	26 460 273

4   Experiments

In our experiments, we first computed PageRank on the introduced link graphs. We then measured the pairwise rank correlations (Spearman's ρ and Kendall's τ)[13] between these rankings and the reference datasets (of which three are also based on PageRank and two are based on page-view data of Wikipedia). With the resulting correlation scores, we investigated on the following hypotheses:

---

[11] DBpedia abstract extraction – http://git.io/vGZ4J.

[12] DBpedia 2015-04 dump dates – http://wiki.dbpedia.org/services-resources/data sets/dataset-2015-04/dump-dates-dbpedia-2015-04.

[13] Both measures have a value range $[-1, 1]$ and are specifically designed for measuring correlations between ranked lists.

H1 Links in templates are created in a "please fill out this form" manner and rather negatively influence the general estimate of salience that PageRank scores should represent.

H2 Links that are mentioned at the beginning of articles are more often clicked and, therefore, the ATL-RP and ABL rankings correlate stronger with the page-view-based rankings.

H3 The practice of resolving redirects does not strongly impact the final ranking (in accordance to PageRank scores) as redirect pages pass most of their score on to the respective target page.

4.1 PageRank Configuration

We computed PageRank with the following parameters on the introduced link graphs ALL, ATL, ATL-RP, ABL, and TEL: non-normalized, 40 iterations, damping factor 0.85, start value 0.1.

4.2 Reference Datasets

We use the following rankings as reference datasets:

DBpedia PageRank (DBP) The scores of DBpedia PageRank [11] are based on the "DBpedia PageLinks" dataset (i.e., Wikipedia PageLinks as extracted by DEF, redirected). The computation was performed with the same configuration as described in Sect. 4.1. The scores are regularly published as TSV and Turtle files. The Turtle version uses the vRank vocabulary [10]. Since DBpedia version 2015-04, the DBP scores are included in the official DBpedia SPARQL endpoint (see Listing 1.1 for an example query). In this work, we use the following versions of DBP scores based on English Wikipedia: DBpedia 3.8, 3.9, 2014, and 2015-04.

DBpedia PageRank Unredirected (DBP-U) This dataset is computed in the same way as DBP but uses the "DBpedia PageLinks Unredirected" dataset.[14] As the name suggests, Wikipedia redirects are not resolved in this dataset (see Sect. 2.3 for more background on redirects in Wikipedia). We use the 2015-04 version of DBP-U.

SubjectiveEye3D (SUB) Paul Houle aggregated the Wikipedia page views of the years 2008 to 2013 with different normalization factors (particularly considering the dimensions articles, language, and time).[15] As such, SubjectiveEye3D reflects the aggregated chance for a page view of a specific article in the interval years 2008 to 2013. However, similar to unnormalized PageRank, the scores need to be interpreted in relation to each other (i.e., the scores do not reflect a proper probability distribution as they do not add up to one).

---

[14] DBpedia PageLinks Unredirected – http://downloads.dbpedia.org/2015-04/core-i18n/en/page-links-unredirected_en.nt.bz2.

[15] SubjectiveEye3D – https://github.com/paulhoule/telepath/wiki/SubjectiveEye3D.

The Open Wikipedia Ranking (TOWR) The TOWR project is maintained by the Laboratory for Web Algorithmics of the Università degli Studi di Milano. It provides Wikipedia rankings in accordance to different ranking methods in a Web interface[16] for direct comparison. They provide the following measures:[17]

TOWR-PR PageRank computed on the Wikipedia link graph with the parallel Gauß-Seidel method [7] of the LAW[18] library.

TOWR-H Harmonic centrality as introduced in [3] computed on the Wikipedia link graph.

TOWR-I Indegree, ranks Wikipedia pages in accordance to their number of incoming links.

TOWR-PV Page views, ranks Wikipedia pages in accordance to "the number of page views in the last year"[19].

The two page-views-based rankings (i.e., SUB and TOWR-PV) serve as a reference in order to compare the different graph-based rankings. We show the mutual overlap of entities covered by the individual rankings in Table 2.

Table 2. Number of mutually covered entities (the colors are used for better readability and comprise no further meaning).

	TOTAL	DBP 3.8	DBP 3.9	DBP 2014	DBP 2015-04	DBP-U 2015-04	ALL	ATL	ATL-RP	ABL	TEL	TOWR-PR	TOWR-H	TOWR-I	TOWR-PV	SUB
TOTAL	23035755	17082708	18172871	19437352	20473313	20473371	18493968	17846024	17846024	12319754	5028217	4853042	4853042	4853042	4853042	6211717
DBP 3.8	17082708	17082708	16553638	16084755	15814436	15814433	14501459	14119610	14119610	10236803	4086481	4082009	4082009	4082009	4082009	4899380
DBP 3.9	18172871	16553638	18172871	17528557	17183483	17183460	15682785	15241442	15241442	10880926	4339961	4316452	4316452	4316452	4316452	5234094
DBP 2014	19437352	16084755	17528557	19437352	18923198	18923126	17151451	16613563	16613563	11639177	4676614	4612952	4612952	4612952	4612952	5193106
DBP 2015-04	20473313	15814436	17183483	18923198	20473313	20473209	18479125	17833498	17833498	12310229	5026674	4781197	4781197	4781197	4781197	5235341
DBP-U 2015-04	20473371	15814433	17183460	18923126	20473209	20473371	18479281	17833616	17833616	12310235	5026723	4781197	4781197	4781197	4781197	5235318
ALL	18493968	14501459	15682785	17151451	18479125	18479281	18493968	17845902	17845902	12311648	5028094	4780590	4780590	4780590	4780590	4936935
ATL	17846024	14119610	15241442	16613563	17833498	17833616	17845902	17846024	17846024	12311477	4382197	4779031	4779031	4779031	4779031	4936085
ATL-RP	17846024	14119610	15241442	16613563	17833498	17833616	17845902	17846024	17846024	12311477	4382197	4779031	4779031	4779031	4779031	4936085
ABL	12319754	10236803	10880926	11639177	12310229	12310235	12311648	12311477	12311477	12319754	4062460	4739103	4739103	4739103	4739103	4425820
TEL	5028217	4086481	4339961	4676614	5026674	5026723	5028094	4382197	4382197	4062460	5028217	3320432	3320432	3320432	3320432	2913541
TOWR-PR	4853042	4082009	4316452	4612952	4781197	4781197	4780590	4779031	4779031	4739103	3320432	4853042	4853042	4853042	4853042	3986482
TOWR-H	4853042	4082009	4316452	4612952	4781197	4781197	4780590	4779031	4779031	4739103	3320432	4853042	4853042	4853042	4853042	3986482
TOWR-I	4853042	4082009	4316452	4612952	4781197	4781197	4780590	4779031	4779031	4739103	3320432	4853042	4853042	4853042	4853042	3986482
TOWR-PV	4853042	4082009	4316452	4612952	4781197	4781197	4780590	4779031	4779031	4739103	3320432	4853042	4853042	4853042	4853042	3986482
SUB	6211717	4899380	5234094	5193106	5235318	5235318	4936935	4936085	4936085	4425820	2913541	3986482	3986482	3986482	3986482	6211717

Legend 30000000   0

4.3 Results

We used MATLAB for computing the pairwise Spearman's ρ and Kendall's τ correlation scores. The Kendall's τ rank correlation measure has $\mathcal{O}(n^2)$ complexity and takes a significant amount of time for large matrices. In order to speed this up, we sampled the data matrix by a random selection of 1M rows for

---

[16] The Open Wikipedia Ranking – http://wikirank.di.unimi.it/.

[17] For their 2015 edition (that we analyze), the link-graph-based measures are applied on an English Wikipedia extract of April 3, 2015. Links in infoboxes were not considered.

[18] LAW – http://law.di.unimi.it/.

[19] Source: http://wikirank-2015.di.unimi.it/more.html.

Kendall's τ. The pairwise correlation scores of ρ and τ are reported in Tables 3 and 4 respectively. The results are generally as expected: For example, the pageview-based rankings correlate strongest with each other. The DDP rankings correlate strongest with the respective neighboring DBP versions. Also DBP-U 2015-04 and ALL have a very strong correlation (these rankings should be equal).

Table 3. Correlation: Spearman's ρ (the colors are used for better readability and comprise no further meaning).

	DBP 3.8	DBP 3.9	DBP 2014	DBP 2015-04	DBP-U 2015-04	ALL	ATL	ATL-RP	ABL	TEL	TOWR-PR	TOWR-H	TOWR-I	TOWR-PV	SUB	Legend
DBP 3.8	1.000	0.965	0.930	0.885	0.696	0.689	0.692	0.646	0.672	0.295	0.832	0.736	0.777	0.624	0.541	1.000
DBP 3.9	0.965	1.000	0.960	0.910	0.707	0.699	0.701	0.653	0.685	0.289	0.872	0.768	0.810	0.638	0.537	0.500
DBP 2014	0.930	0.960	1.000	0.941	0.719	0.709	0.712	0.661	0.700	0.278	0.904	0.796	0.836	0.648	0.502	0.000
DBP 2015-04	0.885	0.910	0.941	1.000	0.771	0.756	0.758	0.708	0.770	0.164	0.772	0.697	0.723	0.654	0.551	
DBP-U 2015-04	0.696	0.707	0.719	0.771	1.000	1.000	0.985	0.945	0.792	0.344	0.773	0.695	0.726	0.657	0.582	
ALL	0.689	0.699	0.709	0.756	1.000	1.000	0.985	0.945	0.788	0.346	0.782	0.707	0.731	0.661	0.565	
ATL	0.692	0.701	0.712	0.758	0.985	0.985	1.000	0.958	0.797	0.294	0.792	0.711	0.732	0.658	0.551	
ATL-RP	0.646	0.653	0.661	0.708	0.945	0.945	0.958	1.000	0.794	0.315	0.794	0.714	0.736	0.646	0.642	
ABL	0.672	0.685	0.700	0.770	0.792	0.788	0.797	0.794	1.000	0.263	0.542	0.441	0.535	0.499	0.455	
TEL	0.295	0.289	0.278	0.164	0.344	0.346	0.294	0.315	0.263	1.000	0.487	0.425	0.522	0.419	0.407	
TOWR-PR	0.832	0.872	0.904	0.772	0.773	0.782	0.792	0.794	0.542	0.487	1.000	0.859	0.889	0.845	0.593	
TOWR-H	0.736	0.768	0.796	0.697	0.695	0.707	0.711	0.714	0.441	0.425	0.859	1.000	0.809	0.677	0.614	
TOWR-I	0.777	0.810	0.836	0.723	0.726	0.731	0.732	0.736	0.535	0.522	0.889	0.809	1.000	0.668	0.616	
TOWR-PV	0.624	0.638	0.648	0.654	0.657	0.661	0.658	0.646	0.499	0.419	0.845	0.877	0.668	1.000	0.857	
SUB	0.541	0.537	0.502	0.551	0.582	0.565	0.551	0.642	0.455	0.407	0.593	0.614	0.616	0.857	1.000	

H1 seems to be supported by the data as the TEL PageRank scores correlate worst with any other ranking. However, ATL does not correlate better with SUB and TOWR-PV than ALL. This indicates that the reason for the bad correlation might not be due to the "bad semantics of links in the infobox". With random samples on ATL—which produced similar results—we found that the computed PageRank values of TEL are mostly affected by the low total link count (see Table 1). With respect to the initial example, the PageRank score of "Carl Linnaeus" is reduced to 217 in ATL. However, this subjective perception of improvement can not be generalized (with the used measures).

On a side note: we assume that computing PageRank on DBpedia's RDF data would produce similar scores as TEL because DBpedia extracts its semantic relations mostly from Wikipedia's infoboxes.

Indicators for **H2** are the scores of ABL and ATL-RP. However, similar to TEL, ABL does not produce enough links for a strong ranking. ATL-RP, in contrast, produces the strongest correlation with SUB. The improvement of ATL-RP comparred to ATL is clearly visible. This is an indication that—indeed—articles that are linked at the beginning of a page are more often clicked. This is supported by related findings of Dimitrov et al. [5] where actual HTTP referrer data was analyzed.

With respect to **H3**, we expected DBP-U 2015-04 and DBP 2015-04 to correlate much stronger than the results suggest. As a reason, we found that DEF does not implement the full workflow of Fig. 1: although it introduces a link $A \to C$ and removes the link $A \to B$, it does not remove the link $B \to C$. As such, the article B occurs in the final entity set with the lowest PageRank score of 0.15 (as it has no incoming links). In contrast, in DBP-U 2015-04, these pages often accumulate PageRank scores of 1000 and above. If B would not occur in the final ranking of DBP 2015-04, it would not be considered by the rank correlation measures. This explains the comparatively weak correlation between the redirected and unredirected datasets.

Further Observations. Another surprising result is the rather weak correlation of TOWR-PR with all the other PageRank-based rankings. As the Wikipedia dump date of DBpedia 2015-04 (that we also used for our measures, see Sect. 3) is only two months apart from the dump date used by TOWR, we expected much stronger correlations here. This is amplified by the observation that TOWR-PR correlates stronger with older DBP versions. However, Table 2 already suggests a clear difference with respect to the number of covered entities. Therefore, we assume that the preprocessing of the link graph performed by TOWR induces this bias. This is also supported by the strong correlations between the link-graph-based TOWR measures (i.e., TOWR-PR, TOWR-H, and TOWR-I) visible in Tables 3 and 4.

Table 4. Correlation: Kendall's τ on a sample of 1 000 000 (the colors are used for better readability and comprise no further meaning).

	DBP 3.8	DBP 3.9	DBP 2014	DBP 2015-04	DBP-U 2015-04	ALL	ATL	ATL-RP	ABL	TEL	TOWR-PR	TOWR-H	TOWR-I	TOWR-PV	SUB	Legend
DBP 3.8	1.000	0.931	0.879	0.798	0.611	0.606	0.604	0.548	0.571	0.209	0.695	0.569	0.625	0.455	0.383	1.000
DBP 3.9	0.931	1.000	0.924	0.826	0.627	0.620	0.618	0.556	0.583	0.205	0.740	0.598	0.658	0.464	0.379	0.500
DBP 2014	0.879	0.924	1.000	0.862	0.647	0.637	0.633	0.565	0.598	0.199	0.785	0.623	0.686	0.471	0.354	0.000
DBP 2015-04	0.798	0.826	0.862	1.000	0.761	0.743	0.725	0.632	0.689	0.116	0.615	0.524	0.563	0.473	0.392	
DBP-U 2015-04	0.611	0.627	0.647	0.761	1.000	0.990	0.948	0.837	0.680	0.254	0.615	0.521	0.585	0.474	0.413	
ALL	0.606	0.620	0.637	0.743	0.990	1.000	0.951	0.839	0.675	0.256	0.623	0.532	0.569	0.478	0.400	
ATL	0.604	0.618	0.633	0.725	0.948	0.951	1.000	0.859	0.686	0.207	0.642	0.538	0.572	0.476	0.389	
ATL-RP	0.548	0.556	0.565	0.632	0.837	0.839	0.859	1.000	0.689	0.222	0.633	0.540	0.573	0.464	0.463	
ABL	0.571	0.583	0.598	0.689	0.680	0.675	0.686	0.689	1.000	0.198	0.405	0.321	0.408	0.363	0.328	
TEL	0.209	0.205	0.199	0.116	0.254	0.256	0.207	0.222	0.198	1.000	0.360	0.313	0.397	0.304	0.294	
TOWR-PR	0.695	0.740	0.785	0.615	0.615	0.623	0.642	0.633	0.405	0.360	1.000	0.687	0.743	0.467	0.425	
TOWR-H	0.569	0.598	0.623	0.524	0.521	0.532	0.538	0.540	0.321	0.313	0.687	1.000	0.647	0.494	0.443	
TOWR-I	0.625	0.658	0.686	0.563	0.565	0.569	0.572	0.573	0.408	0.397	0.743	0.647	1.000	0.500	0.457	
TOWR-PV	0.455	0.464	0.471	0.473	0.474	0.478	0.476	0.464	0.363	0.304	0.467	0.494	0.500	1.000	0.695	
SUB	0.383	0.379	0.354	0.392	0.413	0.400	0.389	0.463	0.328	0.294	0.425	0.443	0.457	0.695	1.000	

In addition to ATL-RP, also the link-graph-based TOWR measures exhibit a stronger correlation with SUB than the other PageRank-based measures. However, with respect to Table 2 it becomes clear that their overlap with SUB is 949 603 entities less than the one of ATL-RP (or -19% relative to the overlap of ATL-RP and SUB). With this difference, the correlation scores are not directly comparable.

4.4 Conclusions

Whether links from templates are excluded or included in the input link graph does not impact strongly on the quality of rankings produced by PageRank. WLRank on articles produces best results with respect to the correlation to page-view-based rankings. In general, although there is a strong correlation, we assume that link and page-view-based rankings are complementary. This is supported by Table 5 which contains the top-50 scores of SUB, DBP 2015-04, and ATL-RP: The PageRank-based measures are strongly influenced by articles that relate to locations (e.g., countries, languages, etc.) as they are highly interlinked and referenced by a very high fraction of Wikipedia articles. In contrast, the page-view-based ranking of SubjectiveEye3D covers topics that are frequently accessed and mostly relate to pop culture and important historical figures or events. We assume that a strong and more objective ranking of entities is most likely achieved by combining link-structure and page-view-based rankings on Wikipedia. For applications that deal with NLP, we recommend to use the unredirected version of DBpedia PageRank.

5   Related Work

There are two common types of Wikipedia rankings: one is based on measures on the link graph, the other is based on consumption (e.g., page views). In the following, we briefly introduce the state of the art in both Wikipedia ranking methods.

Measures on the Wikipedia link graph: The work of Eom et al. [6] investigates on the difference between 24 language editions of Wikipedia with PageRank, 2DRank, and CheiRank rankings. The analysis focuses on the rankings of the top-100 persons in each language edition. We consider this analysis as seminal work for investigation on mining cultural differences with Wikipedia rankings. This is an interesting topic as different cultures often use the same language edition of Wikipedia (e.g., United Kingdom and the United States use English). Similarly, the work of Lages et al. provide rankings of universities of the world in [8]. Again, 24 language editions were analyzed with PageRank, 2DRank, and CheiRank. PageRank is shown to be efficient in producing similar rankings like the "Academic Ranking of World Universities (ARWU)" (that is provided yearly by the Shanghai Jiao Tong University). The Open Wikipedia Ranking (TOWR) also applies different graph measures on the Wikipedia link graph (see Sect. 4.2).

The above approaches vary the applied graph measures (PageRank, 2DRank, CheiRank, indegree, harmonic centrality) but do not vary the link extraction methods. In this paper, we experiment with both, different input graphs and a combination of a new weighted input graph and WLRank.

Wikipedia consumption patterns: The official page view statistics of various Wikipedia projects are publicly available as dumps[20] or as a Web API[21]. Our work on this paper was mainly influenced and motivated by an initial experiment that was performed by Paul Houle: in the Github project documentation of SubjectiveEye3D (see Sect. 4.2 for more details on SubjectiveEye3D), he reports

---

[20] Page view statistics for Wikimedia projects – https://dumps.wikimedia.org/other/pagecounts-raw/.

[21] Wikipedia Pageview API – https://wikitech.wikimedia.org/wiki/Analytics/PageviewAPI.

Table 5. The top-50 rankings of SubjectiveEye3D (< 0.3, above are: Wiki, HTTP 404, Main Page, How, SDSS), DBP 2015-04, and ATL-RP.

	SUB	DBP 2015-04	ATL-RP
1	YouTube	Category:Living people	United States
2	Searching	United States	World War II
3	Facebook	List of sovereign states	France
4	United States	Animal	United Kingdom
5	Undefined	France	Race and ethnicity in the United States Census
6	Lists of deaths by year	United Kingdom	Germany
7	Wikipedia	World War II	Canada
8	The Beatles	Germany	Association football
9	Barack Obama	Canada	Iran
10	Web search engine	India	India
11	Google	Iran	England
12	Michael Jackson	Association football	Latin
13	Sex	England	Australia
14	Lady Gaga	Australia	Russia
15	World War II	Arthropod	China
16	United Kingdom	Insect	Italy
17	Eminem	Russia	Japan
18	Lil Wayne	Japan	Village
19	Adolf Hitler	China	Moth
20	India	Italy	World War I
21	Justin Bieber	English language	Romanize
22	How I Met Your Mother	Poland	Spain
23	The Big Bang Theory	London	Romanization
24	World War I	Spain	Europe
25	Miley Cyrus	New York City	Romania
26	Glee (TV series)	Catholic Church	Soviet Union
27	Favicon	World War I	London
28	Canada	Bakhsh	English language
29	Sex position	Latin	Poland
30	Kim Kardashian	Village	New York City
31	Australia	Counties of Iran	Catholic Church
32	Rihanna	Provinces of Iran	Brazil
33	Steve Jobs	Lepidoptera	Netherlands
34	Selena Gomez	California	Greek language
35	Internet Movie Database	Brazil	Category:Unprintworthy redirects
36	Sexual intercourse	Romania	Scotland
37	Harry Potter	Europe	Sweden
38	Japan	Soviet Union	California
39	New York City	Chordate	Species
40	Human penis size	Netherlands	French language
41	Germany	New York	Mexico
42	Masturbation	Administrative divisions of Iran	Genus
43	September 11 attacks	Iran Standard Time	United States Census Bureau
44	Game of Thrones	Mexico	Turkey
45	Tupac Shakur	Voivodeship (Poland)	New Zealand
46	1	Sweden	Census
47	Naruto	Powiat	Middle Ages
48	Vagina	Gmina	Paris
49	Pornography	Moth	Communes of France
50	House (TV series)	Departments of France	Switzerland

about Spearman and Kendall rank correlations between SubjectiveEye3D and our published PageRank computations [11].[22] His results are similar to our computations. In a recent work, Dimitrov et al. introduce a study on the link traversal behavior of users within Wikipedia with respect to the positions of the followed links [5]. The authors conclude that a great fraction of clicked links can be found in the top part of the articles.

Comparing ranks on Wikipedia is an important topic and with our contribution we want to emphasize the need for considering the features "link graph" and "page views" in combination.

6   Summary and Future Work

In this work, we compared different input graphs for the PageRank algorithm, the impact on the scores, and the correlation to page-view-based rankings. The main findings can be summarized as follows:

1. Removing template links has no general influence on the PageRank scores.
2. The results of WLRank with respect to the relative position of a link indicate a better correlation to page-view-based rankings than other PageRank methods.
3. If redirects are resolved, it should be done in a complete manner as, otherwise, entities get assigned artificially low scores. We recommend using an unredirected dataset for applications in the NLP context.

Currently, we use the link datasets and the PageRank scores in our work on entity summarization [12,13]. However, there are many applications that can make use of objective rankings of entities. Therefore, we plan to investigate further on the combination of page-view-based rankings and link-graph-based ones. In effect, for humans, rankings of entities are subjective and it is a hard task to approximate "a general notion of importance".

Acknowledgments. The authors would like to thank Thimo Britsch for his contributions on the first versions of the SiteLinkExtractor tool. They also would like to thank Paul Houle and Sebastiano Vigna for their pointers and insights. The research leading to these results has received funding from the European Union Seventh Framework Programme (FP7/2007–2013) under grant agreement no. 611346 and by the German Federal Ministry of Education and Research (BMBF) within the Software Campus project "SumOn" (grant no. 01IS12051).

---

[22] Paul Houle on the correlation between DBP and SUB – https://github.com/paulhoule/telepath/wiki/Correlation-of-Subjective-Importance-Scores.

References

1. Auer, S., Bizer, C., Kobilarov, G., Lehmann, J., Cyganiak, R., Ives, Z.G.: DBpedia: a nucleus for a web of open data. In: Aberer, K., et al. (eds.) ASWC 2007 and ISWC 2007. LNCS, vol. 4825, pp. 722–735. Springer, Heidelberg (2007)
2. Baeza-Yates, R., Davis E.: Web page ranking using link attributes. In: Proceedings of the 13th International World Wide Web Conference on Alternate Track Papers & Amp; Posters, WWW Alt. 2004, pp. 328–329. ACM, New York (2004)
3. Boldi, P., Vigna, S.: Axioms for centrality. Internet Math. **10**(3–4), 222–262 (2014)
4. Brin, S., Page, L.: The Anatomy of a large-scale hypertextual web search engine. In: Proceedings of the Seventh International Conference on World Wide Web 7, pp. 107–117. Elsevier Science Publishers B. V, Amsterdam (1998)
5. Dimitrov, D., Singer, P., Lemmerich, F., Strohmaier, M.: Visual positions of links and clicks on Wikipedia. In: Proceedings of the 25th International Conference Companion on World Wide Web, WWW 2016 Companion, pp. 27–28. International World Wide Web Conferences Steering Committee (2016)
6. Eom, Y.-H., Aragn, P., Laniado, D., Kaltenbrunner, A., Vigna, S., Shepelyansky, D.L.: Interactions of cultures and top people of Wikipedia from ranking of 24 language editions. PLoS ONE **10**(3), 1–27 (2015)
7. Kohlschütter, C., Chirita, P.-A., Nejdl, W.: Efficient parallel computation of pagerank. In: Lalmas, M., MacFarlane, A., Rüger, S.M., Tombros, A., Tsikrika, T., Yavlinsky, A. (eds.) ECIR 2006. LNCS, vol. 3936, pp. 241–252. Springer, Heidelberg (2006)
8. Lages, J., Patt, A., Shepelyansky, D.L.: Wikipedia ranking of world universities. Eur. Phys. J. B **89**(3), 69 (2016)
9. von Linné, C., Salvius, L., Linnaei, C.: Systema naturae per regna tria naturae: secundum classes, ordines, genera, species, cum characteribus, differentiis, synonymis, locis., volume v. 1. Impensis Direct. Laurentii Salvii, Holmiae (1758)
10. Roa-Valverde, A., Thalhammer, A., Toma, I., Sicilia, M.-A.: Towards a formal model for sharing and reusing ranking computations. In: Proceedings of the 6th International WS on Ranking in Databases in conjunction with VLDB 2012 (2012)
11. Thalhammer, A.: DBpedia pagerank dataset (2016). http://people.aifb.kit.edu/ath#DBpedia_PageRank
12. Thalhammer, A., Lasierra, N., Rettinger, A.: LinkSUM: using link analysis to summarize entity data. In: Bozzon, A., Cudré-Mauroux, P., Pautasso, C. (eds.) ICWE 2016. LNCS, vol. 9671, pp. 244–261. Springer, Heidelberg (2016). doi:10.1007/978-3-319-38791-8_14
13. Thalhammer, A., Rettinger, A.: Browsing DBpedia entities with summaries. In: Presutti, V., Blomqvist, E., Troncy, R., Sack, H., Papadakis, I., Tordai, A. (eds.) ESWC 2014. LNCS, vol. 8798, pp. 511–515. Springer, Heidelberg (2014). doi:10.1007/978-3-319-11955-7_76

LDQ: 3rd Workshop on Linked Data Quality

Increasing Quality of Austrian Open Data by Linking Them to Linked Data Sources: Lessons Learned

Tomáš Knap[1,2(✉)]

[1] Faculty of Mathematics and Physics, Charles University in Prague,
Malostranské nám. 25, 118 00 Praha 1, Czech Republic
knap@ksi.mff.cuni.cz

[2] Semantic Web Company, Mariahilfer Straße 70/8, 1070 Vienna, Austria
t.knap@semantic-web.at

Abstract. One of the goals of the ADEQUATe project is to improve the quality of the (tabular) open data being published at two Austrian open data portals by leveraging these tabular data to Linked Data, i.e., (1) classifying columns using Linked Data vocabularies, (2) linking cell values against Linked Data entities, and (3) discovering relations in the data by searching for evidences of such relations among Linked Data sources. Integrating data at Austrian data portals with existing Linked (Open) Data sources allows to, e.g., increase data completeness and reveal discrepancies in the data. In this paper, we describe lessons learned from using TableMiner+, an algorithm for (semi)automatic leveraging of tabular data to Linked Data. In particular, we evaluate TableMiner+'s ability to (1) classify columns of the tabular data and (2) link (disambiguate) cell values against Linked Data entities in Freebase. The lessons learned described in this paper are relevant not only for the goals of the ADEQUATe project, but also for other data publishers and wranglers who need to increase quality of open data by (semi)automatically interlinking them to Linked (Open) Data entities.

Keywords: Open Data · Linked Data · Data quality · Data linking · Data integration · Entity disambiguation

1 Introduction

The advent of Linked Data [1] accelerates the evolution of the Web into an exponentially growing information space where the unprecedented volume of data offers information consumers a level of information integration that has up to now not been possible. Consumers can now mashup and readily integrate information for use in a myriad of alternative end uses.

This work has been supported in part by the Austrian Research Promotion Agency (FFG) under the project ADEQUATe (grant no. 849982) and by Charles University project P4.

H. Sack et al. (Eds.): ESWC 2016 Satellite Events, LNCS 9989, pp. 243–254, 2016.
DOI: 10.1007/978-3-319-47602-5_42

In the recent days, governmental organizations publish their data as open data (most typically as CSV files). To fully exploit the potential of such data, the publication process should be improved, so that data are published as Linked Open Data. By leveraging open data to Linked Data, we increase usefulness of the data by providing global identifiers for things and we enrich the data with links to external sources.

To leverage CSV files to Linked Data[1], it is necessary to (1) classify CSV columns based on its content and context against existing knowledge bases (2) assign RDF terms (HTTP URLs, blank nodes and literals) to the particular cell values according to Linked Data principles (HTTP URL identifiers may be reused from one of the existing knowledge bases), (3) discover relations between columns based on the evidence for the relations in the existing knowledge bases, and (4) convert CSV data to RDF data properly using data types, language tags, well-known Linked Data vocabularies, etc.

To introduce an illustrative example of leveraging CSV files to Linked Data, if the published CSV file would contain names of the movies in the first column and names of the directors of these movies in the second column, the leveraging of CSV files to Linked Data should automatically (1) classify first and second column as containing instances of classes 'Movie' and 'Director' respectively, (2) convert cell values in the 'movies' and 'directors' columns to HTTP URL resources, e.g., instead of using 'Matrix' as the name of the movie, URL 'http://www.freebase.com/m/02116f' may be used pointing to Freebase knowledge base[2] and standing for 'Matrix' movie with further attributes of that movie and links to further resources, and (3) discover relations between columns, such as relation 'isDirectedBy' between first and second column[3].

In this paper, we focus on the CSV files available at two Austrian data portals – http://www.data.gv.at and http://www.opendataportal.at. The first one is the official national Austrian data portal, with lots of datasets published by the Austrian government. Our goal is not to find a solution, which automatically leverages tabular data to Linked Data, as this is really challenging and we are aware of that, but our goal is to help data wranglers to convert tabular data to Linked Data by suggesting them (1) concepts classifying the columns and (2) entities the cell values may be disambiguated to. To realize these steps, we evaluate TableMiner+, an algorithm for (semi)automatic leveraging of tabular data to Linked Data. By successfully classifying columns and disambiguating cell values, we immediately increase the quality of the data along the *interlinking* quality dimension [9].

The main contributions of this paper are lessons learned from evaluating TableMiner+ to classify columns and disambiguate cell values in CSV files obtained from the national Austrian open data portal. In [10], they also

---

[1] By leveraging the data we mean improving the way how data is published by converting it from CSV files to Linked Data, with all the benefits Linked Data provides [1].
[2] http://freebase.com.
[3] The classes 'Movie' and 'Director' and the relation 'isDirectedBy' mentioned above should be reused from some well know Linked Data vocabulary.

evaluate TableMiner+, nevertheless, (1) they do not evaluate TableMiner+ on top of CSV files and (2) they do not evaluate TableMiner+ on top of governmental open data, containing, e.g., lots of statistical data.

The rest of the paper is organized as follows. Section 2 discusses possible approaches for leveraging tabular data to Linked Data and justifies selection of TableMiner+ as the most promising algorithm for leveraging CSV files to Linked Data. Section 3 evaluates TableMiner+ algorithm on top of the data obtained from the national Austrian data portal. Section 4 summarizes lessons learned and we conclude in Sect. 5.

2   TableMiner+ and Related Work

TableMiner+ [10] is an algorithm for (semi)automatic leveraging of tabular data to Linked Data. TableMiner+ consumes a table as the input. Further, it (1) discovers subject column of the table (the 'primary key' column containing identifiers for the rows), (2) classifies columns of the table to concepts (topics) available in Freebase, (3) links (disambiguates) cell values against Linked Data entities in Freebase, and (4) discovers relations among the columns by trying to find evidence for the relations in Freebase. TableMiner+ uses Freebase as its knowledge base; as the authors in [10] claim, Freebase is currently the largest knowledge base and Linked Data set in the world, containing over 2.4 billion facts about over 43 million topics (e.g., entities, concepts), significantly exceeding other popular knowledge bases such as DBpedia[4] and YAGO [7]. TableMiner+ is available under an open license – Apache License v2.0.

Limaye et al. [3] model table components (e.g. headers of columns, cells) and their interdependence using a probabilistic graphical model, which consists of two components: *variables* that model different table components, and *factors* modeling (1) the compatibility between the variable and each of its candidate and (2) the compatibility between the variables believed to be correlated. For example, given a named entity column, the header of the column is a variable that takes values from a set of candidate concepts; each cell in the column is a variable that takes values from a set of candidate entities. The task of inference amounts to searching for an assignment of values to the variables that maximizes the joint probability [10].

Mulwad et al. [5] argue that computing the joint probability distribution in Limaye's method [3] is very expensive. Built on the earlier work by Syed et al. [8] and Mulwad et al. [4,6], they propose a lightweight semantic message passing algorithm that applies inference to the same kind of graphical model.

When comparing approach of Limaye et al. and Mulwad et al. with TableMiner+ approach, as the authors [10] state, TableMiner+ approach is fundamentally different since it (1) adds a subject column detection algorithm, (2) deals with both named entity columns and literal columns, while Mulwad et al. only handle named entity columns, (3) uses an efficient approach bootstrapped

---

[4] http://dbpedia.org.

by sampled data from the table while Mulwad et al. and also Limaye et al. build a model that approaches the task in an exhaustive way, which is not efficient, (4) uses different methods for scoring and ranking candidate entities, concepts and relations; and (5) models interdependence differently which, if transformed to an equivalent graphical model, would result in fewer factor nodes.

In [2], the authors present an approach for enabling the user-driven semantic mapping of large amounts of tabular data using MediaWiki[5] system. Although we agree that user's feedback is important when judging about the correctness of the suggested concept for classification or suggested entity for disambiguation, and completely automated solutions leveraging tabular data to Linked Data are very challenging, the approach in [2] relies solely on the user-driven mappings, which expects too much effort from the users.

Open Refine[6] with RDF extension[7] provides a service to disambiguate cell values to Linked Data entities, e.g., from DBpedia. Nevertheless, the disambiguation is not interconnected with the classification as in case of, e.g., the TableMiner+ approach introduced in [10], so either a user has to manually specify the concept (class) restricting the candidate entities for disambiguation or all entities are considered during disambiguation, which is inefficient. Furthermore, the disambiguation phase is based just on the comparison of labels, without taking into account the context of the cell – further row cell values, column values, column header, etc.

We decided to use TableMiner+ to leverage CSV data from national Austrian data portal to Linked Data, because it outperforms similar algorithms, such as the one proposed by Mulwad et al. [5] or the algorithm presented in [3] and is available under an open license.

3   Evaluation

In this section, we describe the evaluation of TableMiner+ algorithm on top of CSV files obtained from the national Austrian data portal available at http:// data.gv.at. First we provide basic statistics about the data we use in the evaluation and then we describe evaluation metrics and results obtained during evaluation of (1) subject column detection, (2) classification, and (3) disambiguation. We do not evaluate in this paper the process of discovering binary relations among columns of the input files.

Since the standard distribution of TableMiner+ algorithm[8] expects HTML tables as the input, we extended the algorithm, so that it supports also CSV files as the input[9].

---

[5] http://mediawiki.org.

[6] http://openrefine.org/.

[7] https://github.com/fadmaa/grefine-rdf-extension.

[8] https://github.com/ziqizhang/sti.

[9] https://github.com/odalic/sti.

3.1   Data and Basic Statistics

We evaluated TableMiner+ on top of 753 files out of 1491 CSV files (50.5 %) obtained from the national Austrian data portal http://data.gv.at. The files processed were randomly selected from the files having less than 1 MB in size and having correct non-empty headers for all columns. We processed at most first 1000 rows from every such file. The processed files had in average 8.46 columns and 1.47 named entity columns.

3.2   Subject Column Detection

From all the processed files, we selected those for which TableMiner+ algorithm identified more than one named entity column and for those, we evaluated precision of the subject column detection by comparing the subject column selected by the TableMiner+ algorithm for the given file and the subject column manually annotated as being correct by a human annotator[10].

Results. In 97.15 % of cases, the subject column was properly identified by the TableMiner+ algorithm. There were couple of issues, e.g., considering column with companies, rather than with projects as the subject column in the CSV file containing list of projects. In case of statistical data containing couple of dimensions and measures, every dimension (except of the time dimension) was considered as a correctly identified subject column.

3.3   Classification

In TableMiner+ algorithm, candidate concepts classifying certain column are computed in phases. First, a sample of cells of the processed column is selected, disambiguated and the concepts of the disambiguated entities vote for the initial winning concept classifying the column. Further, all cells within that column are disambiguated, taking into account restrictions given by the initial winning concept, and, afterwards, all disambiguated cells vote once again for the concept classifying the column. If the winning concept classifying the column changes, disambiguation and voting is iterated. Lastly, candidate concepts for the given column are reexamined in the context of other columns and their candidate concepts, which may once again lead to the change of the winning concept suggested by TableMiner+ algorithm for the column. At the end, TableMiner+ algorithm reports the winning concept for every named entity column and also further candidate concepts, together with their scores (winning concept has the highest score).

To evaluate precision of such classification, for each processed file and named entity column, we marked down the candidate concepts for the classification

---

[10] When talking about a human annotator here and further in the text, we always mean a person who has at least university master degree and at least basic knowledge of German language (to understand data within the Austrian portals).

together with the scores computed by TableMiner+ algorithm, sorted by the descending scores. Then we selected candidate concepts having 5 highest scores (since more candidate concepts may have the same score, this may include more than 5 candidate concepts). Afterwards, we selected a random sample of these selected candidate concepts (containing selected candidate concepts for 100 columns) and let annotators to annotate for each file and named entity column the classifications suggested by the TableMiner+ – annotators marked the suggested column classification either with *best*, *good* or *wrong* labels. Label *best* means that the candidate concept is the best concept which may be used in the given situation – it must properly describe the semantics of the classified column and it must be the most specific concept as possible as the goal is to prefer the most specific concepts among all suitable concepts; for example, instead of the concept *location/location*, the concept *location/citytown* is the preferred concept for the column containing list of Austrian cities. Label *good* means that the candidate concept is appropriate (it properly describes the semantics of the cell values in the column), but it is not necessarily the most suitable concept. Label *wrong* means that the candidate concept is inappropriate, it has a different semantics.

Let us denote #*Cols* as the number of columns annotated by annotators. Further, let us define function $top_N(c)$, which is equal to 1 if the candidate concept c annotated as *best* for certain column was also identified by TableMiner+ as a concept having up to N-th highest score, $N \in 1, 2, 3, 4, 5$. If $N = 1$ and $top_1(c) = 1$ for certain concept c, it means that the winning concept suggested by TableMiner+ is the same as the concept annotated as *best* by the annotators. Further, let us define metric $best_N$ which computes the percentage of columns in which the candidate concept c annotated as *best* for certain column was also identified by TableMiner+ as a concept having N-th highest score at worst; divided by total number of annotated named entity columns:

$$best_N = 100 \cdot \sum_c top_N(c)/\#Cols$$

So, for example, $best_1$ denotes the percentage of cases (columns) for which the concept annotated as *best* is also the winning concept suggested by TableMiner+.

The formula above does not penalize situations when more candidate concept share the same score. Since our goal is not to automatically produce Linked Data or column classification from the result of the TableMiner+, but we expect that user is presented with couple of candidate concepts (s)he verifies/selects from, it is not important whether (s)he is presented with 5 or 8 concepts, but it is important to evaluate how often the concept annotated as *best* is among the highest scored concepts.

Results. The winning concepts (Freebase topics) discovered by the TableMiner+ algorithm running on top of all 753 files from the portal which were suggested for at least 20 columns and the number of columns for which these concepts were suggested as winning concepts are depicted in Table 1.

Table 1. The winning concepts (Freebase topics) as discovered by TableMiner+

Freebase concept	Number of columns
location/location	478
music/recording	166
music/single	51
organization/organization	48
people/person	45
music/artist	35
location/statistical_region	34
location/dated_location	26
base/aareas/schema/administrative_area	25
fictional_universe/fictional_character	25
film/film_character	25
business/employer	22
location/citytown	22
music/release_track	22

As we can see, majority of the columns were classified with the Freebase concept *location/location*. Although this is correct in most cases, typically, there is a better (more specific) concept available, such as *location/citytown*. There are also concepts, such as *music/recording* or *film/film_character*, which are in most cases results of the wrong classification due to low evidence for correct concepts during disambiguation of the sample cells.

Selected results of the $best_N$ measure are introduced in Table 2. As we can see, 20 % of concepts annotated as *best* were properly suggested by the TableMiner+ algorithm as the winning concepts; 36 % of concepts annotated as *best* for certain columns were among concepts suggested by TableMiner+ and having highest or second highest score, etc. In other words, there is 76 % probability that the concept annotated as being *best* will appear within candidate concepts suggested by TableMiner+ and having 5th highest score at worst.

Furthermore, in 68 % of the analyzed columns, only concepts annotated as *best* and *good* appear among concepts suggested by TableMiner+ and having 1st, 2nd, or 3rd highest score.

In 24 % of the analyzed columns, all concept candidates suggested by TableMiner+ were wrongly suggested. The reasons for completely wrong suggested
classifications are typically two-fold: (1) low disambiguation recall due to low evidence for the cell values within the Freebase knowledge base or (2) wrong disambiguation due to short named entities having various unintended meanings.

We did not evaluated recall of the concept classification, as there was always a suggested concept classifying the column, although the precision could have been low.

Table 2. Results of the $best_N$ measure

N	$best_N$ (in percentage)
1	20
2	36
3	64
4	74
5	76

3.4 Disambiguation

For selected concepts from Table 1, we computed precision and recall of the entities disambiguation. Precision is calculated as the number of distinct entities (cell values) being correctly linked to Freebase entities divided by the number of all distinct entities (cell values) linked to Freebase (restricted to the given concept). Recall is computed as the number of distinct entities being linked to Freebase divided by number of all distinct entities (restricted to the given concept). To know which entities were correctly linked to Freebase, we again asked annotators to annotate, for the columns classified with the selected concepts, each distinct winning disambiguation of the cell value to Freebase entity – annotators could have marked the winning disambiguated entity either as being *correct* or *wrong*. The disambiguation is *correct* if the disambiguated entity represents correctly the semantics of the cell value. Otherwise, it is marked as *wrong*.

Results. In case of *location/citytown* concept, we analyzed disambiguation of cities in 16 files, where the concept *location/citytown* was suggested as the winning concept by the TableMiner+ algorithm. The precision of the disambiguation was 95.2 %; the recall 88.1 %. We also analyzed other 24 files, where there was a column containing cities and one of the suggested concepts classifying that column (but not the winning concept) was *location/citytown* concept with the score above 1.0. In this case, precision was 99 % and recall 99.8 %, taking into account more than 500 distinct disambiguated entities. It is also worth mentioning that TableMiner+ algorithm properly disambiguates and classifies cell values based on the context of the cell; thus, in case of the column with the cities, the cell value *Grambach* is properly classified as the city and not the river.

We analyzed 23 files where there was a column containing districts of Austria classified with the winning concept *location/location*. The precision was 38.3 % and recall 100 %. The precision is lower because in this case, more than half of the districts (e.g. *Leibnitz, Leoben*) were classified as cities. The reason why these columns were classified at the end with the rather generic concept *location/location* and not with a more appropriate *location/administrative_division* is that some values within that column were wrongly disambiguated to cities and voted for *location/citytown*, some were disambiguated correctly to districts and voted for the best concept *location/admi nistrative_division*, and, since both these types of entities also belong to the concept *location/location*, the concept *location/location* was chosen as the winning one.

Concept *base/aareas/schema/administrative_area* has high precision 88 % and 100 % recall, but there were only 17 distinct districts of Linz processed.

Concept *organization/organization* has reasonable precision for columns holding schools – it links faculties to the proper universities with precision 75 % and recall 81 %. For other types of organizations, such as pharmacies, hospitals, etc., disambiguation does not work properly, because there are no corresponding entities to be linked in Freebase.

Disambiguation of *people/person* concept has very low precision. The reason for that is that vast majority of people are not in the knowledge base. For the same reason, also the precision of the concept *business/employer* is very low.

4 Lessons Learned

There is a high correlation between precision of the disambiguation and classification, which is caused by the fact that initial candidate concepts for the classification of a column are based on the votes of the disambiguated entities for the selected sample set of cells.

If the recall of the disambiguation is low (not much entities are disambiguated), it does not make sense to classify the column, as it will be in most cases misleading. In these cases, it is better to report that there is not enough evidence for the classification, rather than trying to classify the column somehow, because this ends up by suggesting completely irrelevant concepts, which confuses users.

Row context used by TableMiner+ algorithm proofed its usefulness in many situation. For example, it allowed to properly disambiguate commonly named cities having more than one matching entities in Freebase, i.e., the cities were properly disambiguated w.r.t. to the countries to which they belong.

If the cell values to be disambiguated are too short (e.g., abbreviations) and the precision of the subject column disambiguation, defining the context for these abbreviations, is low, it does not make sense to disambiguate these short cell values as the precision of such disambiguation will be low.

Classification/disambiguation in TableMiner+ has higher precision when the processed tabular data have subject column, which is further described by other columns, thus, classification/disambiguation may use reasonable row context. In case of statistical data, which merely involves measurements and dimension identifiers, the row context is not that beneficial and the precision of the classification/disambiguation is lower.

In many cases, the generic knowledge base, such as Freebase, is not sufficient as it does not include all needed information, e.g., it does not include information about all schools, hospitals, playgrounds, etc., in the country's states/regions/cities. So apart from generic knowledge bases, such as Freebase, also more focused knowledge bases should be used or constructed upfront.

TableMiner+ algorithm should use knowledge bases defining hierarchies of concepts within the knowledge base, so that we can avoid situations when more generic concepts are denoted as winning concepts because more entities in the given columns vote for such generic concepts. Using hierarchy of concepts would increase precision and improve performance of the classification/disambiguation algorithm.

4.1 Contributions to Data Quality

Paper [9] provides a survey of Linked Data quality assessment dimensions and metrics. In this section, we discuss how successful classification and disambiguation conducted by TableMiner+ contribute towards higher quality of the resulting Linked Data along the quality dimensions introduced in [9].

Successful classification and disambiguation increase number of links to external (linked) data sources (such as general knowledge bases, e.g. DBpedia, or Freebase), thus, directly increase the quality of the data along the *interlinking* dimension [9]. By having links to external (linked) data sources, it is then possible to improve the quality of the data along the quality assessment dimensions introduced in further subsections.

Completeness. In [9], completeness is defined as "the degree to which all required information is present in a particular dataset". Further, they distinguish:

- Schema completeness: the degree to which the classes and properties of an ontology are represented
- Property completeness: the degree to which values for a specific property are available for objects of a certain type or in general
- Population completeness: the degree to which all real-world objects of a particular type are represented in the dataset
- Interlinking completeness: the degree to which objects in the dataset are interlinked.

By disambiguating cell values to Linked Data entities, TablerMiner+ increases interlinking completness. By running TableMiner+ algorithm, we may also increase property completeness by introducing more facts about the objects from other (linked) data sources, e.g. from DBpedia or Freebase.

Semantic Accuracy. In [9], semantic accuracy is defined as "the degree to which data values correctly represent the real world facts".

It is possible to reveal discrepancies in the data classified/disambiguated by TablerMiner+ by comparing the attributes of the disambiguated entities with the attributes of the data introduced in external (linked) data sources, e.g., in DBpedia or Freebase.

Trustworthiness. In [9], trustworthiness is defined as "the degree to which the information is accepted to be correct, true, real and credible".

It is possible to increasing trustworthiness of the data processed by TableMiner+ by providing further evidence for the data from external (linked) data sources.

Interoperability. In [9], interoperability is defined as "the degree to which the format and structure of the information conforms to previously returned information as well as data from other sources".

Paper [9] distinguishes two metrics for interoperability dimension: (1) re-use of existing terms and (2) re-use of existing vocabularies – to which extent relevant vocabularies are used.

By reusing existing identifiers from external (linked) data sources, we increase re-use of existing terms. Furthermore, by discovering relations in TableMiner+, we also contribute towards re-use of existing vocabularies.

5   Conclusions and Next Steps

We evaluated TableMiner+ algorithm on top of the Austrian open data obtained from the Austrian national open data portal available at http://www.data.gv.at.

We showed that in 76 % of cases the concept annotated by humans as being the *best* in the given situation appears within the candidate concepts suggested by TableMiner+ with 5th highest score at worst. This is a promising result, as our main purpose is to provide to data wranglers not only the winning concepts, but also certain number of alternative concepts.

Classification and disambiguation had very high precision for concept of cities (95 %+) and reasonable precision for certain other concepts, such as districts, states, organizations. Nevertheless, for certain columns/cell values, the precision of the classification/disambiguation was rather low, which was caused either by (1) missing evidence for the disambiguated cell values in the Freebase knowledge base or (2) by trying to disambiguate cell values which have various alternative meanings. We showed that in 24 % cases, the analyzed columns had irrelevant classification, which is rather confusing for users and in these cases it would be better not to produce any classification at all.

Although the first results are promising, we plan to experiment further (1) with different knowledge bases, such as WikiData[11], and (2) also plan to improve TableMiner+ algorithm, so that it behaves, e.g., more conservative in cases of low evidence for the classification/disambiguation.

---

[11] www.wikidata.org.

References

1. Bizer, C., Heath, T., Berners-Lee, T.: Linked Data - the story so far. Int. J. Semant. Web Inf. Syst. **5**(3), 1–22 (2009)
2. Ermilov, I., Auer, S., Stadler, C.: Csv2rdf: user-driven CSV to RDF mass conversion framework. In: Proceedings of the ISEM 2013, 04–06 September 2013, Graz, Austria (2013)
3. Limaye, G., Sarawagi, S., Chakrabarti, S.: Annotating and searching web tables using entities, types and relationships. PVLDB **3**(1), 1338–1347 (2010)
4. Mulwad, V., Finin, T., Joshi, A.: Generating Linked Data by inferring the semantics of tables. In: Proceedings of the First International Workshop on Searching and Integrating New Web Data Sources - Very Large Data Search, Seattle, WA, USA, 2 September 2011, pp. 17–22 (2011)
5. Mulwad, V., Finin, T., Joshi, A.: Semantic message passing for generating Linked Data from tables. In: Alani, H., Kagal, L., Fokoue, A., Groth, P., Biemann, C., Parreira, J.X., Aroyo, L., Noy, N., Welty, C., Janowicz, K. (eds.) ISWC 2013, Part I. LNCS, vol. 8218, pp. 363–378. Springer, Heidelberg (2013). doi:10.1007/978-3-642-41335-3_23
6. Mulwad, V., Finin, T., Syed, Z., Joshi, A.: Using linked data to interpret tables. In: Proceedings of the First International Workshop on Consuming Linked Data, Shanghai, China, 8 November 2010 (2010)
7. Suchanek, F.M., Kasneci, G., Weikum, G.: Yago: a core of semantic knowledge. In: Proceedings of the 16th International Conference on World Wide Web, WWW 2007, pp. 697–706. ACM, New York (2007)
8. Syed, Z., Finin, T., Joshi, A.: Wikipedia as an ontology for describing documents. In: Proceedings of the Second International Conference on Weblogs and Social Media. AAAI Press, March 2008
9. Zaveri, A., Rula, A., Maurino, A., Pietrobon, R., Lehmann, J., Auer, S.: Quality assessment for linked data: a survey. Semant. Web J. **7**, 63–93 (2015)
10. Zhang, Z.: Effective and efficient semantic table interpretation using tableminer+. Semant. Web J. (2016)

Fourth Workshop on Linked Media (LiME-2016)

Anno4j - Idiomatic Access to the W3C Web Annotation Data Model

Emanuel Berndl[(⊠)], Kai Schlegel, Andreas Eisenkolb, Thomas Weißgerber,
and Harald Kosch

Chair of Distributed Information Systems,
University of Passau, Innstraße 43, 94032 Passau, Germany
{emanuel.berndl,kai.schlegel,andreas.eisenkolb,
thomas.weissgerber,harald.kosch}@uni-passau.de

Abstract. The Web Annotation Data Model proposes standardised
RDF structures to form "Web Annotations". These annotations are
used to express metadata information about digital resources and are
designed to be shared, linked, tracked back, as well as searched and discovered across different peers. Although this is an expressive and rich way
to create metadata, there exists a barrier for non-RDF and SPARQL
experts to create and query such information. We propose Anno4j, a
Java-based library, as a solution to this problem. The library supports
an Object-RDF mapping that enables users to generate Web Annotations by creating plain old Java objects - concepts they are familiar with
- while a path-based querying mechanism allows comprehensive information querying. Anno4j follows natural object-oriented idioms including
inheritance, polymorphism, and composition to facilitate the development. While supporting the functionality of the Web Annotation Data
Model, the library is implemented in a modular way, enabling developers
to add enhancements and use case specific model alterations. Features
like plugin functionality, transactions, and input/output methods further
decrease the boundary for non-RDF experts.

Keywords: Semantic Web · Linked Data · Web Annotations · Java ·
Developer tool

1 Introduction

Annotating things in the web is more and more common and a desired feature
in the internet of today. Users support comments to various media, tag people
on pictures, support helpful links to different topics, and so on. Often times, this
is done using RDF, the Resource Description Format [4], the de facto standard
for interlinking resources in the Semantic Web. Therefore, the Web Annotation Data Group[1] recently issued a new version of the Web Annotation Data
Model WADM [8] (derived from the Open Annotation Data Model OADM [7]).

---

[1] https://www.w3.org/annotation/.

© Springer International Publishing AG 2016
H. Sack et al. (Eds.): ESWC 2016 Satellite Events, LNCS 9989, pp. 257–270, 2016.
DOI: 10.1007/978-3-319-47602-5_43

The central concept of their model is the **Annotation**, being used as a way to give further details, description, or information about another "thing", in general digital resources. Examples could be a comment or tag on a web page or image, or a blog post about a news article. One annotation is devised out of three core components: the **body**, which contains the actual content of the annotation, the **target** specifying the "thing" that the annotation is about, and an **annotation node** itself, which joins the body and the target, while supporting provenance information of the whole annotation. This splits the content from its context in a modular way, which fosters the fundamentals of the annotation model: the designed annotations are interoperable and can be used and interpreted at various different locations, offering the possibility to connect information and metadata beyond the boundaries of single enterprises, data silos, and/or applications. The resulting combined knowledge base increases the benefit as well as the degree of information for every participant.

This aligns heavily with the "purpose" of the Semantic Web, with its promising advantages of combined and interlinked data. However, there exists an initial hurdle to make oneself familiar with the Semantic Web technologies as well as Linked Data "rules" [2]. Often times, developers are very skilled in their own respective programming domain, but lack the knowledge of producing RDF and consuming Linked Data over SPARQL.

To overcome some of these shortcomings and lower the barrier of Semantic Web technologies, Anno4j[2] provides a Java library to directly map Java objects to and from the Web Annotation Data Model. The core contributions are as follows:

- An Object-relational-like mapping (ORM) provides idiomatic access to W3C Web Annotation Data Model following natural object-oriented idioms. The use of RDF is enhanced by the advantages and assets of Java.
- Support for use-case specific Web Annotation Data Model alterations and extensions.
- A library built on-top of OpenRDF Alibaba[3], allowing for a broad field of application.
- Path-based query criteria for extensive annotation search functionality.
- Developer-friendly Open-Source Apache V2 license.

The remaining paper is structured as follows. Section 2 briefly lists the core features of the library and Sect. 3 highlights related work in the context of the WADM and its applications. Sections 4, 5, and 6 will give more detailed information about persistence, querying, and insights about the internals of Anno4j. In Sect. 7, convenience and RDF features that have been added to the library are enlisted. Section 8 shows a use case of Anno4j in the MICO[4] project. Finally, Sect. 9 will conclude the paper by depicting future work and a planned roadmap.

---

[2] https://github.com/anno4j/anno4j.

[3] https://bitbucket.org/openrdf/alibaba.

[4] http://www.mico-project.eu/.

2   Overview

Anno4j is a Java-based library, that offers developers the opportunity to easily create and consume annotations conform to the WADM by writing Java POJOs. The framework has been designed in a modular and extensible fashion, allowing users to extend at every feature of Anno4j. The core functionalities of Anno4j are:

- **Persistence**: Simple Java objects provide the basis of persistence and can easily be created and persisted with a given Anno4j object (see Sect. 4).
- **Querying**: A `QueryService` object created by an Anno4j instance can be supported with different query criteria (formalised as LDPath expression) to query and consume respective annotations of the Anno4j database (see Sect. 5).
- **WADM implementations**: Built-in and predefined implementations for all basic classes proposed by the W3C Web Annotation Data Model.
- **Transactional behaviour**: By working with transactions, an Anno4j user can create sets of operations, that can either be fully integrated in the database (commit) or not at all (rollback). This ensures the consistency of the database (see Sect. 7.1).
- **Context awareness**: RDF databases are often using different contexts to divide their data into subgraphs. This feature is also possible in Anno4j, turning RDF triples into quads (see Sect. 7.2).
- **Plugin Extensions**: By supporting a plugin interface, users can define own RDF functions in combination with respective evaluation operators, which then can be used as custom querying criteria in order to even enhance the querying functionality and fine tune it to their particular use-case (see Sect. 7.3).
- **Input and Output**: Anno4j is able to both read and write annotations from and to different standardised serialisations, such as JSON-LD, TURTLE, N3-Triples, RDF/XML, etc. (see Sect. 7.4).

3   Related Work

Some approaches to create and query for Web Annotations already exist, and therefore can be considered related to our topic. Alongside the WADM, the Web Annotation Working Group also issued a protocol, namely the Web Annotation Protocol WAP [6]. The purpose is to provide a standard set of actions which are to be supported by both an annotation client and annotation server in order to cooperate smoothly. This enables the formerly discussed advantages of the WADM to be fully exploited, as it is not just desired to build up pairs of participants, but rather a client and server architecture that allows multiple users to consume the information of single data sources. The protocol makes use of the Linked Data Platform LDP [10] to define their core concept of an annotation container. Supported REST and HTTP functionality offers different methods to create and easily query annotations from a given container and provide information about the container itself. This allows to further familiarise a client with the information or the structure that a given server will support. Listing 1.1 shows an exemplary POST request to create an annotation. The desired annotation content is supported as JSON, in this case a simple annotation that has a `oa:EmbeddedContent` (referring to the URI "http://www.w3.org/ns/oa# EmbeddedContent") body node containing the String (relationship `rdf:value`) "I like this page!". The target of the annotation is the homepage "http://www. example.com/index.html".

Listing 1.1. Example POST request to create an annotation using the WAP protocol

```
1  POST /annotations/ HTTP/1.1
2  Host: example.org
3  Content-Type: application/ld+json
4
5  {
6    "@context": "http://www.w3.org/ns/oa",
7    "@type": "oa:Annotation",
8    "body": {
9      "@type": "oa:EmbeddedContent",
10     "value": "I like this page!"
11   },
12   "target": "http://www.example.com/index.html"
13 }
```

Before the WAP has been fully published, similar approaches have been issued in order to give the annotations of the WADM (or in this case formerly known as the Open Annotation Data Model OA [7]) a new facet of interoperability by making them shareable more easily over the web. In [5] Pyysalo et al. describe a minimal web interface using REST, that allows users to create and share OA annotations. They implement two core components, the "OA Store" as client and the "OA Explorer" as server respectively, that can further be enhanced with two components for validation and format conversion.

In comparison, the REST-based approaches offer some advantages. As it is a protocol, there exists the freedom of implementing only parts of the specification, allowing the WAP-conform server or client to be adapted to a specific use case and thusly be more lightweight. The REST interface enables the server to be used as a platform independently, annotations are queried as well as created using JSON. On the other hand, all RDF instances in Anno4j are present as a POJO, allowing them to be used further in object-oriented ways, without the need of up-front parsing. Anno4j is also able to read and generate different serialisations of the RDF objects if needed.

4   Persistence

Anno4j's persistence implementation follows a very simple mechanism. Associated with a local in-memory store or a supported remote SPARQL endpoint, the persistence features allow users to generate RDF information by creating plain old Java objects - one direction of the ORM. In the following section, this process is illustrated with the example of an image detection, that expresses that "something" has been detected on a given image with a certain confidence. Figure 1 shows the image that will be utilised in the examples, while Fig. 2 depicts an exemplary Web Annotation that states that "Barack Obama" has been detected to "85 %".

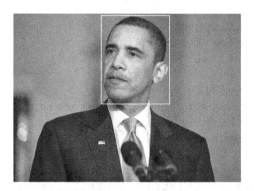

Fig. 1. Picture of Barack Obama. The red rectangle illustrates the area a face was found in

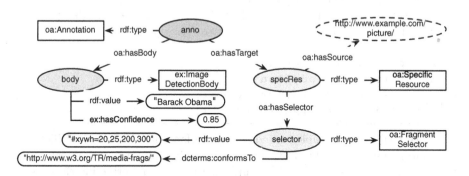

Fig. 2. Resulting RDF graph created with the code shown in Listing 1.2. Round and coloured nodes represent RDF node instances, a rectangle symbolises an RDF class, and rounded rectangles are RDF literals

Listing 1.2 shows a simple workflow example of how to create RDF information with Anno4j Java objects. All required RDF nodes are created independently, using the same Anno4j instance. Various fields are set and the nodes are connected at the end. All associated RDF nodes, relationships, and properties are created automatically and the respective triples are persisted at the registered triplestore. RDF class information is generated automatically when creating a respective instance, relationships and properties are associated through the setters. The resulting triples that are created after the workflow shown in Listing 1.2 can be seen in Fig. 2.

Listing 1.2. Exemplary creation and initialisation of an object

```
1   // Anno4j object initialised
2   Anno4j anno4j = new Anno4j();
3
4   // Create the nodes
5   Annotation annotation =
        anno4j.createObject(Annotation.class);
6
7   ImageDetectionBody body =
        anno4j.createObject(ImageDetectionBody.class);
8   body.setDepicts("Barack Obama");
9   body.setConfidence(0.85);
10
11  FragmentSelector selector =
        anno4j.createObject(FragmentSelector.class);
12  selector.setConformsTo("http://www.w3.org/
13                  TR/media-frags/");
14  selector.setValue("#xywh=1345,25,1050,1325");
15
16  SpecificResource specRes =
        anno4j.createObject(SpecificResource.class);
17  specRes.setSource("http://www.example.com/picture");
18
19  // Connect the nodes
20  annotation.setBody(body);
21  specRes.setSelector(selector);
22  annotation.addTarget(specRes);
```

Every RDF node is implemented as an interface class in Anno4j. When creating an instance of it, the proxy pattern is applied and a proxy object of the respective interface is generated. By implementing the supported interfaces, subclasses can be created and the information is also reflected in the RDF graph. Anno4j already comes with built-in and predefined implementations for most of the classes proposed by the Web Annotation Data Model. This enables developers to start right away with the core functionality of producing annotations. However, use-case specific alterations and enhancements are easy to integrate by extending the persistence layer with their own respective classes.

The main element to do so is the Java annotation @Iri, which is used at interface level of the given Java interface and on each setter/getter pair. The annotation at interface level sets the RDF class (relationship rdf:type) of the respective RDF object, while an assigned @Iri at setter **and** getter method will create the respective predicate attached to the RDF object.

Listing 1.3 shows an example of the implementation of a body interface that is used to convey detected things on a given image. It was already used in the previous example in Listing 1.2. In order to implement the required information of the example shown in Fig. 2, the body node needs to have a specific type (relationship rdf:type with the value ex:ImageDetectionBody) - defined with the @Iri Java Annotation at line 2, and two properties for the detected thing and the confidence about it (properties rdf:value in lines 7 and 10, and ex:hasConfidence in lines 13 and 16 respectively). Note that the interface extends the Body interface, an interface supported for body classes by Anno4j, in order to have the desired behaviour.

Listing 1.3. Exemplary body implementation with one field "depiction"

```
1  // Set the type of the class
2  @Iri(EX.IMAGE_DETECTION_BODY)
3  public interface ImageDetectionBody extends Body {
4
5      // Setters and getters required for the RDF
           predicates
6      @Iri(RDF.VALUE)
7      void setDepicts(String depiction);
8
9      @Iri(RDF.VALUE)
10     String getDepicts();
11
12     @Iri(EX.CONFIDENCE)
13     void setConfidence(Double confidence);
14
15     @Iri(EX.CONFIDENCE)
16     Double getConfidence();
17 }
```

5   Querying

Besides persisting Web Annotations, Anno4j also provides ways to query the annotations or only subparts of them that fit specific needs. On the one hand, Anno4j provides convenient mechanisms to directly query e.g. for all annotation bodies with a particular type or Anno4j Java class. On the other hand, Anno4j offers more expressive ways, using the path-based query language LDPath[5], to define query criteria to reduce the effort for non-SPARQL experts. LDPath, which is similar to XPath, allows a more compact and inline definition of criteria in contrast to the verbose pattern-based query language SPARQL. A

---

[5] http://marmotta.apache.org/ldpath/.

fluent interface API supports readability and comprehensible query definition. A collection of individual criteria defines the desired characteristic of the resulting annotations. Hereby, multiple criteria are combined with a logical AND operation. Helpful functions are supported, that can be used to create key query criteria LDPath expressions via Java methods (such as the type of the body node, the selection features, etc.). This further enhances the usability for beginners and non-SPARQL experts.

Listing 1.4 shows an example using LDPath to define two different criteria for annotations. In this example we are searching for annotations which satisfy the conditions that the annotation body should be a dctypes:StillImage and Barack Obama is depicted on the image. The .execute(Class type) method does not only define the class of the returned objects, it also does define the starting point of the query and LDPath expressions in the RDF graph. If no type is supported, the standard case is to query for oa:Annotation.

Listing 1.4. Anno4j Query Example

```
1  List<Annotation> annotations = queryService
2      .addCriteria("oa:hasBody[is-a dctypes:StillImage]")
3      .addCriteria("oa:hasBody/rdf:value","Barack Obama")
4      .execute(Annotation.class);
```

Although Anno4j uses LDPath as syntax for query criteria, there is no need for a special LDPath-capable RDF endpoint, because LDPath criteria are translated to an equivalent valid SPARQL 1.1 query. This allows developers to reuse generic SPARQL 1.1 endpoints for their use-cases. Besides basic path criteria, Anno4j also supports a wide range of different LDPath condition types[6]:

- Forward and reverse path conditions
- Resolving of namespace abbreviations
- Recursive pathing like OneOrMore(+) or ZeroOrMore(*)
- Comparison methods like equal, greater, or lower for conditions
- Union of multiple paths
- Type or datatype conditions
- Logical combination of conditions
- Custom functions (see Sect. 7.3).

After execution of the translated SPARQL query against the specified endpoint, all query results are automatically transformed to corresponding annotated Java objects. This abstraction layer allows developers to easily work with RDF information in contrast to constructing complex SPARQL queries and parsing the SPARQL results.

6   Internals

At its core, Anno4j builds upon the OpenRDF Alibaba library (former Elmo codebase) which provides simplified RDF store abstractions and combines the

---

[6] For further and detailed description of the LDPath criteria, please refer to the LDPath specification at http://marmotta.apache.org/ldpath/language.html.

flexibility and adaptivity of RDF with the powerful object-oriented programming model of Java. It is able to map Java objects to and from RDF resources in a non-intrusive manner that enables developers to work with resources stored in a SPARQL endpoint. Nested properties of RDF resources are lazy evaluated to avoid unnecessary fetching of unused information for faster and efficient query evaluation. Considering Listing 1.3, lazy evaluation implies that the RDF value for `depicts` is not fetched from the SPARQL endpoint until the `.getDepicts()`-method of the respective Java object is called.

As mentioned before, Anno4j uses LDPath syntax to define query criteria. Internally, LDPath criteria is automatically transformed to valid SPARQL 1.1 syntax. Some LDPath expressions can be directly mapped to similar SPARQL 1.1 property path expressions (e.g. path selection, inverse path selection "^", or alternative path "|"). LDPath expressions which can't be directly mapped to SPARQL 1.1 are translated to similar SPARQL constructs (e.g. datatype or "is-a" tests are mapped to SPARQL `FILTER` constructs). To support extensibility, the translation process follows the Interpreter software pattern. Hence there exists a specific interpreter for each LDPath expression. This allows developers to easily integrate custom LDPath expressions, such as function predicates, test functions, and filters, as well as register a query interpreter which transforms the new query element to valid SPARQL 1.1 to ensure full compatibility with generic SPARQL endpoints.

7 Extensions

The basic functionality of the library Anno4j in order to query and persist RDF via Java POJOs has been covered in Sects. 4 and 5. Next to this, several additional features have been implemented, partly supporting the usability of the library in terms of convenience features as well as some features that give a richer RDF feature support. The following section will give insights into those features, which are namely: transactional behaviour (see Sect. 7.1), subgraphs and contexts (see Sect. 7.2), plugin extensibility (see Sect. 7.3), and input/output functionality (see Sect. 7.4).

7.1 Transactions

The Anno4j library features a transactional behaviour, allowing the user to work in an atomic fashion. By creating sets of actions that either are completely executed (**commit**) or not at all (**rollback**), the database is always at a consistent state. A crash in the midst of a work procedure does not create an unclean or untraceable state of data. The basic behaviour (if no `Transaction` object is used) is set to auto-commit, so every action is persisted at the respective database automatically. A transaction itself has to be **started** and ended, which means a **commit** or **rollback**. Listing 1.5 shows an example that creates, begins, and ends a transaction while showing the possibility to create objects and a Query-Service. This shows, how persistence and querying operations can be added to that respective set of actions.

Listing 1.5. Use of a Transaction in Anno4j.

```
1  Anno4j anno4j = new Anno4j();
2
3  Transaction transaction =
       anno4j.createTransaction();
4  transaction.begin();
5
6  // Create and query using the Transaction object
7  Annotation annotation =
       transaction.createObject(Annotation.class);
8  QueryService qs = transaction.createQueryService();
9
10 transaction.commit(); / transaction.rollback();
```

7.2 Contexts and Subgraphs

A convenience feature of RDF is the use of contexts. Contexts allow users to split their whole RDF graph into smaller contextualised subgraphs. Therefore, RDF triples are turned into so-called **quads**, which have a fourth component after subject, predicate, and object that implements the URI of the subgraph the triple is to be contained in. On the one hand, if no context is defined, the default context is used. On the other hand, context can be utilised in one of two ways in Anno4j:

- Anno4j instance level: Two out of the four possible methods to create an object (`Anno4j.createObject( ... )`) support an additional URI parameter standing for the context. Creating an object this way will insert it in the respective subgraph.
- Transaction level (see Sect. 7.1 for transactions): Every **Transaction** object supports a `.setAllContexts(String uri)` method, which defines the subgraph that the transaction is to write to and read from.

Listing 1.6 shows two examples using a context in Anno4j. Line 1 defines a new URI for a respective subgraph, while line 4 creates an **Annotation** object in that subgraph. Line 7 creates a **Transaction** object and line 8 changes its context to the defined context **uri**.

Listing 1.6. Setting a context for an Anno4j and Transaction object.

```
1  URI uri = new URIImpl("http://www.somePage.com/");
2
3  // Create an Item in the uri context
4  Annotation annotation =
       anno4j.createObject(Annotation.class, uri);
5
6  // Create a Transaction object and define its
       context to uri
```

```
7   Transaction transaction =
       anno4j.createTransaction();
8   transaction.setAllContexts(uri);
```

7.3 Plugin Functionality

By implementing a plugin for Anno4j, users are given the opportunity to add
their own querying logic to the library. This is formalised and implemented by
introducing an LDPath function in combination with the corresponding logic
behind it. Using this, not an actual relationship in the RDF graph is requested,
but rather semantic information between given entities are utilised, like for example the target and its selector in order to evaluate the corresponding function.
When applied, the query logic is evaluated at the point of time the query is
executed. Because of this, the size of the result can be confined beforehand,
rather than "doing the logic by hand" afterwards on a bigger result set. In
order to implement a plugin, the user has to define the LDPath function expression (QueryExtension), as well as the querying logic (QueryEvaluator). An
exemplary expression (taken from SPARQL-MM [3]) can be seen in Listing 1.7.
Integrating that criteria could lead to a result set of only those annotations,
that detected both an elephant and a lion, standing next to each other with the
elephant found left besides the lion.

Listing 1.7. Exemplary plugin expression in an LDPath criteria.

```
1   QueryService qs = anno4j.createQueryService();
2
3   qs.addCriteria("sparqlmm:leftBesides("elephant","lion")");
```

7.4 Input and Output

To improve the usability of the library, a small extension to the ORM has been
implemented. Users can parse their RDF triples formulated in various RDF
serialisations to create the respective Java objects, as well as write their Java
objects as serialised RDF triples. Among the available serialisations are rdf/xml,
ntriples, turtle, n3, jsonld, rdf/json, etc.

In order to read a given RDF annotation (available as Java String), an
ObjectParser object is needed. Its .parse(String annotation, String uri,
RDFFormat format) requires the **annotation** as String, a **uri** for namespacing,
and the supported **format**. It will then return a Java list of the parsed Annotations. Important to note is, that all RDF nodes that are to be parsed need to be
supported as respective Anno4j interfaces. An ObjectParser will keep its parsed
annotations locally, a respective call to the .getAnnotations() method will
return a list of them. Listing 1.8 shows an example reading a turtle annotation.

Listing 1.8. Reading an annotation from a given turtle serialisation.

```
1  String TURTLE = "@prefix oa:
       <http://www.w3.org/ns/oa#> ." +
2  "@prefix ex: <http://www.example.com/ns#> ." +
3  "@prefix dctypes: <http://purl.org/dc/dcmitype/>
       ." +
4  "@prefix rdf:
       <http://www.w3.org/1999/02/22-rdf-syntax-ns#>
       ." +
5
6  "ex:anno1 a oa:Annotation;" +
7  "   oa:hasBody ex:body1;" +
8  "   oa:hasTarget ex:target1.";
9
10 URL url = new URL("http://example.com/");
11
12 ObjectParser objectParser = new ObjectParser();
13 List<Annotation> annotations =
       objectParser.parse(TURTLE, url,
       RDFFormat.TURTLE);
14
15 objectParser.shutdown();
```

To write a given Anno4j Java object to respective RDF serialisations, the ResourceObject interface (which every Anno4j object descends from) supports a .getTriples(RDFFormat format) method, which returns the representation of the object as a Java String in the supported **format**. Listing 1.9 shows an example that writes a given annotation as turtle triples.

Listing 1.9. Writing a given Java item as turtle RDF serialisation.

```
1  Annotation annotation =
       anno4j.createObject(Annotation.class);
2  ...
3  String itemAsTurtle =
       annotation.getTriples(RDFFormat.TURTLE);
```

8   Application

Anno4j was developed within the MICO project [9] and was tailored to the specific project requirements. The MICO project deals with the semantic analysis of multimedia objects in order to create a rich metadata background for the analysed data. This is done to narrow the semantic gap and make the multimedia be more useful to machines. By combining different analysis procedures on the same item, hidden semantics can be found whereby an even wider knowledge base can be created. To do this, the whole analysis process of the MICO project is based on a single instance, called the **MICO platform**. It combines data and metadata storage, an orchestration unit (the MICO broker [1]), recommendation features, a fully integrated persistence API, and the possibility to combine different (local and external) extractors to form workflows and thereby jointly analyse the supported content.

For all the various extraction results with their formats a unified metadata model was necessary. The MICO Metadata Model MMM[7] is an extension to the WADM. The intermediary and final results are persisted in the form of Web Annotations, while the MMM introduces an RDF structure to connect the various result annotations. This allows to combine different analyses of the same input multimedia item, workflow outcomes are combined and provenance can be traced back.

Hence, it was a requirement for the extractor implementer to both persist and query the metadata using (template supported, but still) verbose and complex SPARQL queries. Additionally, when requesting data from the MICO platform, it had to be interpreted "by hand" first. At this point, Anno4j solved a lot of problems at both sides. By replacing the SPARQL queries, the extractor experts could now both persist and query their respective data using POJOs of the programming language they are familiar with and their extractor is written in. Additionally, querying with LDPath expressions was easier to adopt and had less lines of code than the SPARQL queries. Next to that, when querying data from the respective triple store, having POJOs rather than the answer of a SPARQL query, the result data could be used right away and did not require any further up-front parsing or manual object creation.

9 Conclusion

This paper introduced the library Anno4j, which enables Java developers to create and consume RDF annotations. Those annotations are conform to the WADM, allowing them to be shared and exchanged between different locations. Anno4j features simple persistence of RDF objects via Java objects, its querying functionality is based on LDPath, supporting a wide range of combinable path criteria to form a powerful annotation consumption tool. Both features can be extended easily, so developers can fine tune their respective Anno4j instance to their needs.

Anno4j is available under Apache V2 license at Github[8]. Future work on this library will include the attempt to provide the functionality of Anno4j to other programming languages. First proof-of-concepts show positive results for a C++ mapping using Java Native Interfaces. This would allow developers to write their software natively in C++ but still use Anno4j for a convenient creation and querying of Web Annotations. Other extensions like SPARQL-MM querying allow Anno4j to be used in a broader spectrum.

---

[7] http://mico-project.bitbucket.org/vocabs/mmm/2.0/documentation/.

[8] https://github.com/anno4j/anno4j.

The current application of the Anno4j library has led to various lessons learned, as different projects make use of Web Annotations in conjunction with Anno4j to make a more convenient use of the annotations. However, as nearly every application of today is web-oriented, a web facet could lift Anno4j to a broader use case. This requirement is exactly tailored to the WAP specification, so future steps will include a layer on top of Anno4j, allowing it to be used as a WAP-conform server. Additionally, as the querying mechanism of Anno4j in combination with LDPath is manifold, we intend to extend the WAP requirements for our implementation to deliver more comprehensive querying.

Acknowledgments. The presented work was developed within the MICO project partially funded by the EU Seventh Framework Programme, grant agreement number 610480.

References

1. Aichroth, P., Sieland, M., Cuccovillo, L., Köllmer, T.: The mico broker: an orchestration framework for linked data extractors. In: Joint Proceedings of the 4th International Workshop on Linked Media and the 3rd Developers Hackshop (LiME 2016, SemDev 2016), Co-located with 13th Extended Semantic Web Conference ESWC 2016 Heraklion, Crete, Greece, 30 May 2016
2. Bizer, C., Heath, T., Berners-Lee, T.: Linked data-the story so far. In: Semantic Services, Interoperability and Web Applications: Emerging Concepts, pp. 205–227 (2009)
3. Kurz, T., Schlegel, K., Kosch, H.: Enabling access to linked media with SPARQL-MM. In: Proceedings of the 24nd International Conference on World Wide Web (WWW2015) Companion (LIME15) (2015)
4. Manola, F., Miller, E.: RDF primer. W3C Recommendation, W3C, February 2004. http://www.w3.org/TR/2004/REC-rdf-primer-20040210/
5. Pyysalo, S., Campos, J., Cejuela, J.M., Ginter, F., Hakala, K., Li, C., Stenetorp, P., Jensen, L.J.: Sharing annotations better: RESTful open annotation. In: ACL-IJCNLP 2015, p. 91 (2015)
6. Sanderson, R.: WAP web annotation protocol. W3C Working Draft, W3C (2015). https://www.w3.org/TR/annotation-protocol/
7. Sanderson, R., Ciccarese, P., de Sompel, H.V.: OADM open annotation data model. W3C Community Draft, W3C, February 2013. http://www.openannotation.org/spec/core/
8. Sanderson, R., Ciccarese, P., Young, B.: WADM web annotation data model. W3C Working draft, W3C, October 2015. https://www.w3.org/TR/annotation-model/
9. Schlegel, K., Berndl, E., Granitzer, M., Kosch, H., Kurz, T.: A platform for contextual multimedia data: towards a unified metadata model and querying. In: Proceedings of the 15th International Conference on Knowledge Technologies and Data-Driven Business, i-KNOW 2015, pp. 1:1–1:8. ACM, New York (2015). http://doi.acm.org/10.1145/2809563.2809586
10. Speicher, S., Arwe, J., Malhotra, A.: LDP linked data platform 1.0. W3C Recommendation, W3C, February 2015. https://www.w3.org/TR/ldp/

Managing the Evolution and Preservation of the Data Web

Continuous Client-Side Query Evaluation over Dynamic Linked Data

Ruben Taelman[(⊠)], Ruben Verborgh, Pieter Colpaert, and Erik Mannens

imec – Ghent University – IDLab,
Sint-Pietersnieuwstraat 41, 9000 Ghent, Belgium
{ruben.taelman,ruben.verborgh,pieter.colpaert,
erik.mannens}@ugent.be

Abstract. Existing solutions to query dynamic Linked Data sources extend the SPARQL language, and require continuous server processing for each query. Traditional SPARQL endpoints already accept highly expressive queries, so extending these endpoints for time-sensitive queries increases the server cost even further. To make continuous querying over dynamic Linked Data more affordable, we extend the low-cost Triple Pattern Fragments (TPF) interface with support for time-sensitive queries. In this paper, we introduce the TPF Query Streamer that allows clients to evaluate SPARQL queries with continuously updating results. Our experiments indicate that this extension significantly lowers the server complexity, at the expense of an increase in the execution time per query. We prove that by moving the complexity of continuously evaluating queries over dynamic Linked Data to the clients and thus increasing bandwidth usage, the cost at the server side is significantly reduced. Our results show that this solution makes real-time querying more scalable for a large amount of concurrent clients when compared to the alternatives.

Keywords: Linked Data · Linked Data Fragments · SPARQL · Continuous querying · Real-time querying

1 Introduction

As the Web of Data is a *dynamic* dataspace, different results may be returned depending on when a question was asked. The end-user might be interested in seeing the query results update over time, for instance, by re-executing the entire query over and over again ("polling"). This is, however, not very practical, especially if it is unknown beforehand when data will change. An additional problem is that many public (even static) SPARQL query endpoints suffer from a low availability [5]. The unrestricted complexity of SPARQL queries [15] combined with the public character of SPARQL endpoints entails a high server cost, which makes it expensive to host such an interface with high availability. *Dynamic* SPARQL streaming solutions offer combined access to dynamic data streams and static background data through continuously executing queries. Because of this

© Springer International Publishing AG 2016
H. Sack et al. (Eds.): ESWC 2016 Satellite Events, LNCS 9989, pp. 273–289, 2016.
DOI: 10.1007/978-3-319-47602-5_44

continuous querying, the cost for these servers is even higher than with static querying.

In this work, we therefore devise a solution that enables clients to continuously evaluate non-high frequency queries by polling specific fragments of the data. The resulting framework performs this without the server needing to remember any client state. Its mechanism requires the server to *annotate* its data so that the client can efficiently determine when to retrieve fresh data. The generic approach in this paper is applied to the use case of public transit route planning. It can be used in various other domains with continuously updating data, such as smart city dashboards, business intelligence, or sensor networks. This paper extends our earlier work [17] with additional experiments.

In the next section, we discuss related research on which our solution will be based. After that, Sect. 3 gives a general problem statement. In Sect. 4, we present a motivating use case. Section 5 discusses different techniques to represent dynamic data, after which Sect. 6 gives an explanation of our proposed query solution. Next, Sect. 7 shows an overview of our experimental setup and its results. Finally, Sect. 8 discusses the conclusions of this work with further research opportunities.

2   Related Work

In this section, we first explain techniques to perform RDF annotation, which will be used to determine freshness. Then, we zoom in on possible representations of temporal data in RDF. We finish by discussing existing SPARQL streaming extensions and a low-cost (static) Linked Data publication technique.

2.1   RDF Annotations

Annotations allow us to attach metadata to triples. We might for example want to say that a triple is only valid within a certain time interval, or that a triple is only valid in a certain geographical area.

RDF 1.0 [11] allows triple annotation through *reification*. This mechanism uses *subject*, *predicate*, and *object* as predicates, which allow the addition of annotations to such reified RDF triples. The downside of this approach is that one triple is now transformed to three triples, which significantly increases the total amount of triples.

Singleton Properties [14] create unique instances (singletons) of predicates, which then can be used for further specifying that relationship, for example, by adding annotations. New instances of predicates are created by relating them to the old predicate through the `sp:singletonPropertyOf` predicate. While this approach requires fewer triples than reification to represent the same information, it still has the issue of the original triple being lost, because the predicate is changed in this approach.

With RDF 1.1 [6] came *graph* support, which allows triples to be encapsulated into named graphs, which can also be annotated. Graph-based annotation requires fewer triples than both reification and singleton properties when representing the same information. It requires the addition of a fourth element to the triple which transforms it to a quad. This fourth element, the *graph*, can be used to add the annotations to.

2.2 Temporal Data in the RDF Model

Regular RDF triples cannot express the time and space in which the fact they describe is true. In domains where data needs to be represented for certain times or time ranges, these traditional representations should thus be extended. There are two main mechanisms for adding time [9]. *Versioning* will take snapshots of the complete graph every time a change occurs. *Time labeling* will annotate triples with their change time. The latter is believed to be a better approach in the context of RDF, because complete snapshots introduce overhead, especially if only a small part of the graph changes. Gutierrez et al. made a distinction between *point-based* and *interval-based* labeling, which are interchangeable [8]. The former states information about an element at a certain time instant, while the latter states information at all possible times between two time instants.

The same authors introduced a *temporal vocabulary* [8] for the discussed mechanisms, which will be referred to as tmp in the remainder of this document. Its core predicates are:

tmp:interval. This predicate can be used on a subject to make it valid in a certain time interval. The range of this property is a time interval, which is represented by the two mandatory properties tmp:initial and tmp:final.

tmp:instant. Used on subjects to make it valid on a certain time instant as a point-based time representation. The range of this property is xsd:dateTime.

tmp:initial and tmp:final. The domain of these predicates is a time interval. Their range is a xsd:dateTime, and they respectively indicate the start and the end of the interval-based time representation.

Next to these properties, we will also introduce our own predicate tmp:expiration with range xsd:dateTime which indicates that the subject is only valid up until the given time.

2.3 SPARQL Streaming Extensions

Several SPARQL extensions exist that enable querying over data streams. These data streams are traditionaly represented as a monotonically non-decreasing stream of triples that are annotated with their timestamp. These require *continuous processing* [7] of queries because of the constantly changing data.

C-SPARQL [4] is an approach to querying over static and dynamic data. This system requires the client to *register* a query to the server in an extended SPARQL syntax that allows the use of *windows* over dynamic data. This *query registration* [3, 7] must occur by clients to make sure that the streaming-enabled SPARQL endpoint can continuously re-evaluate this query, as opposed to traditional endpoints where the query is evaluated only once. A *window* [2] is a subsection of facts ordered by time so that not all available information has to be taken into account while processing. These windows can have a certain size which indicates the time range and is advanced in time by a *stepsize*. C-SPARQL's execution of queries is based on the combination of a regular SPARQL engine with a *Data Stream Management System* (DSMS) [2]. The internal model of C-SPARQL creates queries that distribute work between the DSMS and the SPARQL engine to respectively process the dynamic and static data.

CQELS [12] is a "white box" approach, as opposed to "black box" approaches like C-SPARQL. This means that CQELS natively implements all query operators without transforming it to another language, removing the overhead of delegating it to another system. The syntax is similar to that of C-SPARQL, also supporting query registration and time windows. According to previous research [12], CQELS performs much better than C-SPARQL for large datasets; for simple queries and small datasets the opposite is true.

2.4 Triple Pattern Fragments

Experiments have shown that more than half of public SPARQL endpoints have an availability of less than 95 % [5]. Any number of clients can send arbitrarily complex SPARQL queries, which could form a bottleneck in endpoints. *Triple Pattern Fragments* (TPF) [18] aim to solve this issue of high interface cost by moving part of the query evaluation to the client, which reduces the server load, at the cost of increased query times and bandwidth. The purposely limited interface only accepts separate triple pattern queries. Clients can use it to evaluate more complex SPARQL queries locally, also over federations of interfaces [18].

3 Problem Statement

In order to lower server load during continuous query evaluation, we move a significant part of the query evaluation from server to client. We annotate dynamic data with their valid time to make it possible for clients to derive an optimal query evaluation frequency.

For this research, we identified the following research questions:

Question 1. Can clients use volatility knowledge to perform more efficient continuous SPARQL query evaluation by polling for data?

Question 2. How does the client and server load of our solution compare to alternatives?

Question 3. How do different time-annotation methods perform in terms of the resulting execution times?

These research questions lead to the following hypotheses:

Hypothesis 1. The proposed framework has a lower server cost than alternatives.

Hypothesis 2. The proposed framework has a higher client cost than streaming-based SPARQL approaches for equivalent queries.

Hypothesis 3. Client-side caching of static data reduces the execution times proportional to the fraction of static triple patterns that are present in the query.

4   Use Case

A guiding use case, based on public transport, will be referred to in the remainder of this paper. When public transport route planning applications return dynamic data, they can account for factors such as train delays as part of a continuously updating route plan. In this use case, different clients need to obtain all train departure information for a certain station. This requires the following concepts:

1. **Departure** (*static*): Unique URI for the departure of a certain train.
2. **Headsign** (*static*): The label of the train showing its destination.
3. **Departure Time** (*static*): The *scheduled* departure time of the train.
4. **Route Label** (*static*): The identifier for the train and its route.
5. **Delay** (*dynamic*): The delay of the train, which can increase through time.
6. **Platform** (*dynamic*): The platform number of the station at which the train will depart, which can be changed through time if delays occur.

Listing 1.1 shows example data in this model. The SPARQL query in Listing 1.2 can retrieve all information using this basic data model.

```
@prefix t: <http://example.org/train/>.
@prefix td: <http://example.org/traindata/>.
td:departure-48 t:delay          "0S"^^xsd:xs:duration;
                t:platform       td:platform-1a;
                t:departureTime "2014-12-05T10:37:00+01:00"^^xsd:
    dateTimeStamp;
                t:headSign       "Ghent";
                t:routeLabel     "IC 1831".
```
Listing 1.1. Train information with static time information according to the basic data model.

```
SELECT ?delay ?platform ?headSign ?routeLabel ?departureTime
WHERE {
    _:id t:delay         ?delay.
    _:id t:platform      ?platform.
    _:id t:departureTime ?departureTime.
    _:id t:headSign      ?headSign.
    _:id t:routeLabel    ?routeLabel.
    FILTER (?departureTime > "2015-12-08T10:20:00"^^xsd:dateTime).
    FILTER (?departureTime < "2015-12-08T11:20:00"^^xsd:dateTime).
}
```
Listing 1.2. The basic SPARQL query for retrieving all upcoming train departure information in a certain station. The two first triple patterns are dynamic, the last three are static.

5   Dynamic Data Representation

Our solution consists of a partial redistribution of query evaluation workload from the server to the client, which requires the client to be able to access the server data. There needs to be a distinction between regular static data and continuously updating dynamic data in the server's dataset. For this, we chose to define a certain temporal range in which these dynamic facts are valid, as a consequence the client will know when the data becomes invalid and has to fetch new data to remain up-to-date. To capture the temporal scope of data triples, we annotate this data with time. In this section, we discuss two different types of time labeling, and different methods to annotate this data.

5.1   Time Labeling Types

We use interval-based labeling to indicate the *start and endpoint* of the period during which triples are valid. Point-based labeling is used to indicate the *expiration time*.

With expiration times, we only save the latest version of a given fact in a dataset, assuming that the old version can be removed when a newer one arrives. These expiration times provide enough information to determine when a certain fact becomes invalid in time. We use time intervals for storing multiple versions of the same fact, i.e., for maintaining a history of facts. These time intervals must indicate a start- and endtime for making it possible to distinguish between different versions of a certain fact. These intervals cannot overlap in time for the same facts. When data is volatile, consecutive interval-based facts will accumulate quickly. Without techniques to aggregate or remove old data, datasets will quickly grow, which can cause increasingly slower query executions. This problem does not exist with expiration times because in this approach we decided to only save the latest version of a fact, so this volatility will not have any effect on the dataset size.

5.2   Methods for Time Annotation

The two time labeling types introduced in the last section can be annotated on triples in different ways. In Sect. 2.1 we discussed several methods for RDF annotation. We will apply time labels to triples using the singleton properties, graphs and implicit graphs annotation techniques.

Singleton Properties. *Singleton properties* annotation is done by creating a singleton property for the predicate of each dynamic triple. Each of these singleton properties can then be annotated with its time annotation, being either a time interval or expiration times.

Graphs. To time-annotate triples using *graphs*, we can encapsulate triples inside contexts, and annotate each context graph with a time annotation.

Implicit Graphs. A TPF interface gives a unique URI to each fragment corresponding to a triple pattern, including patterns without variables, i.e., actual triples. Since Triple Pattern Fragments [18] are the basis of our solution, we can interpret each fragment as a graph. We will refer to these as *implicit graphs*. This URI can then be used as graph identifier for this triple for adding time information. For example, the URI for the triple <s> <p> <o> on the TPF interface located at http://example.org/dataset/ is http://example.org/dataset?subject=s&predicate=p&object=o.

The choice of time annotation method for publishing temporal data will also depend on its capability to *group* time labels. If certain dynamic triples have identical time labels, these annotations can be shared to further reduce the required amount of triples if we are using singleton properies or graphs. When we would have three train delay triples which are valid for the same time interval using graph annotation, these three triples can be placed in the same graph. This will make sure they refer to the same time interval without having to replicate this annotation two times more. In the case of implicit graph annotation, this grouping of triples is not possible, because each triple has a unique graph identifier determined by the interface. This would be possible if these different identifiers are linked to each other with for example `sameAs` relationships that our query engine takes into account, which would introduce further overhead.

We will execute our use case for each of these annotation methods. In practise, an annotation method must be chosen depending on the requirements and available technologies. If we have a datastore that supports quads, graph-based annotation is the best choice because of it requires the least amount of triples. If our datastore does not support quads, we can use singleton properties. If we have a TPF-like interface at which our data is hosted, we can use implicit graphs as annotation technique, if however many of those triples can be grouped under the same time label, singleton properties are a better alternative because the latter has grouping support.

6 Query Engine

TPF query evaluation involves server and client software, because the client actively takes part in the query evaluation, as opposed to traditional SPARQL endpoints where the server does all of the work. Our solution allows users to send a normal SPARQL query to the local query engine which autonomously detects the dynamic parts of the query and continuously sends back results from that query to the user. In this section, we discuss the architecture of our proposed solution and the most important algorithms that were used to implement this.

6.1 Architecture

Our solution must be able to handle regular SPARQL 1.1 queries, detect the dynamic parts, and produce continuously updating results for non-high frequency queries. To achieve this, we chose to build an extra software layer on top of the existing TPF client that supports each discussed labeling type and annotation method and is capable of doing dynamic query transformation and result streaming. At the TPF server, dynamic data must be annotated with time depending on the used combination of labeling type and method. The server expects dynamic data to be pushed to the platform by an external process with varying data. In the case of graph-based annotation, we have to extend the TPF server implementation, so that it supports quads. This dynamic data should be pushed to the platform by an external process with varying data.

Figure 1 shows an overview of the architecture for this extra layer on top of the TPF client, which will be called the TPF *Query Streamer* from now on. The left-hand side shows the *User* that can send a regular SPARQL query to the TPF Query Streamer entry-point and receives a stream of query results. The system can execute queries through the local *Basic Graph Iterator*, which is part of the TPF client and executes queries against a TPF server.

The TPF Query Streamer consists of six major components. First, there is the *Rewriter* module which is executed only once at the start of the query streaming loop. This module is able to transform the original input query into a *static* and a *dynamic query* which will respectively retrieve the static background data and the time-annotated changing data. This transformation happens by querying metadata of the triple patterns against the entry-point through the local TPF client. The *Streamer* module takes this dynamic query, executes it and forwards its results to the *Time Filter*. The *Time Filter* checks the time annotation for each of the results and rejects those that are not valid for the current time. The minimal expiration time of all these results is then determined and used as a delayed call to the *Streamer* module to continue with the *streaming loop*, which is determined by the repeated invocation of the *Streamer* module. This minimal expiration time will make sure that when at least one of the results expire, a

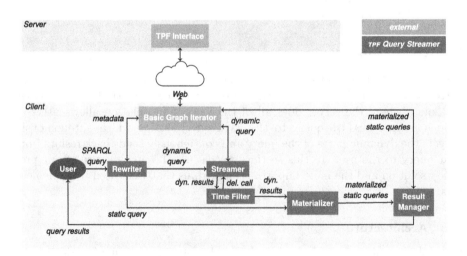

Fig. 1. Overview of the proposed client-server architecture

new set of results will be fetched as part of the next query iteration. The filtered dynamic results will be passed on to the *Materializer* which is responsible for creating *materialized static queries*. This is a transformation of the *static query* with the dynamic results filled in. These *materialized static queries* are passed to the *Result Manager* which is able to cache these queries. Finally, the *Result Manager* retrieves previous *materialized static query* results from the local cache or executes this query for the first time and stores its results in the cache. These results are then sent to the client who had initiated continuous query.

6.2   Algorithms

Query Rewriting. As mentioned in the previous section, the *Rewriter* module performs a preprocessing step that can transform a regular SPARQL 1.1 query into a static and dynamic query. A first step in this transformation is to detect which triple patterns inside the original query refer to static triples and which refer to dynamic triples. We detect this by making a separate query for each of the triple patterns and transforming each of them to a dynamic query. An example of such a transformation can be found in Listing 1.3. We then evaluate each of these transformed queries and assume a triple pattern is *dynamic* if its corresponding query has at least one result. Another step before the actual query splitting is the conversion of blank nodes to variables. We will end up with one static query and one dynamic query, in case these graphs were originally connected, they still need to be connected after the query splitting. This connection is only possible with variables that are visible, meaning that these variables need to be part of the SELECT clause. However, a variable can also be anonymous and not visible: these are blank nodes. To make sure that we take into account blank nodes that connect the static and dynamic graph, these nodes have to be converted to variables, while maintaining their semantics. After this step, we iterate over each triple pattern of the original query and assign them to either the static or the dynamic query depending on whether or not the pattern is respectively static or dynamic. This assignment must maintain the hierarchical structure of the original query, in some cases this causes triple patterns to be present in the dynamic query when using complex operators like UNION to maintain correct query semantics. An example of this query transformation for our basic query from Listing 1.2 can be found in Listings 1.4 and 1.5.

Query Materialization. The *Materializer* module is responsible for creating *materialized static queries* from the static query and the current dynamic query results. This is done by filling in each dynamic result into the static query variables. It is possible that multiple results are returned from the dynamic query evaluation, which is the same amount of materialized static queries that can be derived. Assuming that we, for example, find the following single dynamic query result from the dynamic query in Listing 1.5: {?id $\mapsto$ <http://example.org/train#train4815>, ?delay $\mapsto$ "P10S"^^xsd:duration} then we can derive the materialized static query by

```
SELECT ?s ?p ?o ?time WHERE {
    GRAPH ?g0 { ?s ?p ?o }
    ?g0 tmp:expiration ?time
}
```

Listing 1.3. Dynamic SPARQL query for the triple pattern ?s ?p ?o for graph-based annotation with expiration times.

```
SELECT ?id ?headSign ?routeLabel ?departureTime
WHERE {
    ?id t:departureTime ?departureTime.
    ?id t:headSign       ?headSign.
    ?id t:routeLabel     ?routeLabel.
    FILTER (?departureTime > "2015-12-08T10:20:00"^^xsd:dateTime).
    FILTER (?departureTime < "2015-12-08T11:20:00"^^xsd:dateTime).
}
```

Listing 1.4. Static SPARQL query which has been derived from the basic SPARQL query from Listing 1.2 by the *Rewriter* module.

```
SELECT ?id ?delay ?platform ?final0 ?final1
WHERE {
    GRAPH ?g0 { ?id t:delay ?delay. }
    ?g0 tmp:expiration ?final0.
    GRAPH ?g1 { ?id t:platform ?platform. }
    ?g1 tmp:expiration ?final1.
}
```

Listing 1.5. Dynamic SPARQL query which has been derived from the basic SPARQL query from Listing 1.2 by the *Rewriter* module. Graph-based annotation is used with expiration times.

```
SELECT ?headSign ?routeLabel ?departureTime
WHERE {
    <http://example.org/train#train4815> t:departureTime ?departureTime.
    <http://example.org/train#train4815> t:headSign       ?headSign.
    <http://example.org/train#train4815> t:routeLabel     ?routeLabel.
    FILTER (?departureTime > "2015-12-08T10:20:00"^^xsd:dateTime).
    FILTER (?departureTime < "2015-12-08T11:20:00"^^xsd:dateTime).
}
```

Listing 1.6. Materialized static SPARQL query derived by filling in the dynamic query results into the static query from Listing 1.6.

filling in these two variables into the static query from Listing 1.4, the resulting query can be found in Listing 1.6.

Caching. The *Result manager* is the last step in the streaming loop for returning the materialized static query results of one time instance. This module is responsible for either getting results for given queries from its cache, or fetching the results from the TPF client. First, an identifier will be determined for each materialized static query. This identifier will serve as a key to cache static data and should correctly and uniquely identify static results based on dynamic results. This is equivalent to saying that this identifier should be the *connection* between the static and dynamic graphs. This connection is the intersection of the variables present in the WHERE clause of the static and dynamic queries. Since the dynamic query results are already available at this point, these variables all have values, so this cache identifier can be represented by these variable results. The graph connection between the static and dynamic queries from Listings 1.4 and 1.5 is ?id. The cache identifier for a result where ?id is "train:4815" is for example "?id=train:4815".

7   Evaluation

In order to validate our hypotheses from Sect. 3, we set up an experiment to measure the impact of our proposed redistribution of workload between the client and server by simultaneously executing a set of queries against a server using our proposed solution. We repeat this experiment for two state-of-the-art solutions: C-SPARQL and CQELS.

To test the client and server performance, our experiment consisted of one server and ten physical clients. Each of these clients can execute from one to twenty unique concurrent queries based on the use case from Sect. 4. The data for this experiment was derived from real-world Belgian railway data using the iRail API[1]. This results in a series of 10 to 200 concurrent query executions. This setup was used to test the client and server performance of different SPARQL streaming approaches.

For comparing the efficiency of different time annotation methods and for measuring the effectiveness of our client-side cache, we measured the execution times of the query for our use case from Sect. 4. This measurement was done for different annotation methods, once with the cache and once without the cache. For discovering the evolution of the query evaluation efficiency through time, the measurements were done over each query stream iteration of the query.

The discussed architecture was implemented[2] in JavaScript using Node.js to allow for easy communication with the existing TPF client.

The tests[3] were executed on the Virtual Wall (generation 2) environment from iMinds [10]. Each machine had two Hexacore Intel E5645 (2.4 GHz) CPUS with 24 GB RAM and was running Ubuntu 12.04 LTS. For CQELS, we used version 1.0.1 of the engine [13]. For C-SPARQL, this was version 0.9 [16]. The dataset for this use case consisted of about 300 static triples, and around 200 dynamic triples that were created and removed each ten seconds. Even this relatively small dataset size already reveals important differences in server and client cost, as we will discuss in the paragraphs below.

---

[1] https://hello.irail.be/api/1-0/.

[2] The source code for this implementation is available at https://github.com/LinkedDataFragments/QueryStreamer.js/tree/eswc2016.

[3] The code used to run these experiments with the relevant queries can be found at https://github.com/rubensworks/TPFStreamingQueryExecutor-experiments/.

Server Cost. The server performance results from our main experiment can be seen in Fig. 2a. This plot shows an increasing CPU usage for C-SPARQL and CQELS for higher numbers of concurrent query executions. On the other hand, our solution never reaches more than one percent of server CPU usage. Figure 3a shows a detailed view on the measurements in the case of 200 simultaneous query executions: the CPU peaks for the alternative approaches are much higher and more frequent than for our solution.

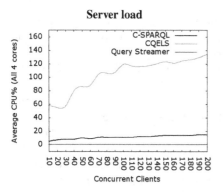 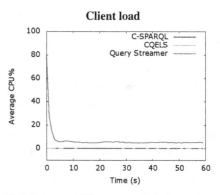

Fig. a: The server CPU usage of our solution proves to be influenced less by the number of clients.

Fig. b: In the case of 200 concurrent clients, client CPU usage initially is high after which it converges to about 5%. The usage for C-SPARQL and CQELS is almost non-existing.

Fig. 2. Average server and client CPU usage for one query stream for C-SPARQL, CQELS and the proposed solution. Our solution effectively moves complexity from the server to the client.

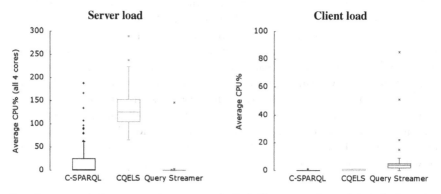

Fig. a: Server CPU peaks for C-SPARQL and CQELS compared to our solution.

Fig. b: Client CPU usage for our solution is significantly higher.

Fig. 3. Detailed view on all server and client CPU measurements for C-SPARQL, CQELS and the solution presented in this work for 200 simultaneous query evaluations against the server.

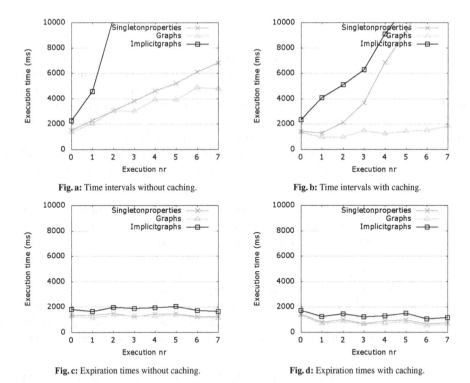

Fig. a: Time intervals without caching.

Fig. b: Time intervals with caching.

Fig. c: Expiration times without caching.

Fig. d: Expiration times with caching.

Fig. 4. Executions times for the three different types of dynamic data representation for several subsequent streaming requests. The figures show a mostly linear increase when using time intervals and constant execution times for annotation using expiration times. In general, caching results in lower execution times. They also reveal that the graph approach has the lowest execution times.

Client Cost. The results for the average CPU usage across the duration of the query evaluation of all clients that sent queries to the server in our main experiment can be seen in Figs. 2b and 3b. The clients that were sending C-SPARQL and CQELS queries to the server had a client CPU usage of nearly zero percent for the whole duration of the query evaluation. The clients using the client-side TPF Query Streamer solution that was presented in this work had an initial CPU peak reaching about 80 %, which dropped to about 5 % after 4 s.

Annotation Methods. The execution times for the different annotation methods, once with and once without cache can be seen in Fig. 4. The three annotation methods have about the same relative performance in all figures, but the execution times are generally lower in the case where the client-side cache was used, except for the first query iteration. The execution times for expiration time annotation when no cache is used are constant, while the execution times with caching slightly decrease over time.

8   Conclusions

In this paper, we researched a solution for querying over dynamic data with a low server cost, by continuously polling the data based on volatility information. In this section, we draw conclusions from our evaluation results to give an answer to the research questions and hypotheses we defined in Sect. 3. First, the server and client costs for our solution will be compared with the alternatives. After that, the effect of our client-side cache will be explained. Next, we will discuss the effect of time annotation on the amount of requests to be sent, after which the performance of our solution will be shown and the effects of the annotation methods.

Server Cost. The results from Sect. 7 confirm Hypothesis 1, in which we wanted to know if we could lower the server cost when compared to C-SPARQL and CQELS. Not only is the server cost for our solution more than ten times lower on average when compared to the alternatives, this cost also increases much slower for a growing number of simultaneous clients. This makes our proposed solution more scalable for the server. Another disadvantage of C-SPARQL and CQELS is the fact that the server load for a large number of concurrent clients varies significantly, as can be seen in Fig. 3a This makes it hard to scale the required processing powers for servers using these technologies. Our solution has a low and more constant CPU usage.

Client Cost. The results for the client load measurements from Sect. 7 confirm Hypothesis 2, which stated that our solution increases the client's processing need. The required client processing power using our solution is clearly much higher than for C-SPARQL and CQELS. This is because we redistributed the required processing power from the server to the client. In our solution, it is the client that has to do most of the work for evaluating queries, which puts less load on the server. The load on the client still remains around 5 % for the largest part of the query evaluation as shown in Fig. 2b. Only during the first few seconds, the query engines CPU usage peaks, which is because of the processor-intensive rewriting step that needs to be done once at the start of each dynamic query evaluation.

Caching. We can also confirm Hypothesis 3 about the positive effect of caching from the results in Sect. 7. Our caching solution has a positive effect on the execution times. In an optimal scenario for our use case, caching would lead to an execution time reduction of 60 % because three of the five triple patterns in the query for our use case from Sect. 4 are dynamic. For our results, this caching leads to an average reduction of 56 % which is close to the optimal case. Since we are working with dynamic data, some required background-data is bound to overlap, in these cases it is advantageous to have a client-side caching solution so that these redundant requests for static data can be avoided. The longer our query evaluation runs, the more static data the cache accumulates, so the bigger the chance that there are cache hits when background data is needed in a certain query iteration. Future research should indicate what the limits of such a client-side cache for static data are, and whether or not it is advantageous to reuse this cache for different queries.

Request Reduction. By annotating dynamic data with a time annotation, we successfully reduced the amount of required requests for polling-based SPARQL querying to a minimum, which answers Research Question 1 about the question if clients can use volatility knowdledge to perform continuous querying. Because now, the client can derive the exact moment at which the data can change on the server, and this will be used to shedule a new query execution on the server. In future research, it is still possible to reduce the amount of requests our client engine needs to send through a better caching strategy, which could for example also temporarily cache dynamic data which changes at different frequencies. We can also look into differential data transmission by only sending data to the client that has been changed since the last time the client has requested a specific resource.

Performance. For answering Research Question 2, the performance of our solution compared to alternatives, we compared our solution with two state-of-the-art approaches for dynamic SPARQL querying. Our solution significantly reduces the required server processing per client, this complexity is mostly moved to the client. This comparison shows that our technique allows data providers to offer dynamic data which can be used to continuously evaluate dynamic queries with a low server cost. Our low-cost publication technique for dynamic data is useful when the number of potential simultaneous clients is large. When this data is needed for only a small number of clients in a closed off environment and query evaluation must happen fast, traditional approaches like CQELS or C-SPARQL are advised. These are only two possible points on the *Linked Data Fragments* axis [18], depending on the publication requirements, combinations of these approaches can be used.

Annotation Methods. In Research Question 3, we wanted to know how the different annotation methods influenced the execution times. From the results in Sect. 7, we can conclude that graph-based annotation results in the lowest execution times. It can also be seen that annotation with time intervals has the problem of continuously increasing execution times, because of the continuously growing dataset. Time interval annotation can be desired if we for example want to maintain the history of certain facts, as opposed to just having the last version of facts using expiration times. In future work, we will investigate alternative techniques to support time interval annotation without the continuously increasing execution times.

In this work, the frequency at which our queries are updated is purely datadriven using time intervals or expiration times. In the future it might be interesting, to provide a control to the user to change this frequency, if for example this user only desires query updates at a lower frequency than the data actually changes.

In future work, it is important to test this approach with a larger variety of use cases. The time annotation mechanisms we use are generic enough to transform all static facts to dynamic data for any number of triples. The City-Bench [1] RSP engine benchmark can for example be used to evaluate these different cases based on city sensor data. These tests must be scaled (both in terms of clients as in terms of dataset size), so that the maximum number of concurrent requests can be determined, with respect to the dataset size.

References

1. Ali, M.I., Gao, F., Mileo, A.: CityBench: a configurable benchmark to evaluate RSP engines using smart city datasets. In: Arenas, M., et al. (eds.) ISWC 2015. LNCS, vol. 9367, pp. 374–389. Springer, Heidelberg (2015). doi:10.1007/978-3-319-25010-6_25
2. Arasu, A., Babcock, B., Babu, S., Cieslewicz, J., Datar, M., Ito, K., Motwani, R., Srivastava, U., Widom, J.: STREAM: the Stanford data stream management system. Book Chapter (2004)
3. Barbieri, D., Braga, D., Ceri, S., Della Valle, E., Grossniklaus, M.: Stream Reasoning: where we got so far. In: Proceedings of the NeFoRS2010 Workshop (2010)
4. Barbieri, D.F., Braga, D., Ceri, S., Valle, E.D., Grossniklaus, M.: Querying RDF streams with C-SPARQL. SIGMOD Rec. **39**(1), 20–26 (2010)
5. Buil-Aranda, C., Hogan, A., Umbrich, J., Vandenbussche, P.-Y.: SPARQL web-querying infrastructure: ready for action? In: Alani, H., et al. (eds.) ISWC 2013, Part II. LNCS, vol. 8219, pp. 277–293. Springer, Heidelberg (2013)
6. Cyganiak, R., Wood, D., Lanthaler, M.: RDF 1.1: Concepts and abstract syntax. Recommendation, W3C, February 2014. http://www.w3.org/TR/2014/REC-rdf11-concepts-20140225/
7. Della Valle, E., Ceri, S., van Harmelen, F., Fensel, D.: It's a streaming world! Reasoning upon rapidly changing information. IEEE Intell. Syst. **24**(6), 83–89 (2009)
8. Gutierrez, C., Hurtado, C., Vaisman, A.: Introducing time into RDF. IEEE Trans. Knowl. Data Eng. **19**(2), 207–218 (2007)
9. Gutierrez, C., Hurtado, C.A., Vaisman, A.A.: Temporal RDF. In: Gómez-Pérez, A., Euzenat, J. (eds.) ESWC 2005. LNCS, vol. 3532, pp. 93–107. Springer, Heidelberg (2005)
10. iLab.t, iMinds: Virtual Wall: wired networks and applications. http://ilabt.iminds.be/virtualwall
11. Klyne, G., Carroll, J.J.: Resource Description Framework (RDF): Concepts and abstract syntax. Recommendation W3C, February 2004. http://www.w3.org/TR/2004/REC-rdf-concepts-20040210/
12. Le-Phuoc, D., Dao-Tran, M., Xavier Parreira, J., Hauswirth, M.: A native and adaptive approach for unified processing of linked streams and linked data. In: Aroyo, L., Welty, C., Alani, H., Taylor, J., Bernstein, A., Kagal, L., Noy, N., Blomqvist, E. (eds.) ISWC 2011, Part I. LNCS, vol. 7031, pp. 370–388. Springer, Heidelberg (2011)
13. Levan, C.: CQELS engine: instructions on experimenting CQELS. https://code.google.com/p/cqels/wiki/CQELS_engine
14. Nguyen, V., Bodenreider, O., Sheth, A.: Don't like RDF reification? Making statements about statements using singleton property. In: Proceedings of the 23rd International Conference on World Wide Web, WWW 2014, New York, NY, USA, pp. 759–770 (2014)

15. Pérez, J., Arenas, M., Gutierrez, C.: Semantics and complexity of SPARQL. In: Cruz, I., Decker, S., Allemang, D., Preist, C., Schwabe, D., Mika, P., Uschold, M., Aroyo, L.M. (eds.) ISWC 2006. LNCS, vol. 4273, pp. 30–43. Springer, Heidelberg (2006)

16. StreamReasoning: Continuous SPARQL (C-SPARQL) ready to go pack. http://streamreasoning.org/download

17. Taelman, R., Verborgh, R., Colpaert, P., Mannens, E., Van de Walle, R.: Continuously updating query results over real-time Linked Data. In: Proceedings of the 2nd Workshop on Managing the Evolution and Preservation of the Data Web, May 2016

18. Verborgh, R., Vander Sande, M., Hartig, O., Van Herwegen, J., De Vocht, L., De Meester, B., Haesendonck, G., Colpaert, P.: Triple pattern fragments: a low-cost knowledge graph interface for the web. J. Web Semant. **37–38**, 184–206 (2016)

PROFILES'16: 3rd International Workshop on Dataset PROFIling and fEderated Search for Linked Data

Assessing Trust with PageRank
in the Web of Data

José M. Giménez-García[1]([✉]), Harsh Thakkar[2], and Antoine Zimmermann[3]

[1] Univ Lyon, UJM-Saint-Étienne, CNRS,
Laboratoire Hubert Curien UMR 5516, 42023 Saint-Étienne, France
`jose.gimenez.garcia@univ-st-etienne.fr`
[2] Enterprise Information Systems Lab, University of Bonn, Bonn, Germany
`hthakkar@uni-bonn.de`
[3] Univ Lyon, MINES Saint-Étienne, CNRS,
Laboratoire Hubert Curien UMR 5516, 42023 Saint-Étienne, France
`antoine.zimmermann@emse.fr`

Abstract. While a number of quality metrics have been successfully proposed for datasets in the Web of Data, there is a lack of trust metrics that can be computed for any given dataset. We argue that reuse of data can be seen as an act of trust. In the Semantic Web environment, datasets regularly include terms from other sources, and each of these connections express a degree of trust on that source. However, determining *what* is a dataset in this context is not straightforward. We study the concepts of dataset and dataset link, to finally use the concept of Pay-Level Domain to differentiate datasets, and consider usage of external terms as connections among them. Using these connections we compute the PageRank value for each dataset, and examine the influence of ignoring predicates for computation. This process has been performed for more than 300 datasets, extracted from the LOD Laundromat. The results show that reuse of a dataset is not correlated with its size, and provide some insight on the limitations of the approach and ways to improve its efficacy.

Keywords: Linked data · Trust · Reuse · Interlinking · PageRank · Metric · Assessment

1 Introduction

The WDAqua project[1] aims at advancing the state of the art in data-driven question answering, with a special focus on the Web of Data. The Web of Data comprises thousands of datasets about varied topics, interrelated among them, which contain large quantities of relevant data to answer a question. Nonetheless, in an environment of information published independently by many different actors, data veracity is usually uncertain [17,19], and there is always the risk

---

[1] http://wdaqua.informatik.uni-bonn.de/.

© Springer International Publishing AG 2016
H. Sack et al. (Eds.): ESWC 2016 Satellite Events, LNCS 9989, pp. 293–307, 2016.
DOI: 10.1007/978-3-319-47602-5_45

of consuming misleading data. While some quality metrics have been proposed that can help to identify good datasets [5], there is a lack of trust metrics to provide a confidence on the veracity of the data [23].

In this context, we argue that actual usage of data can be seen as an act of trust. In this paper we focus on reuse of resources by other datasets as a usage metric. We consider reuse of a resource of a dataset by any other given dataset as an outlink from the later to the former. Under this purview, we can compute the PageRank [18] value of each dataset and rank them according to their reuse. PageRank has been successfully used to obtain trust metrics on individual triples [2]. In order to obtain a good measure of reuse, we perform the process on a large scale. We make use of the tools provided by the LOD Laundromat [20] to go beyond LOD Cloud, and process more than 38 billion triples, distributed in more than 600 thousand documents. The LOD Laundromat provides data from data dumps collected from the Internet, so it is not limited to dereferenceable linked data. However, what is regarded as a dataset is an important issue when dealing with data dumps. We make use of the concept of Pay-Level Domain (or PLD, also known as Top-Private Domain) to draw a distinction between datasets, and consider the influence of ignoring predicates when extracting outlinks. We perform a grouping of the triples in datasets according to their PLD and compute their PageRank values as a first measure of trust. Finally, we discuss the results and limitations of the approach, suggesting improvements for future work.

This document is organized as follows: in Sect. 2, we first discuss the relation of trust and popularity in the Web of Data, what should be considered a dataset in our context in order to clarify the problem we address, and finally present the LOD Laundromat; Sect. 3 describes the experiments and results, which we further discuss; Sect. 4 presents relevant related work; finally, we provide some conclusions and directions for future work in Sect. 5.

2   Ranking the Web of Data

2.1   PageRank, Reuse, and Trust in the Web of Data

We would like to assess trust in datasets by measuring their popularity based on the reuse of resources from a dataset in another dataset. To do this, we rely on the PageRank algorithm [18]. PageRank is the original algorithm developed by Page et al. that Google uses to rank their search results. It takes advantage of the graph structure of the web, considering each link from one page as a "vote" from the source to the destination. Using the links, the importance of a page is propagated across the graph, dividing the value of a page among its outlinks. This process is repeated until convergence is reached. The final result of PageRank corresponds to a stationary distribution, where each page value amounts to the probability for a random surfer to be at any moment in the page.

PageRank is meant to measure popularity (*i.e.*, "human interest and attention") on web pages. However, we argue that reuse of resources in the Semantic Web has a slightly different meaning. When there is a link from one web page to another, it does not mean necessarily that the author considers the linked page a trustworthy fact (it could be even linking something the author is criticizing). However, resources are reused to express facts in the author's dataset, which implicitly means that the author trusts that the resource is correct. This is supported by the analysis of predicates used for linking datasets by Schmachtenberg et al. [22], where the top used predicates are used to express statements about identity or relatedness (owl:sameAs, rdfs:seeAlso, skos:exactMatch, skos:closeMatch), authorship (using Doublin Core vocabulary), and social relations (using foaf and sioc vocabularies).

To compute PageRank in a set of datasets, it is first necessary to define what is considered a dataset and what is a link between datasets. RDF graphs, although formally defined as a set of triples, can be seen as directed multigraphs in which predicates play the role of arcs. This view suggests that if a triple contains a resource of dataset A as subject, and a resource of dataset B as object, it can be seen as a link from dataset A to dataset B. However, the links formed by arcs in an RDF graph are irrelevant to the notion of dataset linking. In fact, only the presence of hyperlinks suffices to indicate a link between one source and destination, therefore any HTTP IRI in an RDF graph can be seen as a link. So the question is, what it means that a resource belongs to a dataset, and to what dataset a hyperlink "points to". A naïve approach would be to consider that any IRI existing in a dataset belongs to the dataset and thus, that links connect two datasets having one same resource. However, this would imply, for instance, that any triple anywhere that uses a DBpedia IRI is considered to be linked to from the DBpedia dataset. As a result, any dataset that reuses a DBpedia IRI would increase their PageRank according to this definition.

Alternatively, we could take advantage of the linked data principles which stipulate that IRIs should be addresses pointing to a location on the Web. Again, one could naïvely assume that the location that the address points to is what defines the dataset, that is, the document retrieved when one gets the resource using the HTTP protocol. However, this would lead us, for instance, to define each DBpedia article as an individual dataset.

A second possibility would be to use the domain part of the URL, so datasets are grouped by the same publisher. This approach is taken by Ding and Finin [6] to characterize data in the Semantic Web. This way, it would be easy to determine what dataset is being linked to. Such approach would work well if all datasets were accessible from dereferenceable IRIs. However, there are large portions of the Web of Data that provide access to data dumps only [9,16]. In this case, the domain of the dump does not necessarily match the domain of the individual IRIs found in the dataset. As an example, the DBpedia dumps are found at http://downloads.dbpedia.org/ while all DBpedia IRIs start with http://dbpedia.org/.

The last approach, is to use the concept of *PLD*, *i.e.*, the subdomain component of a URL followed by a public suffix, to identify a dataset. Then, datasets are grouped not necessarily by the same publisher, but by the same publisher authority. This approach has already been used by other works [15,22]. As an example, if a file found at http://download.dbpedia.org/ contains the following triple:

```
<http://dbpedia.org/wiki/Europe>
    <http://www.w3.org/2002/07/owl#sameAs>
        <http://sws.geonames.org/6255148/>
```

we consider that the dataset having the PLD dbpedia.org is linking to the dataset with PLD geonames.org. It is important to notice that the source of the link (dbpedia.org) is obtained from the URL of the document that contains the triple (http://download.dbpedia.org/), not from the subject of the example. This approach enables us to extract outlinks from datasets published in dumps, and therefore access the majority of accessible semantic web data.

Definition 1 (Dataset). *A dataset is a non empty collection of triples that can be retrieved from sources accessible at a URL having a common Pay-Level Domain. The PLD identifies the dataset.*

In the previous example, we see that the predicate IRI is linking to the standard OWL vocabulary. It is very likely that predicates in general will be linking to vocabularies that are extensively reused. However, our intent is to evaluate trust on actual data that can be used to answer questions, and not vocabularies used to describe the data. We predict that extracting outlinks from predicates will lead to higher values for datasets containing only vocabularies. For this reason, we perform the same experiment with and without taking predicates into consideration.

Definition 2 (Dataset link). *There exists a link from a dataset A to a dataset B if and only if there exists a triple in a file at a location having the PLD that identifies A in which the PLD of its subject, its object, or both matches the PLD that identifies B.*

This definition is in line with the PageRank algorithm [18] where the number of links between the same two nodes is irrelevant. Note that since datasets must be non empty, links to PLDs that do not host RDF have to be ignored.

2.2 The LOD Laundromat and Frank

The LOD cloud[2], and in general Linked Open Data, contains a wide variety of formats, publishing schemes, errors, that make it difficult to perform a large-scale evaluation. Yet, to be accurate, our study requires to be comprehensive. Fortunately, the LOD Laundromat [1,21] makes this data available by gathering dataset dumps from the Web, including archived data. LOD Laundromat cleans the data by fixing syntactic errors and removing duplicates, and then makes it available through download (either as gzipped N-Triples or N-Quads, or HDT [10] files), a SPARQL endpoint, and Triple Pattern Fragments [24].

---

[2] http://lod-cloud.net/.

Using the LOD Laundromat is also a better solution than trying to use documents dereferenced by URIs, because most of datasets available online are data dumps [9,16], thus not accessible by dereferencing.

Frank [20] is a command-line tool which serves as an interface of the LOD Laundromat, and makes it easy to run evaluations against very large numbers of datasets.

3   Experiments and Results

The process to compute PageRank involves the following steps, detailed further below and illustrated in Fig. 1. The code and results are provided online.[3]

1. Extracting the document list from LOD Laundromat.
2. Parsing the content of each document to extract the outlinks.
3. Consolidating the results
4. Computing PageRank

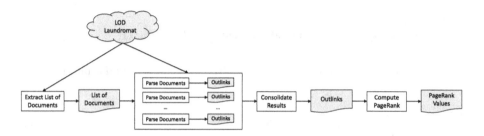

Fig. 1. Outlink extraction and PageRank computation workflow

3.1   Extracting the Document List from LOD Laundromat

We use the Frank command line tool [20] to obtain a snapshot of the contents of the LOD Laundromat. While the output of Frank can be directly pipelined to our process, the next step is performed in parallel in several machines. For this reason, we need that every machine reads the exact same input. An update in the contents of the LOD Laundromat during the next process could have impacted the results in that case. We retrieve the list of documents in the LOD Laundromat with the following command.

```
$ frank documents > documents.dat
```

This command retrieves a list of pairs *(downloadURL-resourceURL)*, where the first is the URL to download the gzipped datasets, and the second the resource identifier in the LOD Laundromat ontology. At the moment of the experiments, it retrieved 649,855 documents.

---

[3] https://github.com/jm-gimenez-garcia/LODRank.

3.2   Parsing the Content of Each Document to Extract the Outlinks

A prototype tool[4] has been developed to stream the contents of the documents end extract the outlinks. This tool reads the list of pairs *(downloadURL-resourceURL)* from the standard input, and accepts two optional parameters for partial processing: *Step* and *Start*. The first one tells how many lines the process reads in every iteration, processing the last one, while the second denotes what line to use for the first input. For each line processed, it queries the SPARQL endpoint to retrieve the URL where that datasets was crawled. This information can be found in the LOD Laundromat ontology connected to the resource, in the case the document was crawled as a single file, or connected to the archive that contains the document, if it was crawled compressed in a compressed file, possibly along other documents. In the first case, we retrieve the URL with Query 1, in the second case we retrieve the URL using Query 2, where %s is substituted by the *resourceURL*. The Pay-Level Domain is then extracted and stored. This will be considered as the identifier of the dataset.

```
SELECT ?url
WHERE {<%s> <http://lodlaundromat.org/ontology/url> ?url}
```

Query 1. Query to retrieve crawled URL of a non-archived document

```
SELECT ?url
WHERE {
        ?archive <http://lodlaundromat.org/ontology/containsEntry> <%s> .
        ?archive <http://lodlaundromat.org/ontology/url> ?url
}
```

Query 2. Query to retrieve crawled URL of an archived document

Then, the gzipped file is streamed from the *downloadURL* and parsed the triples. The subject and object (in case it is a URI) are extracted the Pay-Level Domain and compared against their dataset PLD. If they have a valid PLD and is different from their dataset's Pay-Level Domain, the pair *(datasetPLD-resourcePLD)* is stored as an outlink for the dataset. The output of each dataset is stored in a different file, which will be appended more pairs if a different document is identified as the same dataset.

---

[4] https://github.com/jm-gimenez-garcia/LODRank/tree/master/src/com/chemi2g/lodrank/outlink_extractor.

This process makes use of Apache Jena[5] v3.0.1 to query the SPARQL end-point of the LOD Laundromat and Google Guava[6] v19.0 to extract the Pay-Level Domain of the datasets.

In the experiments the process was launched in parallel in 8 virtual machines using Google Cloud Platform[7] free trial resources, each one processing a different subset of the list downloaded in the previous step. A statistical description of the results of each process, with and without considering predicates, is detailed in Table 1. "Documents" correspond to the number of dump files in the LOD Laundromat, while "Datasets" are the number of PLDs that the process is dealing with. There can be an overlap in the datasets of several processes, so the total number of datasets is not equal to the sum. We can see that the number of triples processed by each process is not proportional to the number of documents processed.

Table 1. Data extracted from the LOD Laundromat by each process

Process	Documents	Triples	Datasets (w. p.)	Datasets (w/o. p.)
1	81,220	3,994,446,393	135	121
2	81,226	3,742,870,561	137	118
3	83,422	4,146,249,367	140	127
4	81,225	3,376,784,600	135	120
5	81,225	3,623,413,245	142	120
6	88,198	3,377,773,585	131	116
7	81,226	4,132,960,522	137	115
8	89,781	3,911,917,919	134	123

3.3 Consolidating the Results

Once the outlinks have been extracted, the different files have to be appended and duplicates removed using a simple tool.[8] In the experiments, the data from each virtual machine was downloaded in a separate folder of a unique machine. Then files with the same name in each folder were concatenated and we removed the duplicates. The total number of datasets after consolidating the results is 412 when considering predicates, and 319 when not. The result was again concatenated in a single file.

---

[5] https://jena.apache.org/.
[6] https://github.com/google/guava.
[7] https://cloud.google.com/.
[8] https://github.com/jm-gimenez-garcia/LODRank/tree/master/src/com/chemi2g/lodrank/duplicate_remover.

3.4   Computing PageRank

For PageRank computation we make use of the igraph R package [4]. The ordered
PageRank values for all datasets can be seen in Figs. 2 and 3, with a logarithmic
scale. The complete list of results is published online.[9] We can see that in both
cases the top-ranked dataset is very much higher than the rest, then the slope
becomes more regular until it reaches a plateau at the end, with a minimum
value shared by several datasets that have no inlinks at all. Tables 2 and 3 show
the 10 highest ranked datasets.

Discussion. Here we provide additional information about the datasets, especially the top-ranked ones, in order to understand how ranking correlates with
other statistical values, such as number of triples, number of documents. We also
discuss how our own choices influenced the results.

The datasets appearing on the top 10 list are generally not surprising, with
the only exception of holygoat.co.uk, the only domain in the top 10 owned by
an individual person, Richard Newman, a computer scientist who wrote several
ontologies in the early days of the Semantic Web. This is even more remarkable
considering that the dataset has only 7 inlinks. The reason is that rdfs.org, the
domain of sioc ontology for instance, includes resources from holygoat.co.uk.
Because this dataset has only 2 outlinks, half of its PakeRank score is forwarded
to holygoat.co.uk, which accounts for 96 % of its PageRank value.

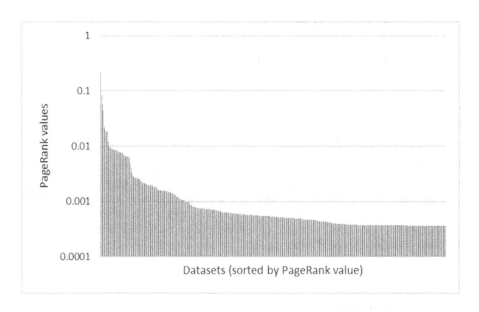

Fig. 2. PageRank values sorted from higher to lower, with predicates

---

[9] https://github.com/jm-gimenez-garcia/LODRank/tree/master/results.

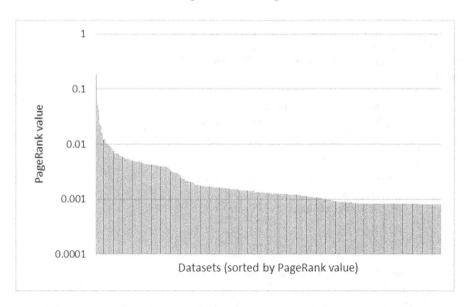

Fig. 3. PageRank values sorted from higher to lower, without predicates
Table 2. PageRank values for the top 10 datasets, with predicates

Rank	Dataset	PageRank value	Inlinks	Outlinks
1	w3.org	0.224806691	411	32
2	purl.org	0.085548846	278	64
3	lodlaundromat.org	0.056963209	188	1
4	xmlns.com	0.045452453	219	3
5	schema.org	0.023239532	32	1
6	creativecommons.org	0.020496922	106	2
7	dbpedia.org	0.018894825	118	160
8	rdfs.org	0.018738995	108	5
9	ogp.me	0.018442606	37	4
10	usefulinc.com	0.012066847	26	4

As predicted, when including predicates the first positions incorporate more datasets about vocabularies. When removing the predicates, w3.org, xmlns.com, schema.org, and ogp.me no longer appear in the top positions, and datasets with factual data move upwards. lodlaundromat.org seems to appear when considering predicates because the LOD Laundromat adds information about the cleaning process when processing the data. While not an optimum solution (considering that purl.org and rdfs.org, hosts of well known ontologies, are still in the top positions), ignoring the predicates proves to be a simple but useful technique.

We used two queries, (Query 3 and Query 4), to obtain the number of documents and triples for each PLD, from the LOD Laundromat.

```
PREFIX llo: <http://lodlaundromat.org/ontology/>
PREFIX ll: <http://lodlaundromat.org/resource/>
SELECT (COUNT(DISTINCT ?resource) AS ?count)
WHERE {
      {
          ?resource llo:url ?url
          FILTER regex(?url,"[^/\\.]*\\.?%s/", "")
      }
      UNION
      {
          ?archive llo:containsEntry ?resource ;
              llo:url ?url
          FILTER regex(?url, "[^/\\.]*\\.?%s/", "")
      }
}
```

Query 3. Query to retrieve the number of documents per dataset

Table 3. PageRank values for the top 10 datasets, without predicates

Rank	Dataset	PageRank value	Inlinks	Outlinks
1	purl.org	0.185304616	181	50
2	creativecommons.org	0.051625742	93	1
3	dbpedia.org	0.04234706	104	119
4	rdfs.org	0.023497322	73	2
5	geonames.org	0.02127494	59	6
6	loc.gov	0.016137225	33	8
7	fao.org	0.012392539	27	8
8	europa.eu	0.012182709	30	13
9	holygoat.co.uk	0.012179038	7	1
10	data.gov.uk	0.010364034	19	11

The result of the queries are given in Table 4 for all the datasets that appear in the 10 top of both experiments.

As we can see, popularity is not at all correlated with the size of the datasets. Indeed, a number of the top ten datasets have less that 200 triples, while dbpedia.org and europa.eu both have billions of triples.

The enormously high page rank of purl.org should be mitigated by the fact that purl.org does not actually host any data. It is a redirecting service that many data publishers are using. This result highlights a drawback in our heuristic for identifying datasets: the PLD is not always referring to a single dataset. To overcome this particular case, we could consider the PLD of the URL of the

```
PREFIX llo: <http://lodlaundromat.org/ontology/>
PREFIX ll: <http://lodlaundromat.org/resource/>
SELECT (COUNT(DISTINCT ?resource) AS ?count) (SUM(?triples) as ?sum)
WHERE {
        {
            ?resource llo:url ?url ;
                llo:triples ?triples
            FILTER (?triples > 0)
                FILTER regex(?url, "[^/\\.]*\\.?%s/", "")
        }
        UNION
        {
            ?archive llo:containsEntry ?resource ;
                llo:url ?url .
            ?resource llo:triples ?triples
            FILTER (?triples > 0)
                FILTER regex(?url, "[^/\\.]*\\.?%s/", "")
        }
}
```

Query 4. Query to retrieve the number of documents with triples and number of triples

Table 4. Documents and triples per dataset in LOD Laundromat

Dataset	Rank	Documents	Documents with triples	Triples
w3.org	1/-	413	256	1,973,715
purl.org	2/1	9,166	9,073	254,548,441
lodlaundromat.org	3/-	68	1	4
xmlns.com	4/-	4	4	1895
schema.org	5/-	1	1	1
creativecommons.org	6/2	1	1	117
dbpedia.org	7/3	1,888	1,752	1,257,930,891
rdfs.org	8/4	6	6	1,808
ogp.me	9/274	68	1	231
usefulinc.com	10/-	2	2	1398
geonames.org	20/5	6	4	9,762
loc.gov	29/6	16	12	263,653,979
fao.org	36/7	17	11	48366
europa.eu	12/8	7734	7,705	3,414,066,228
holygoat.co.uk	37/9	1	1	95
data.gov.uk	42/10	157	88	51,401,490

document obtained after dereferencing the IRI, in the same way as Hogan et al. [15] do for the general case.

Another possible drawback of the approach is that triples with `rdf:type` in predicate position have their object pointing to a class in an ontology. This is in contradiction with our remark in Sect. 2 where we say that we want to rank instance data rather than terminological knowledge. This can have a major impact the results since purl.org is most often used to redirect to vocabularies more than datasets, and rdfs.org only hosts ontologies.

4   Related Work

The authors of Semantic Web Search Engine (SWSE [15]) strongly advocate that the use of a ranking mechanism is very crucial for prioritizing data elements in the search process. Their work is inspired by the Google PageRank algorithm, which treats hyperlinks to other pages as a positive score. The PageRank algorithm is targeted for hyperlink documents and its adaptation to the LOD is however non-trivial, as we have seen. They point out that the primary reason for this is that LOD datasets may not have direct hyperlinks to other datasets but rather in most cases make use of implicit links to other web pages via the re-use of dereferenceable URIs. Here the unit of search becomes the entity and not the document itself. The authors briefly re-introduce the concept of naming authority, from their previous work [13] in order to rank structured data from an open distributed environment. They assume that the naming authority should match the Pay-level domain such that computing PageRank is performed on a naming authority graph where the nodes are PLDs. Their intuition therefore is in accordance with our reasoning from Sect. 2. They have discussed and contrasted the interpretation of naming authorities on a document level (e.g. http://www.danbri.org/foaf.rdf) and a PLD level (danbri.org). Also, they make use of a generalization for the method discussed in the paper [8] for ranking entities and carry out links analysis on the PLD abstraction layer.

The authors of Swoogle [7] develop OntoRank algorithm in order to rank documents. OntoRank, a variation of Google PageRank, is an iterative algorithm for calculating the ranks for documents built on references to terms (i.e., classes and properties) which are defined in other documents.

In the paper [3], the authors calculate the rank of entities (or as they call them objects) based on the logarithm of the number of documents where that particular object is mentioned.

In their work [11] present LinkQA, an extensible data quality assessment framework for assessing the quality of linked data mappings using the network measures. For this, they assess the degree of interlinking of datasets using five network measures, out of which two network measures are specifically designed for Linked Data (namely, Open Same-As chains and Description Richness) and the other three standard network measures (namely, degree, centrality, and the clustering coefficient) in order to assess variation in the quality of the overall linked data with respect to a certain set of links.

In [2], PageRank is used to compute a measure that is in turn associated to individual statements in datasets for the purpose of incorporating trust in reasoning. Therefore, as in our own approach, they consider that PageRank is an indication of trustworthiness. However, they only compute PageRank on a per document basis, and report on the PageRank values of the top 10 documents obtained from their web crawl.

5 Conclusion and Future Work

Data-driven question answering, the aim of project WDAqua mentioned in the introduction of this paper, requires quality data in which one can trust. Our aim has been to provide insight on how a trust measure can be based on dataset interlinking. To that end, we consider Pay-Level Domains as identifiers of unique datasets and compute PageRank on them. Our results show that the design choices greatly affect the results. Whether taking into account or not predicates for outlink extraction impacts how vocabularies are ranked, and the choice of PLD as definition of dataset is arguable, as some PLDs group many data dumps. In order to improve this, we could associate well known datasets to IRI patterns, such as it.dbpedia.org for the Italian version of DBpedia.

In addition, we also intend to explore further applications of PageRank that may be useful for question answering. User interaction that provides trust values in a number of dataset could be used to compute PageRank values with those datasets as a teleport set, as suggested by Gyöngyi et al. [12]. Also, Topic-Sensitive PageRank [14] could help a question-answering system to select different datasets when a question is identified to belong to a specific topic.

Finally, this work is part of a broader objective that we want to pursue: to ascertain the relationship between the perceived trust on a dataset and its objective quality. We will explore this area in a future work where other data reuse metrics will be considered and compared against different quality metrics.

Acknowledgments. This project is supported by funding received from the European Unions Horizon 2020 research and innovation program under the Marie Skłodowska-Curie grant agreement No 642795. We would like to thank Elena Simperl, whose idea jumpstarted the project that lead to this article, and also Elena Demidova, Kemele Endris, and Christoph Lange for the useful discussions related to it.

References

1. Beek, W., Rietveld, L., Bazoobandi, H.R., Wielemaker, J., Schlobach, S.: LOD laundromat: a uniform way of publishing other people's dirty data. In: Mika, P., et al. (eds.) ISWC 2014, Part I. LNCS, vol. 8796, pp. 213–228. Springer, Heidelberg (2014)
2. Bonatti, P.A., Hogan, A., Polleres, A., Sauro, L.: Robust and scalable Linked Data reasoning incorporating provenance and trust annotations. Web Semant.: Sci. Serv. Agents World Wide Web **9**(2), 165–201 (2011)

3. Cheng, G., Qu, Y.: Searching linked objects with falcons: approach, implementation and evaluation. Int. J. Semant. Web Inf. Syst. (IJSWIS) **5**(3), 49–70 (2009)

4. Csardi, G., Nepusz, T.: The igraph software package for complex network research. InterJ. Complex Syst. **1695**(5), 1–9 (2006)

5. Debattista, J., Londoño, S., Lange, C., Auer, S.: Quality assessment of linked datasets using probabilistic approximation. In: Gandon, F., Sabou, M., Sack, H., d'Amato, C., Cudré-Mauroux, P., Zimmermann, A. (eds.) ESWC 2015. LNCS, vol. 9088, pp. 221–236. Springer, Heidelberg (2015)

6. Ding, L., Finin, T.W.: Characterizing the semantic web on the web. In: Cruz, I., Decker, S., Allemang, D., Preist, C., Schwabe, D., Mika, P., Uschold, M., Aroyo, L.M. (eds.) ISWC 2006. LNCS, vol. 4273, pp. 242–257. Springer, Heidelberg (2006)

7. Ding, L., Finin, T., Joshi, A., Pan, R., Cost, R.S., Peng, Y., Reddivari, P., Doshi, V., Sachs, J.: Swoogle: a search and metadata engine for the semantic web. In: Proceedings of the Thirteenth ACM International Conference on Information and Knowledge Management, pp. 652–659. ACM (2004)

8. Ding, L., Pan, R., Finin, T.W., Joshi, A., Peng, Y., Kolari, P.: Finding and ranking knowledge on the semantic web. In: Gil, Y., Motta, E., Benjamins, V.R., Musen, M.A. (eds.) ISWC 2005. LNCS, vol. 3729, pp. 156–170. Springer, Heidelberg (2005)

9. Ermilov, I., Martin, M., Lehmann, J., Auer, S.: Linked open data statistics: collection and exploitation. In: Mouromtsev, D., Klinov, P. (eds.) KESW 2013. CCIS, vol. 394, pp. 242–249. Springer, Heidelberg (2013)

10. Fernández, J.D., Martínez-Prieto, M.A., Gutiérrez, C., Polleres, A., Arias, M.: Binary RDF representation for publication and exchange (HDT). Web Semant. Sci. Serv. Agents World Wide Web **19**, 22–41 (2013)

11. Guéret, C., Groth, P., Stadler, C., Lehmann, J.: Assessing linked data mappings using network measures. In: Simperl, E., Cimiano, P., Polleres, A., Corcho, O., Presutti, V. (eds.) ESWC 2012. LNCS, vol. 7295, pp. 87–102. Springer, Heidelberg (2012)

12. Gyöngyi, Z., Garcia-Molina, H., Pedersen, J.: Combating web spam with TrustRank. In: Proceedings of the Thirtieth International Conference on Very large data bases, vol. 30, pp. 576–587. VLDB Endowment (2004)

13. Harth, A., Kinsella, S., Decker, S.: Using naming authority to rank data and ontologies for web search. In: Bernstein, A., Karger, D.R., Heath, T., Feigenbaum, L., Maynard, D., Motta, E., Thirunarayan, K. (eds.) ISWC 2009. LNCS, vol. 5823, pp. 277–292. Springer, Heidelberg (2009)

14. Haveliwala, T.H.: Topic-sensitive pagerank: a context-sensitive ranking algorithm for web search. IEEE Trans. Knowl. Data Eng. **15**(4), 784–796 (2003)

15. Hogan, A., Harth, A., Umbrich, J., Kinsella, S., Polleres, A., Decker, S.: Searching and browsing Linked Data with SWSE: the semantic web search engine. Web Semant.: Sci. Serv. Agents World Wide Web **9**(4), 365–401 (2011)

16. Hogan, A., Umbrich, J., Harth, A., Cyganiak, R., Polleres, A., Decker, S.: An empirical survey of Linked Data conformance. Web Semant.: Sci. Serv. Agents World Wide Web **14**, 14–44 (2012)

17. Liu, S., d'Aquin, M., Motta, E.: Towards linked data fact validation through measuring consensus. In: Proceedings of the 2nd Workshop on Linked Data Quality Co-located with 12th Extended Semantic Web Conference (ESWC 2015). CEUR Workshop Proceedings, vol. 1376. CEUR-WS.org (2015)

18. Page, L., Brin, S., Motwani, R., Winograd, T.: The PageRank citation ranking: bringing order to the web (1999)

19. Paulheim, H., Bizer, C.: Improving the quality of linked data using statistical distributions. Int. J. Semant. Web Inf. Syst. (IJSWIS) **10**(2), 63–86 (2014)

20. Rietveld, L., Beek, W., Schlobach, S.: LOD lab: experiments at LOD scale. In: Arenas, M., et al. (eds.) ISWC 2015. LNCS, vol. 9367, pp. 339–355. Springer, Heidelberg (2015)
21. Rietveld, L., Verborgh, R., Beek, W., Vander Sande, M., Schlobach, S.: Linked data-as-a-service: the semantic web redeployed. In: Gandon, F., Sabou, M., Sack, H., d'Amato, C., Cudré-Mauroux, P., Zimmermann, A. (eds.) ESWC 2015. LNCS, vol. 9088, pp. 471–487. Springer, Heidelberg (2015)
22. Schmachtenberg, M., Bizer, C., Paulheim, H.: Adoption of the linked data best practices in different topical domains. In: Mika, P., et al. (eds.) ISWC 2014, Part I. LNCS, vol. 8796, pp. 245–260. Springer, Heidelberg (2014)
23. Thakkar, H., Endris, K.M., Giménez-García, J.M., Debattista, J., Lange, C., Auer, S.: Are linked datasets fit for open-domain question answering? A quality assessment. In: Proceedings of the 6th International Conference on Web Intelligence, Mining and Semantics, WIMS 2016, p. 19. ACM (2016)
24. Verborgh, R., Vander Sande, M., Colpaert, P., Coppens, S., Mannens, E., Van de Walle, R.: Web-scale querying through linked data fragments. In: Proceedings of the Workshop on Linked Data on the Web Co-located with the 23rd International World Wide Web Conference (WWW 2014). CEUR Workshop Proceedings, vol. 1184 (2014)

Workshop on Extraction and Processing of Rich Semantics from Medical Texts

The Generation of a Corpus for Clinical Sentiment Analysis

Yihan Deng[1], Thierry Declerck[2], Piroska Lendvai[2], and Kerstin Denecke[3(✉)]

[1] Innovation Center Computer Assisted Surgery,
University of Leipzig, Leipzig, Germany
yihan.deng@medizin.uni-leipzig.de
[2] Department of Computational Linguistics,
Saarland University, Saarbrücken, Germany
declerck@dfki.de, piroska.r@gmail.com
[3] Department of Medical Informatics,
Bern University of Applied Sciences, Biel, Switzerland
kerstin.denecke@bfh.ch

Abstract. Clinical care providers express their judgments and observations towards the patient status in clinical narratives. In contrast to sentiment expressions in general domains targeted by language technology, clinical sentiments are influenced by related medical events such as clinical precondition or outcome of a treatment. We argue that patient status in terms of positive, negative and neutral judgements can only suboptimally be judged with generic approaches, and requires specific resources in term of a lexicon and training corpus targeting clinical sentiment. To address this challenge, we manually developed a corpus based on 300 ICU nurse letters derived from a clinical database, and an annotation scheme for clinical sentiment. The paper discusses influence patterns between clinical context and clinical sentiments as well as a semi-automatic method to generate a larger annotated corpus.

1 Introduction

Attention towards opinion and sentiment analysis has been growing over the past 10 years, in which most existing methods and corpora are developed to process and analyze general domain sentiment, most prominently in Web 2.0 texts. Besides, stance detection has evolved as an extension of opinion analysis towards a given topic or object [1,2]. Sentiment analysis and stance detection represent two levels of observation on polarity: sentiment analysis deals with the detection of polarized terms at token level, whereas the detection of stance, i.e., pro, contra, and neutral attitudes towards a target address the discourse level. Information originating from these two levels necessarily influence one another, and deliver important knowledge for several applied tasks such as information retrieval, text summarization and textual entailment.

Neither sentiment analysis nor stance detection have been thoroughly scrutinized in the clinical domain yet – one of the first studies on sentiment in clinical

© Springer International Publishing AG 2016
H. Sack et al. (Eds.): ESWC 2016 Satellite Events, LNCS 9989, pp. 311–324, 2016.
DOI: 10.1007/978-3-319-47602-5_46

documents that we are aware of is from Denecke and Deng [3] –, whereas their utility in automated analysis is undeniable. One particular challenge that clinical sentiment analysis could support is the collaborative decision making process in clinical practice. According to this, a successful diagnosis or treatment is achieved based on the collaboration and knowledge sharing of care providers from different specialities. Since the communication dialogue directly influences the quality of the treatment, misunderstandings or missing out on evidence can increase the risk of operational failure and thus threaten patient safety. Our development scenario takes place exactly in this context, motivated by the importance of automated methods for sentiment analysis in the clinical domain in order to reconstruct patient status based on consolidated judgments from care providers.

In contrast to the definition of sentiment analysis in non-clinical domains, we consider **clinical sentiment** as an event that reflects the patient status, in which the care provider additionally expresses stance towards clinical and social situations (cf. [3,4]). In this paper, we characterize the patient status event and its corresponding context objects in detail, formulating annotation entities that can accommodate sentiment values.

To our knowledge, there is no annotated corpus or sentiment dictionary available for the clinical domain, and we argue that such resources form crucial prerequisites to develop and evaluate methods for processing clinical narratives. The main contributions of this work are:

– introduction of an annotation schema for clinical sentiment,
– creation of an annotated clinical sentiment dataset,
– assessment of the annotation scheme, collecting suggestions for automated corpus generation.

The paper is structured as follows: Sect. 2 summarizes related work in corpus generation and annotation. The annotation scheme is introduced in Sect. 3. In Sect. 4, we describe our method of corpus generation procedure. The results of the annotation study are presented in Sect. 5. At the end of the paper, we discuss methods for larger corpus generation and pointers for future work.

2   Sentiment Resources and Annotation Level for Clinical Sentiments

To process texts of social media and online news, general domain sentiment lexicons (SentiWordNet [5], WordNetAffect [6], etc.) are utilized in the field of language technology, which some researchers have adapted to the medical domain. Goeuriot et al. [7] extended general domain sentiment lexicons with domain terminology from drug reviews to constitute a sentiment lexicon for pharmaceutical evaluation. In more detail, the MPQA lexicon [8] and SentiWordNet were employed as basic sources for getting opinionated terms. However, a more comprehensive study using these two lexicons based on six different data sets of health and clinical texts [3] have already proved the insufficiency of the merged sentiment lexicons with respect to clinical sentiment analysis, noting that clinical practice has a completely different usage of sentiment terminology and emphasis in the expression. Experts are more likely to use patient reaction and clinical outcome to express polarity instead of sentiment terms seen in generic domains (adjective, adverb and part of the noun). Deng presented a merged event/entity level sentiment corpus [9].

Most of the work in sentiment analysis focused on English texts. Recently, some German lexicons [10] and annotated corpora based on a Twitter data set for sentiment analysis have been developed [11]. Besides, a German aspect-oriented sentiment corpus of smartphone app store reviews were developed by Saenger et al. [12]. Their study builds on data-driven mechanisms and conditional random field prediction to build the baseline. In addition, a newly emerged annotation schema for verb-centered sentiment inference has been reported by Klenner et al. [13], discussing sentiment features extracted from or implied by verbs.

Importantly, conventional sentiment corpora typically assign labels to spans of tokens, which does not allow for representing the influence of context on the target event (in our case, patient status). Our annotation effort aims to overcome this limitation and allows to gather influence patterns from context objects towards patient status events.

3   Annotation Scheme

Existing corpora in generic domains feature annotations at document level [14], sentence or phrase level [15] to event level [9]. Our targeted sentiment objects relate to fine-grained sentiment aspects of the patient status and its healthcare context. Given the limitations we discussed above, our work focuses on designing an annotation scheme targeting clinical texts, involving patient status as well as clinical context objects. Our intention is that the scheme should simulate the judgment process between clinical experts, and thus should be able to represent multiple perspectives pertaining to a single patient status.

We regard clinical sentiment as an event that reflects the patient bodily status, coupled with clinical, pharmaceutical, and social objects (i.e., procedures, events, states, phenomena) that influence this status:

$$Patient\,status\,event + set\,of\,(CI|PI|SC)$$

with CI = clinical intervention, PI = pharmaceutical intervention, SC = social connection. The main target of clinical sentiment is the event that represents the change of the patient status (see Fig. 1).

The patient status event can be further categorized into different aspects of the status. The 'concrete', i.e., physical aspect refers to descriptions of body parts, movement, input and output of patients as well as the vital signs. Next to the patient status, another type of annotation target are the context objects, which are divided into three types: clinical intervention, pharmaceutical intervention and social connections. The context objects annotation class provides important context information for the expression of the patient status. The patient event and its context objects are defined below in detail.

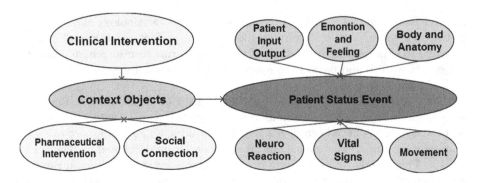

Fig. 1. Annotation schema for patient status event and context objects

Patient status event: Patient reactions, physical condition, emotional state, input and output, movement.

Context objects: (1) Clinical interventions (physical therapy, ventilator, paralyzed, balloon placement), laboratory values. (2) Pharmaceutical intervention (treatment, anesthesia, end of life care). (3) Patient's social connection (relatives' visiting, religious support).

Sentiment polarity is associated to the patient status, while the context objects additionally influence the patient status and the entire polarity outcome. Three sentiment values (positive, negative, neutral) are annotated with respect to the patient status, while the polarity of context objects is only limited to two categories: positive and negative (see Fig. 2). The neutral polarity of context objects is excluded, as it is assumed not to alter the patient status. The polarity of the context object and patient status can overlap or interfere with each other, forming an aggregated sentiment value.

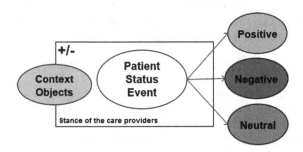

Fig. 2. Patient status event and its context objects

4   Corpus Generation

4.1   Raw Data

Our raw data set consists of 300 nurse letters from MIMIC II database. MIMIC (Multi-parameter Intelligent Monitoring in Intensive Care[1]) is an openly available dataset managed by the MIT Lab for Computational Physiology, comprising anonymized health data associated with more than 40,000 critical care patients. More specifically, we collected nurse letters from this database. A nurse letter is part of a patient record written by nurses on duty while monitoring the patients, containing information such as patient health status and response of the patient to treatments. These documents are written in a relatively free text manner. Acronyms and typos appear very often when describing e.g. vital signs, laboratory tests, medications. In order to ensure that the annotated sentiment will only depend on the treating care provider and the reflected patient status, we have identified a patient cohort with a limited set of diagnoses and treatments, based on the disease classification code (International Classification of Diseases) and procedure code.

4.2   Annotation Guidelines and Examples

The annotation guidelines are derived from the annotation scheme, and pertain to the *patient status event* and the three context objects (*clinical, pharmaceutical, social*). The patient status event can be reflected by up to six types of descriptions, as illustrated in Fig. 1.

As indicated earlier, we regard the polarity of the patient status event as positive, negative and neutral, while the polarity of context object is defined as stance of the care providers, where only two values can be assigned: for and against. Concrete examples of classes, annotation guidelines, and color coding are presented in the following section.

In general, blue marks words referring to the patient status. Red (*negative*), orange (*neutral*) and green(*positive*) reflect the polarity, the yellow color highlights the positive clinical context object whereas brown indicates the negative clinical objects. Silver represents the social connection with positive effect. Purple indicates words referring to social connection with negative effect. Magenta shows positive pharmaceutical context objects, while pink represents the negative pharmaceutical context object.

Patient Status. As illustrated in Fig. 1, the patient status (e.g. movement or emotion) is the basic type of status signs. The patient status concerns the patient's body part, reaction or status changes. Typically, the author of a clinical text directly expresses polarity to show their attitude towards the patient

---

[1] https://physionet.org/mimic2/.

status. The patient status is sometimes self-explained with polarity. Only minimal medical knowledge is required to conduct this kind of annotation. Given the explicit expression of polarity, existing conventional methods are able to detect and analyze these explicit sentiment expressions in clinical texts.

1. Drowsy but easily arousable seems irritated at times nods head Y/N appropriately follows simple commands perla.
   The patient status: Drowsy, arousable, irritated, nods head, follows commands
   Polarity: Easily, at times, appropriately –> positive

2. belly distended firm , abbd wound incision oozing moderate amnt serous secretions
   Patient status: Belly, abbd, secretions
   Polarity: Distended firm –> negative, wound incision in the area of abdomen –>negative, oozing moderate amnt serous –> negative

3. Multiple blisters on the trunk/chest , oozing from the swollen scrotum
   Patient status: blisters, trunk, chest, scrotum
   Polarity: multiple, oozing, swollen –> negative

Clinical Interventions. The objects for clinical interventions, readmission events and pharmaceutical interventions and their relations to the patient status need to be determined with sufficient medical knowledge and experiences. Since the patient status is not presented or only partly presented in the explicit polarity, conventional methods can not extract the implicit sentiments expressed in terms of clinical context objects. The annotators were therefore asked to pick out the clinical context objects and judge their polarity and relation towards the patient status, so that their impact could be determined. Positive and negative clinical events are determined by the annotator separately. Afterwards, the patient status under the influence of the context object has been selected.

1. Pt remains in AFIB(h/o chronic afib) rate initially 100–106 presently down to 90's AFIB with rare PVC noted.
   Patient status: Pt remains
   Clinical object: AFIB 100–106, 90's, premature ventricular contraction (pvc), negated
   Polarity: although it shows "down to" trend, the value 100–106 and 90 –> still negative, rare PVC –> positive

2. Echo done, no ASD or VSD . Found severe biventricular enlargement and biventricular systolic dysfunction , 4+ MR, torn mitral chordae .
   Patient status: Echo, biventricular, mitral chordae
   Clinical object: Atrial septal defect(ASD), Ventricular Septal Defect(VSD), enlargement, systolic dysfunction
   Polarity: Negated ASD and VSD –> positive effect –> positive in this sentence, systolic dysfunction and severe enlargement –> negative effect –> negative patient status

3. pt with ecchymolic area over trunk and arms bilateral
   Patient status: Pt, trunk, arms bilateral
   Clinical object: Ecchymolic
   Polarity: Ecchymolic area renders negative effect towards pt's trunk and arms −> negative
4. Pt remains on ventilator and IHO . Pt received espheageal balloon placement and optimal peep was established
   Patient status: Pt
   Clinical object: Ventilator and IHO, esopheageal balloon placement, peep was established
   Polarity: On ventilator and IHO −> negative effect on patient status, balloon placement successful be conducted and optimal peeps−> positive effect −> positive

Pharmaceutical Interventions. Besides clinical intervention, the pharmaceutical therapy is another important measure to support the entire treatment process in clinical care. There are three main types of usage for the context object in this category. Firstly, the intervention can treat the patient with a certain disease, where recovery is the goal in the long run. Secondly, it can be a medication used in order to support a clinical intervention, such as stabilizing the blood pressure or regulating of the clotting time of blood during the surgery. Thirdly, it may refer to end of life care, where the aim of drug application is to relieve pain. The first two pharmaceutical interventions assist the conduction of therapy. A positive outcome and a potential recovery of patient through usage of medication can be expected, while the latter goal of intervention is only to relieve the pain whereas the negative outcome has already been determined when the medication was applied. As a result, the former two types of pharmaceutical interventions stand for a positive outcome whereas the latter one implies the negative outcome of the patient status.

1. Pt initially paralyzed on cisactricurium-atvian and mso4- , pt 4 twitches out of 4 with med on- cisactricurium to off- cont ativan and mso4- both dec by 1 mg each, pt eyes blinking to stimuli , pupils 4 cm- sluggish rx to light, not moving extremities at this time.
   Patient status: Pt, pt eyes blinking, pupils, rx, extremities
   Pharmaceutical object: Cisactricurium, atvian, mso4-
   Clinical object: paralyzed, stimuli
   Polarity: Cisatracurium-atvian has been applied for relaxation of muscle and assisting breath. The medication is used to support the treatment (positive effect). The status of the patient is as expected. Hence the entire status −> positive
2. continues to be febrile despite Tylenol , but the temp has come down from 39.2 to 38.7 .
   Patient status: Implicit pt, continue, temp

Pharmaceutical object: Tylenol for fever treatment
Polarity: Although the "come down" trend from 39.2 to 38.7 has been confirmed, the body temperature is still higher than normal temperature, Tylenol has a positive effect but the patient status is still –> negative.

3. Fentanyl increased to 300 mcg/hr at 0830 for comfort . CVVHD discontinued . ECMO withdrawn by perfusionist and pt expired .
Patient status: Pt
Pharmaceutical object: Fentanyl
Clinical object: Continuous veno-venous hemodialysis (CVVHD), extracorporeal membrane oxygenation (ECMO)
Polarity: Patient expired after the application of Fentanyl for comfort (negative effect on outcome) CVVHD and ECMO withdrawn –> negative

Social Connections. The third type of objects indicate the patient social relations and relatives' visiting, which shows the social status of the patient. A patient's social connection directly influences the patient status. The activity and implicit sentiments for the appearance of certain social connections entail positive and negative effect towards the patient status.

1. Wife and brother in to visit, discussed with doctor and comforted pt .
Patient status: Pt
Social connections: Brother and wife
Polarity: Come to support and comfort –>Positive
2. Family called in middle of night and in to see pt . Priest called in and last rights given.
Patient status: Pt
Social connection: Family, priest
Polarity: Called in middle of night, last right –> negative
3. wife and daughter into visit and will return this evening daughter upset when leaving
Social connection: Wife, daughter
Polarity: Upset –> negative

5   Annotation Task and Results

5.1   Annotators and Annotation Task

To generate a high-quality corpus, clinical experts were asked to supply the annotations, with whom the annotation schema was jointly improved after initial discussions.

In a pre-annotation phase, all five annotators marked the same 25 nurse letters individually in an iterative process: In a first step, they were asked to label four entities (patient status and three context objects). Then, they had to identify and highlight the polarity terms and the polarity of the context objects (+ *effect* or - *effect*). At last, they were requested to connect the polarity term and

Table 1. Background of the annotators

ID	Knowledge background	Experience	Specialities
A1	Medicine	5 years	Cardiology, surgery
A2	Medicine	6 years	ENT, cardiology
A3	Pharma engineering	6 years	Pathology, targeted drug delivery
A4	Medical informatics	4 years	Text mining, medical ontology
A5	Medical informatics	3.5 years	Knowledge modeling

context objects with the patient status entity to show their relationship. Ehost[2] was the tool used in this annotation task. Annotation was conducted independently by each annotator without discussions. However, annotators were allowed to use the Internet to look up unfamiliar terms. The annotator's background is presented in Table 1: two students of medicine, two medical informatics students and one pharmacy student took part in the annotation. During the annotation, annotators received the annotation guideline and text corpus, which comprises the same texts to be annotated by the five annotators. One introduction seminar was held to explain the annotation guidelines and two rounds of pre-annotation with adjudications were conducted to train the annotators.

After this pre-stage, the annotation guidelines were verified and extended. Subsequently, the remaining corpus was annotated by two annotators who were selected based on good inter-annotator agreement and accordance with the guidelines (annotators A1 and A2 (medical annotators)).

5.2 Annotation Assessment Methods

The objective of annotation assessment was (i) to determine which background is necessary to perform the annotation, and (ii) to assess the recognition rate. The following methods have been applied: the average annotation agreements in the two pre-annotations based on the subset of 25 nurse letters were compared in pairs of annotators. The inter annotator agreement (IAA) was calculated as Kappa statistic value. IAA scores were calculated under exact match criteria (i.e., the annotations should match completely). The recognition of patient status events and context objects from the five annotators were compared with the adjudicated event list (gold standard). The event list was generated through majority voting and discussion among the five annotators. A correct recognition should agree on both the span and polarity ($+/-$ effect) to this span.

During the adjudication, a subset of 25 documents was annotated by all five annotators. The recognition of patient status events and context objects from five annotators are compared with the adjudicated event list (gold standards) (see Table 2). A correct recognition requires the correct annotation of both the span and the polarity ($+/-$ effect) to this span. The recognition rate is *recognized events + polarity/adjudicated events in list*.

---

[2] https://code.google.com/p/ehost/.

The final corpus consists of the consensus of the annotations. Altogether, the 300 documents were annotated with 7,080 patient status events, 2,040 clinical objects, 1,380 pharmaceutical events and 535 social context objects.

5.3   Recognition Rate and Inter-annotator Agreement

The medical annotators (A1 and A2) have achieved a high average recognition rate in all three categories. The pharma-annotator has recognized the highest percentage of pharmaceutical interventions. Generally, the social context objects are well detected by all annotator groups.

Table 2. Statistics of event annotation in two adjudications

Class statistics (total number)	A1	A2	A3	A4	A5
Patient status event (118)	109	106	83	118	79
Clinical context object (34)	33	30	9	17	7
Pharmaceutical context object (23)	20	21	22	15	10
Social context object (9)	9	8	9	9	8

As can be seen in Table 3, the average agreement between medical annotators (A1, A2) have reached 79.8 %. Annotations of A5 received the lowest agreement with others, only 22–34.2 %. The other pairs of annotators had an overlap of around 50 %. Further, the medical annotators (A1 and A2) have achieved a high average recognition rate in all three categories. The pharmacy-annotator A3 has recognized the highest percentage of pharmaceutical interventions. Generally, the social context objects are well detected by all annotator groups.

Table 3. Two-way average inter annotator agreement (Kappa) in two adjudications (all selected classes)

Annotators	A5	A4	A3	A2
A1	0.342	0.494	0.493	0.798
A2	0.291	0.511	0.532	
A3	0.22	0.36		
A4	0.24			

In summary, the statistics show that annotators with medical education background reached a high agreement and recognized more clinical objects than other annotator groups. Moreover, medical annotators can interpret the polarity and its influences on the patient status in a more professional way. For annotators A4 and A5, the other annotations were difficult given the missing medical background.

6   Discussions

6.1   Annotation Errors

Most of the errors that occurred during the annotations are missed classes. This type of error occurred at all annotators irrespective of knowledge backgrounds. We hypothesize that it is mainly due to lack of concentration or understanding of the annotation guidelines.

The second error group pertains to the determination of boundaries of entities Since our goal is to use this corpus for testing sentiment analysis methods, assigning the correct boundary for each class is vitally important, whereby only the target events should be labeled. With respect to the latter, we recognized inconsistencies, e.g. some patient status events such as (pt, abd) appeared several times in one document, but the annotators have only labeled the first occurrence. With respect to the former, some annotators simply marked all the tokens around the target span, often including stop words, conjunctions or other terms without meaning.

Moreover, acronyms with different meanings have proven to be problematic for our annotators. For example, the acronym *BS* can either refer to *breath sounds* or to *blood sugar* in different contexts. Our annotators have only recognized one of these meanings, which caused mistakes in polarity determination. After two rounds of pre-annotation and reversioning, these kinds of disagreements have been reduced.

6.2   Feedback from Annotators

We collected feedback from the annotators during the annotation process for future consideration. Disagreement among annotators was mainly caused by different event recognitions, or idiosyncratic interpretation of the annotation guidelines. For medical annotators, if they have identified the clinical objects, they were typically able to identify the correct polarity and relations between the patient status and context objects, whereas this has proven to be the most difficult part for the annotators with other backgrounds. A non-medical annotator could easily identify the span for context objects, but could not determine the polarity towards the patient status. The low recognition rates for context objects of non-medical annotators presented in Table 2 are mainly due to this.

Another special situation observed by annotators are complex influence patterns that cannot be described by the current annotation schema. For the current study, it was assumed that the context object is going to influence the polarity of patient status, whereby the annotation scope is limited to the sentence level. However, the influences accumulated from multiple objects may each have their impact on the patient status, especially for patients with multiple diseases. Sentiment may therefore need to be determined by the evidences from several annotation levels.

Further difficulties are caused by using unstandardized values or units for medical measurements and unclear precondition of the value pairs in text. For example, *ABG 7.38/49/71/30* is the value for arterial blood gas, but the writing convention of the numeric value is unclear to the annotators. Since the ranges of the value are quite similar, it is impossible to differentiate it without additional hints. Besides, some values are noted with different units such as *mmol/L* or *mg/dl*. These kinds of ambiguities have made the judgment of polarity difficult. A more concrete interpretation should be provided to reach a better understanding about the medical concepts.

Sometimes, the annotation of trends has also rendered a high disagreement between annotators. For example, the phrase *MAP 55 -70, MAP increased* should be labeled as positive outcome for the patient status, if the status is judged according to the trend of value. The value of MAP 55 on its own needs to be considered negative. However, some of the annotators have annotated these values separately, while other just labeled the negative boundary of these both values. Again, this resulted in inconsistent annotations.

6.3 Towards Larger Sentiment Corpora

The annotated text does not contain complete sentences, but phrases with abbreviations (e.g. "Pt" refers to "Patient"). They are rather enumerations of patient status events, their polarity and context objects. These characteristics hamper the automatic processing or annotation as it could be realized using machine learning. The effectiveness of machine learning based methods relies on the size of the training corpus. The manual method can however only reach a limited corpus size. There are mainly two types of methods that can be put to use for corpus extension: data-driven and rule-based [16]. Rule-based approaches are mainly based on a lexicon, ontology or a dictionary, while the data-driven methods are exploiting unsupervised machine learning methods applied to an unannotated text corpus. The text snippets with similiar syntactic and semantic structure will be extracted and grouped together. A hybrid method combining both, rule-based and data-driven methods, could even achieve more reliable results. More specifically, after extending our corpus of 300 documents with manually generated annotations, it could work as a bootstrapping data set for unsupervised learning. In this way, a larger text corpus can be developed. For instance, the corpus can be enlarged gradually through the distributional similarity comparison between manually annotated corpus and unannotated text with the same cohort distribution.

7   Conclusions

In this paper, we introduced an annotation scheme for clinical sentiment and reported an annotation study that exploited the scheme for annotating clinical documents. We argued that existing resources have been developed for generic texts and are not suitable for representing the influence of special clinical context objects on our targeted event: patient status sentiment.

To fill this gap, we generated an annotated corpus of clinical documents that enabled analysis of relations in terms of influence patterns between patient status and context objects. Several terminology lists and acronym lists with corresponding polarities were established, constituting the first step towards a sentiment lexicon for the clinical domain.

Furthermore, the annotation and validation of 300 nurse letters have been prepared[3]. Our annotation scheme and annotated corpus have unveiled important characteristics of clinical sentiment.

Annotation mistakes and inconsistencies have been identified based on annotator feedback and analysing the major disagreements between annotators leading us to a revised annotation scheme and guidelines. As next step, the corpus will be used to generate a clinical sentiment lexicon semi-automatically and it can also be employed to develop and to evaluate methods for sentiment analysis considering the clinical context objects.

References

1. Sridhar, D., Getoor, L., Walker, M.: Collective stance classification of posts in online debate forums. In: ACL 2014, p. 109 (2014)
2. Somasundaran, S., Wiebe, J.: Recognizing stances in online debates. In: Proceedings of the Joint Conference of the 47th Annual Meeting of the ACL, the 4th International Joint Conference on Natural Language Processing of the AFNLP: Volume 1 - Volume 1, ACL 2009, Stroudsburg, PA, USA, pp. 226–234. Association for Computational Linguistics (2011)
3. Denecke, K., Deng, Y.: Sentiment analysis in medical settings. Artif. Intell. Med. **64**(1), 17–27 (2015)
4. Deng, Y., Stoehr, M., Denecke, K.: Retrieving attitudes: sentiment analysis from clinical narratives. In: Proceedings of the Medical Information Retrieval Workshop at SIGIR Co-located with the 37th Annual International ACM SIGIR Conference (ACM SIGIR 2014), Gold Coast, Australia, 11 July 2014, pp. 12–15 (2014)
5. Baccianella, S., Esuli, A., Sebastiani, F.: Sentiwordnet 3.0: an enhanced lexical resource for sentiment analysis and opinion mining. In: LREC (2010)
6. Strapparava, C., Valitutti, A.: Wordnet affect: an affective extension of wordnet. In: Proceedings of the Fourth International Conference on Language Resources and Evaluation, LREC 2004, Lisbon, Portugal, 26–28 May 2004 (2004)
7. Goeuriot, L., Na, J.-C., Kyaing, W.Y.M., Khoo, C., Chang, Y.-K., Theng, Y.-L., Kim, J.-J.: Sentiment lexicons for health-related opinion mining. In: Proceedings of the 2nd ACM SIGHIT International Health Informatics Symposium, IHI 2012, pp. 219–226, New York, NY, USA. ACM (2012)
8. Wilson, T., Wiebe, J., Hoffmann, P.: Recognizing contextual polarity in phrase-level sentiment analysis. In: Proceedings of the Conference on Human Language Technology and Empirical Methods in Natural Language Processing, HLT 2005, pp. 347–354, Stroudsburg, PA, USA. Association for Computational Linguistics (2005)

---

[3] The corpus is available on request.

9. Deng, L., Wiebe, J.: MPQA 3.0: an entity/event-level sentiment corpus. In: NAACL HLT 2015, The 2015 Conference of the North American Chapter of the Association for Computational Linguistics: Human Language Technologies, Denver, Colorado, USA, 31 May–5 June 2015, pp. 1323–1328 (2015)

10. Emerson, G., Declerck, T.: SentiMerge: combining sentiment lexicons in a Bayesian framework. In: Proceedings of the Workshop on Lexical and Grammatical Resources for Language Processing. The COLING 2014 Organizing Committe, August 2014

11. Sidarenka, U.: PotTS: the potsdam twitter sentiment corpus. In: Calzolari, N., et al. (eds.) Proceedings of the Ninth International Conference on Language Resources and Evaluation (LREC 2016), Portoro++, Slovenia, European Language Resources Association (ELRA), May 2016

12. Snger, M., Leser, U., Kemmerer, S., Adolphs, P., Klinger, R.: SCARE - the sentiment corpus of app. reviews with fine-grained annotations in German. In: Proceedings of the Tenth International Conference on Language Resources and Evaluation (LREC 2016), Portoro v z, Slovenia, May 2016. European Language Resources Association (ELRA) (2016)

13. Klenner, M., Amsler, M.: Sentiframes: a resource for verb-centered German sentiment inference. In: Calzolari, N., (Conference Chair), Choukri, K., Declerck, T., Grobelnik, M., Maegaard, B., Mariani, J., Moreno, A., Odijk, J., Piperidis, S. (eds.) Proceedings of the Tenth International Conference on Language Resources and Evaluation (LREC 2016), Paris, France. European Language Resources Association (ELRA), May 2016

14. Pang, B., Lee, L., Vaithyanathan, S.: Thumbs up? Sentiment classification using machine learning techniques. CoRR, cs.CL/0205070 (2002)

15. McDonald, R.T., Hannan, K., Neylon, T., Wells, M., Reynar, J.C.: Structured models for fine-to-coarse sentiment analysis. In: ACL 2007, Proceedings of the 45th Annual Meeting of the Association for Computational Linguistics, Prague, Czech Republic, 23–30 June 2007 (2007)

16. Manishina, E., Jabaian, B., Huet, S., Lefevre, F.: Automatic corpus extension for data-driven natural language generation. In: Calzolari, N. (Conference Chair), Choukri, K., Declerck, T., Goggi, S., Grobelnik, M., Maegaard, B., Mariani, J., Mazo, H., Moreno, A., Odijk, J., Piperidis, S. (eds.) Proceedings of the Tenth International Conference on Language Resources and Evaluation (LREC 2016), Paris, France. European Language Resources Association (ELRA), May 2016

SALAD – Services and Applications over Linked APIs and Data

Modeling and Composing Uncertain Web Resources

Pierre De Vettor$^{(\boxtimes)}$, Michaël Mrissa, and Djamal Benslimane

Université de Lyon, CNRS LIRIS, UMR5205, 69622 Lyon, France
{pierre.de-vettor,michael.mrissa,djamal.benslimane}@liris.cnrs.fr

Abstract. Nowadays, huge quantities of data are produced and published on the Web, coming from individuals, connected objects, and organizations. Uncertainty happens when combining data from different sources that contain heterogeneous, contradictory, or incomplete information. Today, there is still a lack of solutions in order to represent uncertainty that appears on the Web. In this paper, we introduce the concept of uncertain RESTful resource and propose a model and an algebra to interpret such resources.

Keywords: Data uncertainty · RESTful resource · Hypertext composition · Data combination

1 Introduction

Nowadays, individuals, organizations, and connected objects produce and publish a huge amount of data on the Web [12], through APIs and public endpoints [15], which is then combined into mashups [4] to produce high valuable new data. In this context, data uncertainty may occur as data comes from heterogeneous, contradictory, or incomplete sources [11]. In this case, there is a chance that each data source provides different information, which may be correct under some circumstances, and incorrect under others. Instead of choosing a unique version, yet arbitrary, of information, we believe users should be given the whole spectrum of possibilities to describe an entity.

The main objective of this paper is to propose a theoretical framework for describing, manipulating, and exposing uncertain data on the Web. We present a model to define and interpret **uncertain Web resources**. We define an interpretation model and an algebra to compute uncertainty in the context of classical hypertext navigation and in the context of data query evaluation. The paper is structured as follows: Sect. 2 describes our uncertainty model and interpretation. Section 3 explains how we interpret query evaluation in this uncertainty-aware context. Section 4 presents our implementation details and evaluation. Section 5 presents other approaches that handle uncertainty. Finally, Sect. 6 concludes and presents some future work.

© Springer International Publishing AG 2016
H. Sack et al. (Eds.): ESWC 2016 Satellite Events, LNCS 9989, pp. 327–341, 2016.
DOI: 10.1007/978-3-319-47602-5_47

2   Uncertain Web Resources

In order to understand the notion of uncertain resource, it is first required to remind some background knowledge on underlying definitions. Uncertain resources are Web resources, and according to the principles of REST [10], a Web resource is an entity or object, identified by an URI. A resource can provide a single object, a set of sub-resources or an abstract notion, e.g. a concept from an ontology. In order to clarify Web resources, we propose the following notation, and define a resource R as follows:

$$R =< uri_r, rep_r >$$

where uri_r is the URI that identifies this and only this resource and rep_r is the representation of the resource R in the server. A resource can be accessed with the HTTP methods ($RFC\ 2616$[1]), allowing to read (GET), update or create ($POST$), delete ($DELETE$) resources on a domain. In this section, we overcome the definition of Web resources in order to handle data uncertainty.

2.1   Definition

The semantics of uncertain Web resources can be explained based on the theory of possible worlds [17]. In our view, an uncertain resource has several possible representations which can potentially and individually be interpreted as true. These possibilities can be interpreted as a set of possible worlds ($PW_1,..., PW_n$) with a probability $prob(PW_i)$. We call them **possible Webs**, and inside these possible Webs, data is considered as certain.

In order to define the notion of uncertain resource, we rely on several assumptions: **(1)** Due to the REST principles, several representation of one URI (i.e. one resource) cannot coexist, so the possible representations of a resource must be mutually exclusive. **(2)** In order to deal with uncertain resources, rather than with uncertain data, assume that in the statement inside a given possible representation, every piece of data is considered certain. By extension, if a resource property has several possible values, it should appear into separate representations. **(3)** For the sake of simplicity, in this approach we consider that each possible representation of a given resource is represented according to the same model.

Based on the definition of Web resources [10], a Web resource is an entity or object, identified by an URI, accessible via HTTP methods. We define an uncertain Web resource $\widetilde{R}$ as follows:

$$\widetilde{R} =< uri_r, \{< rep_i, P_i > | i \in [1, n], \sum_{i=1}^{n} (P_i) \leq 1\} >$$

where rep_i are the possible representations of $\widetilde{R}$. Since multiple representations of a resource cannot coexist at the same URI, these representations are mutually exclusive, and we have $P_i \in]0; 1]$. Probabilities are not part of representations, they are meta-data provided by the server.

---

[1] http://www.w3.org/Protocols/rfc2616/rfc2616-sec9.html.

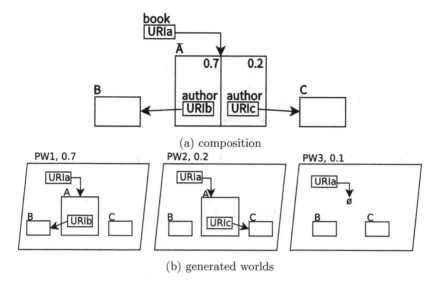

(a) composition

(b) generated worlds

Fig. 1. Uncertain resource example 1

As an example, Fig. 1a shows that the two possible representations of our book resource generate three Webs in which representations are certain. We rely on the popular uncertain database model *Block-Independent Disjoint* (BID) [6] to define the following: *every resource is independent, and each URI identifies a unique resource, whose representation are disjoint, i.e., only one representation is true at a time.* Our model specifies that (1) possible resource representations are disjoint and (2) resource interpretations are independent from each other. Figure 1a shows how we interpret uncertain resources as a set of probable representations with a probability (number in upper right), generating possible Webs in which this representation is true and unique.

2.2   Precision on Unknown Representations

Having $\sum_{i=1}^{n} (P_i) \leq 1$ indicates that a part of the resource is considered as unknown. In this paper, unknown resource representations are noted $\emptyset$. The resource can specify that other representations may exist but their actual content is left unknown. This kind of behavior can happen according to different factors, as an example, the provider could have left a part private or protected, for privacy reasons. Another possibility is that the resource no longer exist in the server, but the URI still points at it. Finally, a provider could have planned to create the resource, but had not already made it. This way, the URI points to something that does not yet exist.

In our example, shown in Fig. 1b, the possible Webs $PW1$, $PW2$, and $PW3$ are generated from the probable possible representations of the uncertain resource A. In possible Web $PW1$, resource A has one representation which contains a link to B; resource C exists but is not connected to A. In possible Web $PW3$, the uncertain unknown resource $\tilde{A}$ has no existing representation. Technically, a GET request over such a resource leads to an HTTP error, such as a 404 not found or a 500 internal server error.

2.3 Programmatic Representation of Uncertain Resources

In order to provide a way to handle uncertain resources, we proposed a formalism to physically represent them. Our mecanisms for programmatically representing an uncertain resource include all the possible representations and their associated probabilities.

Listings 1.1 and 1.2 shows the JSON representations of some uncertain resources.

```
[ {p:0.6,  r:{ title :"Les miserables",
              date:"3 Avr 1862",
              author:"http://dbpedia.org/VHugo"}},
  {p:0.4,  r:{ title :"Les miserables",
              date:"30 Mar 1862",
              author:"http://dbpedia.org/VictorHugo"}} ]
```

Listing 1.1. JSON representation of an uncertain book resource

```
[ {p:0.7,  r:{ name:"Hugo V.",
              birth: 1802,
              city :"http://city/besancon"}},
  {p:0.2,  r:{ name:"Victor HUGO",
              birth: 1802,
              author:"http://city/paris"}} ]
```

Listing 1.2. JSON representation of an uncertain author resource

2.4 HTTP Request over Uncertain Resources

In this subsection, we introduce the notion of uncertainty-aware client, which is a client who is able to manipulate uncertain resources. In order to respect the Web principles, and to adapt to every client, we rely on *content negotiation*. Content negotiation is an HTTP mechanism that allows to serve different *versions* of the same resource representation (i.e., at the same URI), to fit with the client. In this paper, we make a difference between classical and uncertainty-aware GET requests. We propose the notation $\widetilde{GET}$ to describes a GET request from an uncertain-aware client. Let $\tilde{R}$ be an uncertain resource deployed at uri_r, we defined the following expected behaviors:

$$\widetilde{GET}(uri_r) := \{< rep_1, P_1 >, \ldots, < rep_n, P_n >\}$$

In case, where the client performs a $\widetilde{GET}$ request over a certain resource, the response will provide the representation with a probability of 1. In our approach $\widetilde{GET}$ is **not** defining a new HTTP method. $\widetilde{GET}$ acts as a standard GET with a specific HTTP header which we define in Sect. 4 as $X - Accept - Uncertain : true$. We choose to define a specific header to avoid interference with the standardized usage of the *accept* header. Indeed, the *Accept* header is the classical header for content negotiation, as it is used to specify an expected mime-type for the resource representation. The good practice is then to specify an adhoc specific header to respect the HTTP standards (see $RFC7231^2$).

2.5   A Certain Representation of an Uncertain Resource

In this approach, using content negociation provides us with a solution that allows to manipulate uncertain resources as classical certain resources. Doing so, a client who does not know how to process uncertain resources, or does not care about uncertainty representation, will still be able to receive a certain (but arbitrary) version of the resource representation. In this case, we choose to provide the most certain representation of the resource.

On top of that, it is also important to inform the client about the uncertain nature of the resource it is accessing. We provide two additionnal headers to enhance this certain representation with uncertain capabilities. On the one hand, we inform the client that the resource has an uncertain capacity, and is able to accept uncertain requests, by providing a $x - accept - uncertain$ in the response header. Additionnaly, we also provide the probability of the arbitrary selected representation, through the $x - uncertainty - value$. Listing 1.3 shows an example of certain (i.e., classical) GET response (only headers) over an uncertain resource.

```
curl −I http:// liris .cnrs. fr /~pdevetto/uncert/index.php/df/paper/89

HTTP/1.1 200 OK
Date: Fri, 12 Jun 2016 13:35:00 GMT
Server: Apache/2.4.7 (Ubuntu)
X−Accept−Uncertain: 1
X−Uncertainty−Value: 0.225
Content−Type: application/json+ld
```

Listing 1.3. Enrichment of certain GET response over uncertain resources

NB: providers could also choose another method to define the given certain representation of an uncertain resource. This is only an approach we advocate in this paper. We only provide one possibility to do it.

2.6   Composing Uncertain Web Resources

In a composition of Web resources, each combinaton of possible resource representations generates a new possible Web PW_x, whose probability is computed as follows:

---

2 https://tools.ietf.org/html/rfc7231#section-5.3.2.

$$P(PW_x) = \prod_{i \in [1,n]} \Big(prob(rep_i) \Big)$$

where $rep_i \in Card(PW_x)$, and $Card(PW_x)$ being the representations involved in PW_x. The probability of the unknown representations of a resource R_a is computed as follows: $prob(rep_a^x) = 1 - \sum_{i=1}^{n} prob(rep_a^i)$ where rep_a^i are the different representations of resource R_a. Figure 2a shows a more complex example, where resources are certain and uncertain, generating the possible Webs shown in Fig. 2b. As an example, the probability of possible Webs PW_4 is $prob(PW_4) = prob(A^2) \times prob(C^1) \times prob(H) \times prob(E) = 0.2 \times 0.5 \times 1 \times 1 = 0.1$. In the next section, we describe how to interpret and compute a query in an uncertain composition.

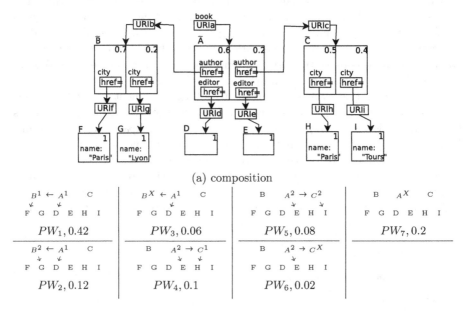

(a) composition

(b) generated Webs

Fig. 2. Uncertain resource example 2

2.7   Particular Composition Cases

Our previous examples show scenario based compositions which we could use to answer our query. The distributed state of resource-based applications, associated with our lightweight model, allow more complex compositions. Figure 3 shows some examples of complex resource orchestration where uncertainty appears, and to which our model could easily adapt. These examples include loop, redundancies and differences in models while navigating through hypertext.

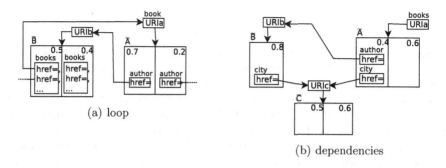

(a) loop

(b) dependencies

Fig. 3. Uncertain resource examples

In these examples, heterogeneities can appear but are handled by our algorithm presented in the next section.

Figure 3a shows a situation where it may exist a loop in the request path. Our algorithm only dereferences the resource once, protecting us from looping infinitely through hypertext path. In Fig. 3b, the resource composition presents a duplicate resource. The important specificity here is related to this duplicate resource, and is handled by our model, which specifies that a resource only have one representation in a possible Web.

2.8   Uncertain Data vs. Uncertain Resources

In this section, we introduced the concept of uncertain Web resources, presented a model and an algebra to compute the probability of uncertain resource composition. One common question about this work, is: "How do you handle resource where representation also contains uncertain data?".

It is sometimes difficult to make a difference between uncertain data and uncertain resources. Let a data set, containing a list of scientific articles, each of them having a list of authors URIs and a publication date, as presented in Table 1.

Table 1. Example of uncertain data set

Paper	Authors		Date	
#1	AuthorA & authorB	0.6	*April2016	1
	AuthorA & authorC	0.4		
#2	AuthorD & authorE	0.7	March 2015	0.6
	AuthorD &authorE & authorF	0.2	April 2015	0.4
	AuthorC & authorE	0.1		
#3	AuthorF & authorG	1	September 2015	0.7
			October 2015	0.2

In this example, each tabular cell can have several values each of them applied with a probability. Representing this uncertainty in resource description, would break the semantic of atomic resources, since our possible representations need to be able to be manipulated as certain representations in the possible Web they exist in.

In order to represent this uncertainty, what we propose, to fit with our model, is to generate each possible representation for each resource, and include all these representations in the resource. This way, we respect our model, and only manipulate finite resource representations. We then obtain a result set of resource representations, where we could compute probabilities by applying a product, of the data pieces involved. Our uncertain data set, generates a set of possible representations for each article as shown in Table 2.

From this example, we show that our results stays relevant and safe. As an example, paper #2 data have generated 6 possible representation whose probability sum is $0.42 + 0.28 + 0.12 + 0.08 + 0.06 + 0.04 = 1$. Each generated representation is a possibility of truth. They are all mutually exclusive.

Table 2. Generate possible representation from uncertain data set

Paper	Authors	Date	Computed
#1	AuthorA & authorB	April 2016	0.6 * 1 = 0.6
	AuthorA & authorC	April 2016	0.4 * 1 = 0.4
#2	authorD & authorE	March 2015	0.7 * 0.6 = 0.42
	AuthorD & authorE	April 2015	0.7 * 0.4 = 0.28
	AuthorD & authorE & authorF	March 2015	0.2 * 0.6 = 0.12
	AuthorD & authorE & authorF	April 2015	0.2 * 0.4 = 0.08
	AuthorC & authorE	March 2015	0.1 * 0.6 = 0.06
	AuthorC & authorE	April 2015	0.1 * 0.4 = 0.04
#3	AuthorF & authorG	September 2015	1 * 0.7 = 0.7
	AuthorF & authorG	October 2015	1 * 0.2 = 0.2

NB: This transformation method to generate uncertain resource from uncertain data set only work when data fields are independent, and data values are mutually exclusive, as our uncertain Web resource model specifies it. In case where there is a dependency between the different data parts, or where values can coexist, a different algebra applies. This is not part of the scope of this paper, and can be solved with help from concepts from related existing approaches such as [8,16].

Complexity of Our Method. Although our method generates a lot of possible representations, it does not specifically increase the complexity of the following approach to interpret hypermedia queries. The important notion we present in the following, to prevent our algorithms for exponential growth in term of execution time, is the reduction step. At each stage of the process, we aggregate duplicates URIs and addition the probabilities. Doing so, it is easier to create several possible representations, since duplicate values will be grouped after probability computation.

3 Query as Resource Paths: Definition and Assessment

In this section, we present our approach to aggregate data from uncertain resources thanks to hypertext navigation. Formally, we define a data query as an ordered set of resource requests, following the same path through the different generated possible Webs. Each Web will provide a unique result, which are then aggregated. Generating each of these possible Webs, i.e., combining and storing each combination in memory to compute the query in each one, is a time and memory-consuming task. There is a need for an approach that allows to aggregate these results directly without having to generate the possible Webs.

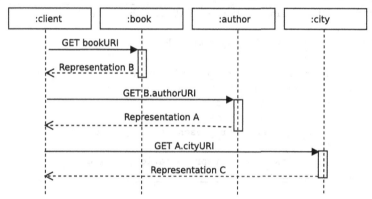

Fig. 4. Query answering in RESTful compositions

Following our example scenario, the query *What is the date and city of birth of the writer of the book "Les Miserables"?* The execution of this path in a classical RESTful composition is detailed in Fig. 4.

In order to follow this path, we must assume that resource representations specify the necessary semantics about their contents. As an example, when searching for the author of a book resource, the **author** functional property is required in order to complete. In our scenario, the semantically enhanced representations we manipulate are represented in the JSON-LD format [14]. When dealing with uncertain resources, we follow our query path through the possible resource representations. This navigation creates a possibility tree pattern, where branches are possible Webs associated with their probability. Figure 5 shows the tree pattern created from our book scenario.

We propose an algorithm, cf. *Algorithm 1*, to compute resulting probabilities without possible Web generation. This algorithm implements an operator, which we call GET_p, who follows a stage-by-stage routing inside the possibility tree.

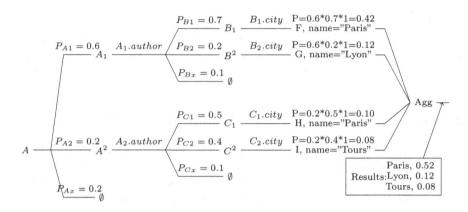

Fig. 5. Generating tree pattern while navigating resources

Algorithm 1. GETp Algorithm

1: **procedure** GETP(input_uris : list of (URI,proba) couple)
2:     results ← List()
3:     **for all** (URI_i,prob_i) ∈ input_uris **do**
4:         $\widetilde{R}$ ← $\widetilde{GET}$(URI_i)
5:         **for all** (representation, prob_r) ∈ $\widetilde{R}$ **do**
6:             //Compute current probability
7:             prob_c ← prob_i * prob_r
8:             **if** representation ∉ results **then**
9:                 results.add(< representation, prob_c >)
10:            **else**
11:                results.update(representation, prob_c)
         return results

GET_p takes as input a list of URIs from an nth stage of the tree, and returns the possible resource representations from the $(n+1)^{th}$ stage. The GET_p operator executes the necessary sequence of HTTP requests over the given URIs, applies the probability formula and returns the set of representation-probability couples.

As an example, we have a list of author URIs, extracted from possible book representations, each with a probability. GET_p gives us the possibility to retrieve the representation of each authors (with their probabilities) and to apply book probabilities to them. This will produce a set of author representations with *global* probabilities. The mutually exclusive status of representations guarantees a safe composition, which means resulting probabilities are coherent and their sum does not exceed 1. Finally, our computation algorithm, see *Algorithm 2*, uses GET_p to iteratively process through the different stages of the probability tree. According to a query, and the URI of the first resource, our algorithm processes its way through the resource path, using object properties to find its way. In the end, the resulting data set contains all the values with their probabilities.

Algorithm 2. Computation Algorithm

```
 1: procedure COMPUTE( query, URI_0 )
 2:     transform query in lists of properties // the path
 3:     // Make the first call / first URI is certain
 4:     result ← PROCESS_PATH( properties, < URI_0, 1 > )
 5: procedure PROCESS_PATH( properties, input_uris )
 6:     // Retrieve the next set of resources descending the path
 7:     rep ← GETp( input_uris )
 8:     // Stop condition, no more properties = end of the path
 9:     if properties[0] = ∅ then
10:         return rep
11:     else
12:         next_uris ← []
13:         // For each couple (representation, probability)
14:         for all (rep_r, prob_r) ∈ rep do
15:             if rep_r.getprop(properties[0]).type == URI then
16:                 // Get the property and add it to the new list
17:                 next_uris[] ← [rep_r.getprop(properties[0]), prob_r]
18:         properties.remove(0)
19:         return PROCESS_PATH( properties, next_uris )
```

4 Implementation and Evaluation

As introduced before, the principles of our approach rely on REST principles, so we are able to use any HTTP client to access our uncertain resources. What we propose here, is an implementation of the GET_p algorithm as well as an implementation of our evaluation algorithm, which will perform the necessary HTTP requests. As introduced before, we rely on content negotiation to specify that an HTTP client is able to understand uncertainty. Listing 1.4 shows an example of HTTP request, using content negociation to retrieve an uncertain resource.

```
curl --header "X-Accept-Uncertain: true""http://uri/resource"
```

Listing 1.4. Enrichment of certain GET response over uncertain resources

In order to keep our approach reusable, and to allow integration with other RESTful approaches, we implemented the GET_p and $COMPUTE$ algorithms as RESTful services. Service calls are made through POST, and GET retrieves a user-friendly description of the service. Listing 1.5 shows an example of HTTP request to call the GET_p service.

```
curl −X POST "{ domain }/op/getp"
    --data "uri0={ uri0 }&prob0={ prob0 }
            &uri1={ uri1 }&prob1={ prob1 }"
```

Listing 1.5. Enrichment of certain GET response over uncertain resources

We propose a Web interface to execute simple SPARQL queries. Our prototype, resources and scenarios are publicly available for testing at the following URL: http://liris.cnrs.fr/~pdevetto/uncert/index.php.

In order to evaluate our approach, we focus on processing time of our algorithms. For this purpose, we hosted RESTful services serving uncertain Web resources in JSON-LD [14] over linked data dumps from the SWDF corpus (http://data.semanticweb.org), representing ESWC2015, ISWC2013, and WWW2012 conference semantic data (author, proceedings, etc.). We created three different scenarios (use case workflows) involving a different amount of resources and with different graph complexities:

- Starting from an *inproceeding article*, the first workflow retrieves all the *articles* that share the same *keywords*, shown in Listing 1.6.
- The second workflow retrieves all the *articles* written by at least one same *author*.
- Finally, the third workflow retrieves the *authors* that have written at least one *article* with one similar *keyword*.

We executed all the workflows with 30 different inproceedings articles as input data. Comparing uncertain workflow executions with the same workflow in a certain context has no meaning, because the number of HTTP request will grow exponentially.

```
PREFIX dc: <http://purl.org/dc/elements/1.1/>
PREFIX al: <http://liris.cnrs.fr/~pdevetto/uncert/>
PREFIX swrc: <http://swrc.ontoware.org/ontology#>
SELECT ?similararticle WHERE {
    ? article a swrc:InProceedings ;
         dc:subject ?subject .
    ?subject dc: inarticle ? similararticle  .
}
```

Listing 1.6. Query 1: Articles that share the same subject than another article (by ID)

Our *compute* algorithm implementation will transform this query in a list of concepts to extract from resource to resource, creating our path descending through the possibility tree. Our implementation uses the ARC2 SPARQL Parser to extract query concepts. In our evaluation, we evaluate the ratio of network latency in the total execution cost of a workflow. We show that the processing cost of our solution is negligible compared to the network cost. The obtained results show the following: while workflows become more complex, the number of HTTP request grows, and the processing time is more and more negligible, compared to HTTP latency. Under a global execution time of 2 s, processing time is less than 5 %. After 3 s, it never exceeds 1 %. On top of that, as long as input resources defines coherent representations (and correct probabilities), no matter the query, it always generates a safe result set, with relevant values and probabilities.

5   Related Work

Uncertainties have been processed in different contexts, we envision different approaches, handling uncertainty, historically in database world, and more recently in services oriented application. Unfortunately, none of these approach handles the uncertainty that can appear when manipulating resources or when dealing with Restful applications.

5.1   Uncertainty in Databases

Fagin et al. [9] envision data exchanges in presence of uncertain data coming from probabilistic databases. Their approach is a generalization of Dong et al. by-table semantics [8] in which probabilistic matching are generated between tables in order to align fields. The generated results are associated with a probability value. The *Fagin et al.* probabilistic approach relies on creating an arbitrary binary relationship between two countable (finite or countably infinite) probability spaces. *Agrawal et al.* [1] propose a local-as-view data integration approach dealing with sources containing uncertain data. Their approach rely on the concept of containment, which means creating a mediated uncertain schema that must contain both databases. *Cheng et al.* [5] propose an approach assuming that concept values from both schemas can overlap to deal with uncertain matchings.

These works are strong although complex approaches to handle with uncertain/probabilistic data, these approaches has inspired our definition of uncertain Web resource. However, if it applies very well to database, these approach does not fit well when working with Web resources. One solution could be to layer data sources with a database endpoint, but it could not provide a sufficient solution for considering our composition semantics. These approach has lead our definition of uncertain Web resources.

5.2   Uncertainties on the Web

Several approach have been proposed to deal with uncertainty in other contexts than databases, most of the time in order to propose heterogeneous data integration approach.

Lamine Ba et al. [3] propose an approach for data integration, combining data from web sources containing uncertainty and dependencies. Their approach confront and merge diverse information about a same subject from diverse web sources. They model the following data as probabilistic XML [13] to process decisions. *Sarma et. al.* [7] envision what they call pay-as-you-go integration systems, which is related to our smart data architecture. Their system rely on a single point interface to a set of data sources, integration of data being made by creating a mediated schema for the domain. *Pivert and Prade*[16] propose a solution to integrate multiple heterogeneous and autonomous information sources, resolving factual inconsistencies by analyzing the existence of suspects answers in both data sets. Their approach verify the data provided by two data source they want to integrate, if a data piece in second source invalidates a data piece in the first data source, it is considered as a suspect answer. Their approach finally return all the candidate answers to a query, rank-ordered according to their level of reliability.

Finally, *Amdouni et al.* [2] propose an approach to handle the uncertainty of the data returned by data services, which they call uncertain data service. They define uncertainty at three levels, in the context of DaaS services, modelisation, invocation and composition. First of all, they extend the Web services standards to model uncertainty of a service in its own description. This model introduces the notion of uncertain data service, whose can be explained by possible worlds theory [17]. These uncertain data services are defined by their inputs and sets of their probable outputs. It is this set of possible outputs returned by an invocation which can be considered as possible worlds, each of these world being Dependant and having a probability value. They defined two different kinds of invocation of uncertain data services, conventional with certain input and probabilistic where inputs are presented containing uncertain data instances.

These works propose several methods and models to process uncertainty in the context of the Web (XML, services, or semantics), but none of them address the uncertainty that can appear while referencing or browsing information through the Web. This is a very common problem, which is usually skipped or decided arbitrarily by providers. Our approach proposes a relevant and adaptable approach to enhance Web-based applications with uncertainty awareness.

6   Conclusion

In this paper, we address the need for a solution to handle data uncertainty while referencing and navigating resources on the Web. We propose a model for uncertain Web resources, as resources which may have several mutually exclusive representations with probabilities. On top of that, we propose an algebra to interpret and evaluate data query in uncertain resource compositions.

Future work includes opening our approach in order to deal with more complex scenarios, where possible representations could be actual Web resources with URIs. This way, we could construct a model based on hypertext navigation to define a resource according to a set of others, giving a possibility to represent the probable equivalence of resources.

References

1. Agrawal, P., Sarma, A.D., Ullman, J., Widom, J.: Foundations of uncertain-data integration. Proc. VLDB Endow. **3**(1–2), 1080–1090 (2010)
2. Amdouni, S., Barhamgi, M., Benslimane, D., Faiz, R.: Handling uncertainty in data services composition. In: 2014 IEEE International Conference on Services Computing (SCC), pp. 653–660, June 2014
3. Ba, M., Montenez, S., Tang, R., Abdessalem, T.: Integration of web sources under uncertainty and dependencies using probabilistic XML. In: Han, W.-S., Lee, M.L., Muliantara, A., Sanjaya, N.A., Thalheim, B., Zhou, S. (eds.) Database Systems for Advanced Applications. LNCS, vol. 8505, pp. 360–375. Springer, Heidelberg (2014)

4. Benslimane, D., Dustdar, S., Sheth, A.P.: Services mashups: the new generation of web applications. IEEE Internet Comput. **12**(5), 13–15 (2008)
5. Cheng, R., Gong, J., Cheung, D., Cheng, J.: Evaluating probabilistic queries over uncertain matching. In: 2012 IEEE 28th International Conference on Data Engineering (ICDE), pp. 1096–1107, April 2012
6. Dalvi, N., Re, C., Suciu, D.: Queries and materialized views on probabilistic databases. J. Comput. Syst. Sci. **77**(3), 473–490 (2011)
7. Das Sarma, A., Dong, X., Halevy, A.: Bootstrapping pay-as-you-go data integration systems. In: Proceedings of the 2008 ACM SIGMOD International Conference on Management of Data, SIGMOD 2008, pp. 861–874, New York, NY, USA. ACM (2008)
8. Dong, X., Halevy, A.Y., Yu, C.: Data integration with uncertainty. In: Proceedings of the 33rd International Conference on Very Large Data Bases, pp. 687–698. VLDB Endowment (2007)
9. Fagin, R., Kimelfeld, B., Kolaitis, P.G.: Probabilistic data exchange. J. ACM (JACM) **58**(4), 15 (2011)
10. Fielding, R.T., Taylor, R.N.: Principled design of the modern web architecture. In: Proceedings of the 22nd International Conference on Software Engineering, ICSE 2000, pp. 407–416, New York, NY, USA. ACM (2000)
11. Halevy, A., Rajaraman, A., Ordille, J., Data integration: the teenage years. In: Proceedings of the 32Nd International Conference on Very Large Data Bases, VLDB 2006, pp. 9–16. VLDB Endowment (2006)
12. Heath, T., Bizer, C., Data, L.: Evolving the Web into a Global Data Space. Synthesis Lectures on the Semantic Web. Morgan & Claypool Publishers, San Rafael (2011)
13. Kimelfeld, B., Senellart, P.: Probabilistic XML: models and complexity. In: Ma, Z., Yan, L. (eds.) Advances in Probabilistic Databases. STUDFUZZ, vol. 304, pp. 39–66. Springer, Heidelberg (2013)
14. Lanthaler, M., Gutl, C.: On using JSON-LD to create evolvable restful services. In: Proceedings of the Third International Workshop on RESTful Design, WS-REST 2012, pp. 25–32, New York, NY, USA. ACM (2012)
15. Maleshkova, M., Pedrinaci, C., Domingue, J.: Investigating web APIs on the world wide web. In: 2010 IEEE 8th European Conference on Web Services (ECOWS), pp. 107–114, December 2010
16. Pivert, O., Prade, H.: Querying uncertain multiple sources. In: Straccia, U., Calì, A. (eds.) SUM 2014. LNCS, vol. 8720, pp. 286–291. Springer, Heidelberg (2014)
17. Sadri, F.: Modeling uncertainty in databases. In: Proceedings of the Seventh International Conference on Data Engineering, pp. 122–131, April 1991

grlc Makes GitHub Taste Like Linked Data APIs

Albert Meroño-Peñuela[1,2(✉)] and Rinke Hoekstra[1,3]

[1] Department of Computer Science,
Vrije Universiteit Amsterdam, Amsterdam, The Netherlands
{albert.merono,rinke.hoekstra}@vu.nl
[2] Data Archiving and Networked Services, KNAW, Amsterdam, The Netherlands
[3] Faculty of Law, University of Amsterdam, Amsterdam, The Netherlands

Abstract. Building Web APIs on top of SPARQL endpoints is becoming common practice. It enables universal access to the integration favorable data space of Linked Data. In the majority of use cases, users cannot be expected to learn SPARQL to query this data space. Web APIs are the most common way to enable programmatic access to data on the Web. However, the implementation of Web APIs around Linked Data is often a tedious and repetitive process. Recent work speeds up this Linked Data API construction by wrapping it around SPARQL queries, which carry out the API functionality under the hood. Inspired by this development, in this paper we present `grlc`, a lightweight server that takes SPARQL queries curated in GitHub repositories, and translates them to Linked Data APIs on the fly.

Keywords: SPARQL · Git · GitHub · Linked Data APIs · RESTFul

1 Introduction

Despite their known benefits for data integration, the Linked Data technologies of RDF and SPARQL still operate in a niche. There is a gap with what average Web-applications and developers have come to expect. RDF and SPARQL remain relatively unknown to the wider Web community. But they are still a requirement for accessing Linked Data. Both have steep learning curves that many developers refuse to face. The W3C specification of SPARQL 1.1 has a limited adoption even within the Linked Data community [16]. Triggered by requirements of Linked Open Data publishers, such as the UK government,[1] the current best practice solution to this problem is the deployment of a custom Web API on top of a Linked Data source. For instance, as an interface to a SPARQL endpoint. This use of APIs to apply the basic principle of encapsulation, and their deployment in large scale Linked Data applications [4], has proved to solve another problem: the expressiveness of SPARQL allows for highly inefficient or

---

Pronounced as in "garlic", written lowercase with no vowels.

[1] https://github.com/UKGovLD/linked-data-api.

© Springer International Publishing AG 2016
H. Sack et al. (Eds.): ESWC 2016 Satellite Events, LNCS 9989, pp. 342–353, 2016.
DOI: 10.1007/978-3-319-47602-5_48

just computationally expensive queries. Abstracting over curated, 'well behaved' queries allows for more efficient query answering and caching. It is an important step in safeguarding query response times.

However, the deployment and configuration of a Linked Data API is still a cumbersome task. Effective Linked Data APIs require careful management of SPARQL queries, reliable storage, abundant documentation, and the overhead of software maintenance. The latter has been recently addressed by Daga et al. [1], who propose a system that builds API operations automatically by taking a SPARQL query, a short description, and an endpoint location as input. However, the question of how to effectively store and organise such API-translated SPARQL queries remains. As shown in this paper, users require organised APIs that adapt to their existing query curation workflows. Moreover, such APIs might need to coexist in systems where SPARQL queries are already being used by other Linked Data applications. This requires a paradigm shift where not just the data, but the queries themselves also become *first class citizens*.

In several Linked Data projects such as CEDAR [11] and CLARIAH-SDH [6] we followed a practice of storing, curating, and publishing illustrative SPARQL queries of their use cases using GitHub repositories. These queries are then used by various client applications to access Linked Data. An analysis of GitHub shows that this is quite common: it currently hosts at least 5000 SPARQL queries, and potentially many more. A search on 'sparql extension:rq' produces 4987 results; 'query extension:rq' gives 4386 results. The seach for 'rdf language:SPARQL' returns over 46k results, and searching for 'SELECT' queries gives over 280k files, but these include a large number of syntax templates used by Audacity, Virtuoso and ClioPatria. From an e-Science perspective, sharing research questions as concrete queries on GitHub has a huge potential for the reproducibility of research outcomes.

In this paper, we investigate how the current practice of curating queries in open GitHub repositories can be decoupled from the custom built applications that use them to interact with Linked Data. This way we can to lower the costs of (1) constructing APIs for Linked Data and (2) developing applications that interact with Linked Data. Concretely, the contribution of this paper is:

- A mapping between a Swagger RESTful API specification and SPARQL query repositories accessible through the GitHub API;
- a decorator syntax to enrich SPARQL queries in Git repositories with metadata about their intended use (Sect. 3.3); and
- a description of the `grlc` service, that automatically exposes such enriched SPARQL queries in GitHub repositories as Linked Data APIs (Sect. 4)

As for the rest of the paper, we survey relevant related work in Sect. 2, evaluate our approach in two use cases in Sect. 5, and conclude in Sect. 6.

2   Related Work

The organization and management of SPARQL queries is central to the study of their efficiency, nature, and use at improving Linked Data applications. SPARQL query logs have been used to study differences between queries by humans and machines [13]. These logs are also useful to understand semantic relatedness of queried entities [7]. Saleem et al. [14] propose to "create a Linked Dataset describing the SPARQL queries issued to various public SPARQL endpoints". Loizou et al. [9] identify (combinations of) SPARQL constructs that constitute a performance hit, and formulate heuristics for writing optimized queries. To the best of our knowledge, no previous work addresses the use of collaborative code platforms to ease deployment of Web APIs.

The Semantic Web has developed significant work on the relationship between Linked Data and Web Services [3,12]. In [15], authors propose to expose REST APIs as Linked Data. These approaches suggest the use of Linked Data technology on top of Web services. Our work is related to results in the opposite direction, concretely the Linked Data API specification[2] and the W3C Linked Data Platform 1.0 specification, which "describes the use of HTTP for accessing, updating, creating and deleting resources from servers that expose their resources as Linked Data"[3]. Kopecký et al. [8] address the specific issue of writing (updating, creating, deleting) these Linked Data resources via Web APIs. However, our work is more related to providing Linked Data access interfaces that function as SPARQL wrappers, like the OpenPHACTS Discovery Platform for pharmacological data [4], and the BASIL server [1]. These approaches build Linked Data APIs compliant with the Swagger RESTful API specification[4] on top of SPARQL endpoints. Our contribution proposes an additional decoupling of Linked Data APIs with SPARQL query curation infrastructures, in order to lower the costs of building and maintaining such APIs.

3   From GitHub Repositories to Linked Data APIs

Git is becoming increasingly popular for maintaining projects that revolve around other content than code: a wide variety of projects use Git repositories to store data and queries over them. This development can be largely attributed to the popularity of GitHub[5], a cloud-based Git repository hosting service [10]. For example, we use GitHub to store important SPARQL queries and templates for the CEDAR and CLARIAH projects. This has brought two key outcomes. First, it contributes to *better maintainability* of the life cycle of SPARQL queries. By leveraging git features and GitHub's infrastructure, queries become easily reusable (since they get unique, dereferenceable URIs), their provenance better traceable [2], their development (through frictionless branching) less error-prone, and their versioning effortless. Second, it lowers *coupling* between SPARQL queries and applications, by separating their development and maintenance workflows while keeping queries accessible. As a consequence, queries are less frequently hard-coded and retyped. The goal of grlc

---

[2] https://github.com/UKGovLD/linked-data-api.
[3] https://www.w3.org/TR/2015/REC-ldp-20150226/.
[4] https://github.com/OAI/OpenAPI-Specification.
[5] https://github.com/.

is to profit from this the decoupling of applications and queries to streamline the infrastructure for building and exposing Linked Data APIs.

This section investigates how the organisational characteristics of GitHub repositories can be used to build, manage and maintain a Swagger-spec compliant API. First, in Sect. 3.2 we study the requirements of Swagger-compliant APIs and map them to elements of the GitHub API. Since these elements are insufficient for a complete API spec, in Sect. 3.3 we propose to complete it with non-intrusive SPARQL decorators.

3.1 The Swagger Specification and User Interface

The OpenAPI Specification, previously known as the Swagger[6] Specification, is a standard[7] specification for machine-readable data structures that describe, produce, consume, and visualize RESTful Web services. Given such a data structure, a variety of tools can generate API code, documentation, and test cases.

The OpenAPI specification allows for the declarative description of API resources, such as operation names (called *paths*), their human-readable descriptions, available access methods (e.g. HTTP GET, POST), parameters, available output formats, and expected responses. The standard is ongoing work, and there exists abundant documentation on all supported features[8]. Client applications can read these declarative resource descriptions, and consume services without knowledge about their specific implementations. The Swagger UI[9] is an example of such a client application. It reads Swagger service specifications, and produces a web-base user interface that displays the contents of the API and offers a variety of ways of interaction.

3.2 Mapping Swagger and GitHub

We propose to align the metaphor of the *repository* with that of the *API*, since both share abstract notions of organizing files and operations in a way that is meaningful for their users. For this reason, in this section we study a possible mapping between the two. Table 1 shows the mapping between the attribute requirements of the Swagger RESTful API specification (see Sect. 3.1), and how these correspond with either attributes of the GitHub API (repository organisation elements) or attributes of the SPARQL usage decorator (usage metadata elements). The latter are discussed in Sect. 3.3.

3.3 SPARQL Decorators

To complete the mapping of the GitHub API to the Swagger RESTful API specification shown in Table 1, we propose *SPARQL decorators* (or tags) to add metadata in queries as comments. These decorators are used by grlc to build more

---

[6] See http://swagger.io/ and https://github.com/swagger-api.

[7] See https://openapis.org/.

[8] For a complete overview of the specification, see http://swagger.io/specification/.

[9] See https://github.com/swagger-api/swagger-ui.

Table 1. Mappings between the Swagger RESTful API and the GitHub API/SPARQL decorators. Such decorators, and the query itself, are parsed through accessing any file with the extension `.rq` in the repo via GET http://raw.githubusercontent.com/:owner/:repo/master/

Swagger attribute	Scope	Description	Mapping
Swagger version	API	Version number of compliant Swagger RESTful API specification	**Static**: independent of the LDA. Currently version 2.0 of the Swagger RESTful API spec is supported
API version	API	Version number of the API	**GitHub API**: last repo release from the release API through GET `/repos/:owner/:repo/releases/latest`
Title	API	Title of the API	**GitHub API**: name of the repository through GET `/repos/:owner/:repo`
Contact name	API	Author and contact information	**GitHub API**: login name of the repository owner through GET `/repos/:owner/:repo`
Contact URL	API	URL to be followed for additional information	**GitHub API**: link to the HTML page of the repository owner through GET `/repos/:owner/:repo`
License	API	License under which the API is released	**Repository file**: a link to the raw LICENSE file of the repo if it exists; empty otherwise
Host	API	Host name to compose the API calls	**grlc parameter**: supplied host name in grlc's configuration; `localhost` by default
Base path	API	Base path to compose the API calls	**GitHub API**: the string `/:owner/:repo` in GET `/repos/:owner/:repo`
Schemes	API	Supported schemes to compose the API calls	**Static**: `http` is supported
Path name	Operation	Name of the API operation	**GitHub API**: the file name, without the extension, of any `.rq` file found in the repository
Path method	Operation	HTTP method for the operation (GET, POST)	**Static**: GET is supported
Path tags	Operation	Tags under which the operation will be classified	**SPARQL decorator**: the parsed list of the decorator *tags* in `.rq` files
Path description	Operation	Description of the API operation	**SPARQL decorator**: the parsed *description* decorator in `.rq` files
Path parameters	Operation	Parameters of the operation	**SPARQL decorator**: all parameter placeholders parsed in the query (see Sect. 4)
Path responses	Operation	Responses of the operation	**SPARQL decorator**: response codes on success, datatypes of parameters (see Sect. 4)

accurate and descriptive Linked Data APIs (see Sect. 4). We assume SPARQL queries organised as .rq files in git repositories. Each of these files will translate into an API operation. We propose to comment them in the first file lines, with the syntax depicted in the following example[10]:

```
#+ summary: A brief summary of what the query does
#+ method: GET
#+ endpoint: http://example.org/sparql
#+ tags:
#+     - UseCase1
#+     - Awesomeness
```

This indicates the *summary* of the query (which will document the API operation), the http *method* to use (GET, POST, etc.), the *endpoint* to send the query, and the *tags* under which the operation falls in. The latter helps to keep operations organized within the API. In addition, we suggest to include two special files in the repository. The first is a LICENSE file containing the license for the SPARQL queries and the API. The second is the endpoint.txt file, with the URI of a default endpoint to direct all queries of the repository. When parsing the repository (see Sect. 4) the target endpoint will be the one indicated by the #+ endpoint decorator, the endpoint.txt file, or http://dbpedia. org/sparql, in this order of preference.

4   grlc

grlc[11] is a thin gateway that automatically builds complete, well documented, and neatly organized Linked Data APIs on the fly, with no input required from users beyond a GitHub user and repository name. To do so, it implements the GitHub API mappings proposed in Sect. 3.2, and uses the SPARQL decorators described in Sect. 3.3. The public online instance of grlc is located at http:// grlc.io/.

grlc provides three basic operations: *(1)* generates the Swagger spec of a specified GitHub repository; *(2)* generates the Swagger UI (see Sect. 3.1) to provide an interactive user-facing frontend of the API contents; and *(3)* translates http requests to call the operations of the API against a SPARQL endpoint with several parameters. If the GitHub repository at https://github.com/: owner/:repo contains decorated SPARQL queries, grlc uses these, together with organisational repo information from the GitHub API, to build the API interface automatically. Assuming that grlc is running at :host, these operations are available at the following routes:

- http://:host/:owner/:repo/spec: JSON Swagger-compliant specification, using the mappings of Sect. 3
- http://:host/:owner/:repo/api-docs: Swagger-UI, rendered using such mappings, as shown in Fig. 1.

---

[10] Additional examples can be found at https://github.com/CEDAR-project/Queries and https://github.com/CLARIAH/wp4-queries.

[11] Source code at https://github.com/CLARIAH/grlc.

Fig. 1. Screenshot of the Swagger user interface generated by `grlc`

- `http://:host/:owner/:repo/:operation?p_1=v_1...p_n=v_n`:
  http GET request to `:operation` with parameters $p_1, ..., p_n$ taking values $v_1, ..., v_n$.

`grlc` composes the Swagger spec as follows: *(1)* the user requests the URI `http://:host/:owner/:repo/spec`[12] to a host running `grlc`; *(2)* `grlc` issues the http GET request to the GitHub API at `https://api.github.com/repos/:owner/:repo`, using the owner and repo names indicated in the previous step; *(3)* for each `.rq` file described in the response, `grlc` derefences `https://raw.githubusercontent.com/:owner/:repo/master/file.rq` to get the SPARQL file contents; *(3)* `grlc` parses these file contents to extract: *(a)* the values of the decorators (if any), and *(b)* any parameter placeholders in the query; *(4)* `grlc` uses all the gathered data to compose the Swagger spec, and returns it to the client as JSON. The composition of the Swagger UI is analogous: first the JSON spec is composed and, after, it is used to render the Swagger UI template[13].

4.1 Parameter Mapping

It is often useful for SPARQL queries to be parameterized. This happens when a resource in a basic graph pattern (BGP) can take specific values that affect the result of the query. Previous work has investigated how to map these values to parameters provided by the API operations [1,4].

`grlc` is compliant with BASIL's convention for Web API parameters mapping[14]. This means that some "parameter-declared" SPARQL variables of a query are interpreted by `grlc` as parameter placeholders. Hence, operations of the

---

[12] Requested from any `http` compliant client: a Web browser, `curl`, etc.

[13] https://github.com/swagger-api/swagger-ui.

[14] See https://github.com/the-open-university/basil/wiki/SPARQL-variable-name-convention-for-WEB-API-parameters-mapping.

```
1   SELECT (SUM(?pop) AS ?tot) FROM <urn:graph:cedar-mini:release> WHERE {
2           ?obs a qb:Observation.
3           ?obs sdmx-dimension:refArea ?_location_iri.
4           ?obs cedarterms:Kom ?_kom_iri.
5           ?obs cedarterms:population ?pop.
6           ?slice a qb:Slice.
7           ?slice qb:observation ?obs.
8           ?slice sdmx-dimension:refPeriod ?_year_integer.
9           ?obs sdmx-dimension:sex ?_sex_iri.
10          ?obs cedarterms:residenceStatus ?_residenceStatus_iri.
11          FILTER (NOT EXISTS {?obs cedarterms:isTotal ?total }) }
```

Listing 1.1. Example of a parametrized SPARQL query (prefixes ahave been omitted).

```
1   SELECT DISTINCT ?_sex_iri FROM <urn:graph:cedar-mini:release> WHERE {
2           ?obs sdmx-dimension:sex ?_sex_iri. }
```

Listing 1.2. Rewritten query by `grlc` to retrieve plausible values for the parameter ?_sex_iri.

form `http://:host/:owner/:repo/:operation?p_1=v_1...p_n=v_n`, are executed by `grlc` by: *(1)* retrieving the raw SPARQL query from https://raw.githubusercontent.com/:owner/:repo/master/:operation.rq; and *(2)* rewriting this query by replacing the placeholders by the parameter values $v_1, ..., v_n$ supplied in the API request. After this, the query is submitted to the endpoint indicated by the methods described in Sect. 3.3. The endpoint results are forwarded to the client application.

An example of such a parameterized query[15] is shown in Listing 1.1. SPARQL variable names staring with ?_ and ?__ indicate mandatory and optional parameters, respectively. If they end with `iri` or `integer`, they are expected to be mapped to IRIs and literal (integer) values, respectively. These parameters are replaced by the user provided parameter values $v_1, ..., v_n$ by `grlc`'s query rewriting engine.

Parameter Enumerations. In order to guide the user at providing valid parameter values, `grlc` tries to fill the enumeration `get->parameters->enum` attribute of an operation in the Swagger specification. This (optional) attribute contains an array with all possible values that a parameter can take. To generate this enumeration, `grlc` sends an additional SPARQL query to the endpoint, replacing the original BGP by the triple pattern where the parameter appears. For instance, to fill the enumeration of parameter ?_sex_iri in Listing 1.1, `grlc` retrieves its plausible values from the endpoint with the query shown in Listing 1.2. Figure 2 shows an example of how the Swagger UI displays these plausible parameter values.

---

[15] The original query can be found at https://github.com/CEDAR-project/Queries/blob/master/residenceStatus_params.rq.

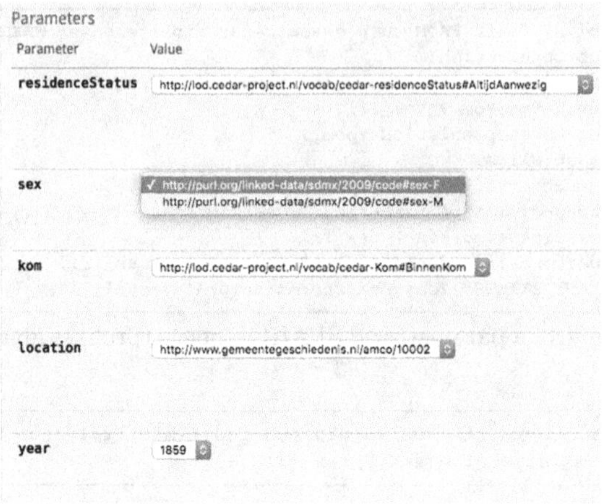

Fig. 2. Screenshot of the Swagger user interface rendering parameter enumerations generated by `grlc`

4.2 Content Negotiation

`grlc` supports content negotiation at two different levels: by *request*, and by *URL*. By request, `grlc` checks the value of the `Accept` header in incoming `http` requests. By URL, `grlc` checks whether a route calling an API operation ends with a trailing `.csv`, `.json` or `.html`.

In both cases, the corresponding `Accept http` header is used in the request to the SPARQL endpoint, delegating support of specific content types to each endpoint. When the response from the server is received, `grlc` sets the `Content-Type` header of the client response to match that received by the endpoint, and therefore it only proxies both requests and responses.

4.3 Caching

Building a Swagger specification retrieving data from the GitHub API can be a costly operation. One GitHub request is needed to retrieve the repository contents; n additional GitHub requests are needed for the n queries in the repository; and m extra endpoint requests per query are required when such query contains m parameters. Thus, the performance of `grlc` can be severely affected by cumulative network latency.

In order to mitigate this cost, `grlc` implements a simple *cache* that maintains the following data structure:

    { <repo_uri> : { 'date' : <date>, 'spec' : <spec> } }

`<repo_uri>` is the URI of a requested GitHub repository; `<date>` is the timestamp at which `grlc` generated the Swagger spec for that repository for the last time; and `<spec>` is the JSON data structure containing the Swagger spec itself.

Every time there is an incoming request to generate a spec for the repository located at `<repo_uri>`, `grlc` checks whether there is a cached spec for `<repo_uri>` in the cache. If there is, `grlc` compares: *(a)* the date at which the repository `<repo_uri>` was updated for the last time, by requesting the value of `pushed_at` to the GitHub API; and *(b)* the date `<date>` of the cached spec. If the cached copy is more recent than the last GitHub update, `grlc`'s cache is up to date and the cached spec can be used instead of generating it from scratch.[16] If there is no `<repo_uri>` in the cache, or its date is older than the last GitHub update, the spec is generated from scratch, as described in Sect. 4, and the cache is updated. This makes `grlc` about 20 times faster to generate API specs, at the cost of building them from scratch when the cache is empty or outdated.

5   Preliminary Evaluation

In this section we evaluate requirements satisfied by `grlc` in two use cases.

Dutch Historical Census Data. The CEDAR project[17] has published the Dutch historical censuses (1795–1971) as 5-star Linked Data [11]. Key queries and templates to interrogate this dataset are available at GitHub[18]. These queries are used in various client applications[19,20]. Before `grlc`, we decided to implement a minimal effort Web API using our own instance of BASIL[21]. However, the queries needed to be retyped in the system, and caused ramifications with respect to the ones in our existing applications. Moreover, it was not possible to mimic the organisation these queries had in the original GitHub repo in the API spec. After `grlc`, we could create this API without interfering with the original applications and queries, effectively reusing them. Furthermore, `grlc` permitted an ecosystem where SPARQL and non-SPARQL savvy applications coexist.

Born Under a Bad Sign. In CLARIAH[22], querying structured humanities data from combined sources is central. This particular use case focuses on validating the hypothesis that prenatal and early-life conditions have a strong impact on socioeconomic and health outcomes later in life, by using 1891 census records of Canada and Sweden. These were converted to Linked Data with QBer [6], and analyzed in the statistical environment R. Before `grlc`, loading the data to be analyzed implied the manual download of a SPARQL query resultset in a file, and then loading this file in R. This was mitigated with the R SPARQL package [5]. However, this resulted in hard-coded, hardly reusable, and difficult to maintain queries. After better organising these queries in a GitHub repository, an API using them became immediately available through `grlc`. As shown in Fig. 3, the R code became *clearer* due to the decoupling with SPARQL; and *shorter*, since a `curl` one-liner calling a `grlc` enabled API operation sufficed to retrieve the data.

---

[16] Note that the cache currently does not track updates to the underlying data. This means that the parameter enumerations 'grlc' generates can become outdated for more dynamic datasets.

[17] http://www.cedar-project.nl/.

[18] https://github.com/CEDAR-project/Queries.

[19] YASGUI-based browsing: http://lod.cedar-project.nl/cedar/data.html.

[20] Drawing historical maps with census data: http://lod.cedar-project.nl/maps/map_CEDAR_women_1899.html.

[21] https://github.com/the-open-university/BASIL.

[22] http://clariah.nl/.

```
46   ## using grlc API call
47   library(RCurl)
48   canada <- getURL("http://grlc.clariah-sdh.eculture.labs.vu.nl/clariah/wp4-
49   canada <- read.csv(textConnection(canada))
50   sweden <- getURL("http://grlc.clariah-sdh.eculture.labs.vu.nl/clariah/wp4-
51   sweden <- read.csv(textConnection(sweden))
52
53   fit_canada_base <- lm(log(hiscam) ~ log(gdppc), data=canada)
54   fit_canada <- lm(log(hiscam) ~ log(gdppc) + I(age^2) + age, data=canada)
55   fit_sweden_base <- lm(log(hiscam) ~ log(gdppc), data=sweden)
56   fit_sweden <- lm(log(hiscam) ~ log(gdppc) + I(age^2) + age, data=sweden)
```

Fig. 3. The use of `grlc` makes Linked Data accessible from any `http` compatible application

6 Conclusion and Future Work

In this paper we presented `grlc`, a novel approach to automatically build Linked Data APIs by using SPARQL queries stored and documented in git repositories. Our approach addresses two pitfalls of current practice in constructing Linked Data APIs: *(1)* the coupling of SPARQL curation workflows and the API infrastructure, which hampers query reuse and forces query retyping and ramifications; and *(2)* the common lack of organisation in Linked Data APIs. `grlc` maps the Swagger specification with GitHub API features and a proposed SPARQL decorator notation, and builds and maintains Linked Data APIs automatically with minimal effort. We argue that this approach enables a better coexistence of SPARQL and non-SPARQL savvy applications, and allows developers to switch their efforts from API infrastructure to applications.

We plan to extend this work in several ways. First, we will support additional repository elements and SPARQL decorators. Second, we will add compatibility with other collaborative coding platforms, like Bitbucket and GitLab, enabling private APIs and authentication. Third, we will investigate ways to map API results pagination and the SPARQL keywords LIMIT and OFFSET. Fourth, we will investigate the use of `grlc`-built APIs for the retrieval of generic endpoint metadata (schema, VoID, etc.) without the need of SPARQL. Finally, we plan to create a `grlc` companion to facilitate the curation of SPARQL queries in Git repositories.

Acknowledgments. This work was funded by the CLARIAH project of the Dutch Science Foundation (NWO) and the Dutch national programme COMMIT. The work on which this paper is based has been supported by the Computational Humanities Programme of the Royal Netherlands Academy of Arts and Sciences. For further information, see http://ehumanities.nl. We want to thank the reviewers for their valuable comments and suggestions.

References

1. Daga, E., Panziera, L., Pedrinaci, C.: A BASILar approach for building web APIs on top of SPARQL endpoints. In: Services and Applications over Linked APIs and Data SALAD2015 (ISWC 2015), vol. 1359, CEUR Workshop Proceedings (2015). http://ceur-ws.org/Vol-1359/

2. De Nies, T., Magliacane, S., Verborgh, R., Coppens, S., Groth, P., Mannens, E., Van de Walle, R.: Git2PROV: Exposing version control system content as W3C PROV. In: Poster and Demo Proceedings of the 12th International Semantic Web Conference (2013). http://www.iswc2013.semanticweb.org/sites/default/files/iswc_demo_32_0.pdf
3. Fielding, R.T.: Architectural styles and the design of network-based software architectures (2000)
4. Groth, P., Loizou, A., Gray, A.J., Goble, C., Harland, L., Pettifer, S.: API-centric linked data integration: the open PHACTS discovery platform case study. Web Semant.: Sci. Serv. Ag. World Wide Web **29**, 12–18 (2014). http://www.sciencedirect.com/science/article/pii/S1570826814000195, life Science and e-Science
5. van Hage, W.R., with contributions from: Tomi Kauppinen, Graeler, B., Davis, C., Hoeksema, J., Ruttenberg, A., Bahls., D.: SPARQL: SPARQL client (2013). http://CRAN.R-project.org/package=SPARQL. Rpackageversion1.15
6. Hoekstra, R., Meroño-Peñuela, A., Dentler, K., Rijpma, A., Zijdeman, R., Zandhuis, I.: An ecosystem for linked humanities data. In: Proceedings of the 1st Workshop on Humanities in the Semantic Web (WHiSe 2016), ESWC 2016 (2016, under review)
7. Huelss, J., Paulheim, H.: What SPARQL query logs tell and do not tell about semantic relatedness in LOD. In: Gandon, F., et al. (eds.) ESWC 2015. LNCS, vol. 9341, pp. 297–308. Springer, Heidelberg (2015). doi:10.1007/978-3-319-25639-9_44
8. Kopecký, J., Pedrinaci, C., Duke, A.: Restful write-oriented API for hyperdata in custom RDF knowledge bases. In: 2011 7th International Conference on Next Generation Web Services Practices (NWeSP), pp. 199–204, October 2011
9. Loizou, A., Angles, R., Groth, P.: On the formulation of performant SPARQL queries. Web Semant.: Sci. Serv. Ag. World Wide Web **31**, 1–26 (2015). http://www.sciencedirect.com/science/article/pii/S1570826814001061
10. McMillan, R.: From collaborative coding to wedding invitations: github is going mainstream. Wired Magazine, 9 February 2013. http://www.wired.com/2013/09/github-for-anything/all
11. Meroño-Peñuela, A., Guéret, C., Ashkpour, A., Schlobach, S.: CEDAR: the Dutch historical censuses as linked open data. Semant. Web - Interoper. Usabil. Appl. (2015, in press)
12. Pedrinaci, C., Domingue, J.: Toward the next wave of services: linked services for the web of data. J. Univ. Comput. Sci. **16**(13), 1694–1719 (2010)
13. Rietveld, L., Hoekstra, R.: Man vs. machine: differences in SPARQL queries. In: Proceedings of the 4th USEWOD Workshop on Usage Analysis and the Web of of Data, ESWC 2014 (2014). http://usewod.org/files/workshops/2014/papers/rietveld_hoekstra_usewod2014.pdf
14. Saleem, M., Ali, M.I., Hogan, A., Mehmood, Q., Ngomo, A.-C.N.: LSQ: the linked SPARQL queries dataset. ISWC 2015. LNCS, vol. 9367, pp. 261–269. Springer, Heidelberg (2015). doi:10.1007/978-3-319-25010-6_15
15. Speiser, S., Harth, A.: Integrating linked data and services with linked data services. In: Antoniou, G., Grobelnik, M., Simperl, E., Parsia, B., Plexousakis, D., De Leenheer, P., Pan, J. (eds.) ESWC 2011, Part I. LNCS, vol. 6643, pp. 170–184. Springer, Heidelberg (2011)
16. Vandenbussche, P.Y., Aranda, C.B., Hogan, A., Umbrich, J.: Monitoring the status of SPARQL endpoints. In: 12th International Semantic Web Conference on Proceedings of the ISWC 2013 Posters and Demonstrations Track (ISWC 2013), pp. 81–84. CEUR-WS (2013)

Semantic Web Technologies in Mobile and Pervasive Environments (SEMPER)

Dem@Home: Ambient Intelligence for Clinical Support of People Living with Dementia

Stelios Andreadis, Thanos G. Stavropoulos[✉], Georgios Meditskos, and Ioannis Kompatsiaris

Center for Research and Technologies - Hellas, Information Technologies Institute, Thessaloniki, Greece
{andreadisst,athstavr,gmeditsk,ikom}@iti.gr

Abstract. With the ever-growing prevalence of dementia, nursing costs are increasing, while the ability to live independently vanishes. Dem@Home is an ambient assisted living framework to support independent living while receiving intelligent clinical care. Dem@Home integrates a variety of ambient and wearable sensors together with sophisticated, interdisciplinary methods of image and semantic analysis. Semantic Web technologies, such as OWL 2, are extensively employed to represent sensor observations and application domain specifics as well as to implement hybrid activity recognition and problem detection. Complete with tailored user interfaces, clinicians are provided with accurate monitoring of multiple life aspects, such as physical activity, sleep, complex daily tasks and clinical problems, leading to adaptive non-pharmaceutical interventions. The method has been already validated for both recognition performance and improvement on a clinical level, in four home pilots.

Keywords: Ambient assisted living · Sensors · Semantic web · Ontologies · Reasoning · Context-awareness · Dementia

1 Introduction

The increase of the average lifespan across the world has been accompanied by an unprecedented upsurge in the occurrence of dementia, with high socio-economic costs, reaching 818 billion US dollars worldwide, in 2015. Nevertheless, its prevalence is increasing as the number of people aged 65 and older with Alzheimer's disease may nearly triple by 2050, from 46.8 million to 131 million people around the world, the majority of which, living in an institution [1].

Assistive technologies could enhance clinicians' diagnosis and decision making, in order to meet individual needs, but also to be used as an objective assessing measure of cognitive status and disease progress of patients. Furthermore, assistive technology is expected to play a critical role in improving patients' quality of life, both on cognitive and physical level, whereas cost is reduced. Drawbacks of current health services are that they often aim to evaluate single needs (e.g. pharmacological treatment) or detect problems solely via interviews, leading to generic interventions by clinicians. However, home remote monitoring of patients is a promising "patient-centered" management

© Springer International Publishing AG 2016
H. Sack et al. (Eds.): ESWC 2016 Satellite Events, LNCS 9989, pp. 357–368, 2016.
DOI: 10.1007/978-3-319-47602-5_49

approach that provides specific and reliable data, enabling the clinicians to monitor patients' daily function and provide adaptive and personalized interventions.

Towards this direction, we propose Dem@Home, a holistic approach for context-aware monitoring and personalized care of dementia at homes, prolonging independent living. To begin with, the system integrates a wide range of sensor modalities and high-level analytics to support accurate monitoring of all aspects of daily life including physical activity, sleep and activities of daily living (ADLs), based on a service-oriented middleware [2]. After integrating them in a uniform knowledge representation format, Dem@Home employs semantic interpretation techniques to infer complex activity recognition from atomic events and highlight clinical problems. Specifically, it follows a hybrid reasoning scheme, using DL reasoning for activity detection and SPARQL to extract clinical problems. Utterly, Dem@Home presents information to applications tailored to clinicians and patients, endorsing technology-aided clinical interventions to improve care. Dem@Home has been deployed and evaluated in four home pilots showing optimistic results with respect to accurate fusion and activity detection and clinical value in care.

The rest of the paper is structured as follows: Sect. 2 presents relevant work, while Sect. 3 gives an overview of the framework. Section 4 elaborates on data analytics, presenting the activity recognition and problem detection capabilities of Dem@Home. Section 5 describes the GUIs supported by the framework to provide feedback to clinical experts or patients, Sect. 6 presents the evaluation results and Sect. 7 concludes the paper.

2   Related Work

Pervasive technology solutions have already been employed in several ambient environments, either homes or clinics, but most of them focus on a single domain to monitor, using only a single or a few devices. Such applications include wandering behavior prevention with geolocation devices, monitoring physical activity, sleep, medication and performance in daily chores [3, 4].

In order to assess cognitive state, activity modelling and recognition appears to be a critical task, common amongst existing assistive technology. OWL has been widely used for modelling human activity semantics, reducing complex activity definitions to the intersection of their constituent parts. In most cases, activity recognition involves the segmentation of data into snapshots of atomic events, fed to the ontology reasoner for classification. Time windows [5] and slices [6] provide background knowledge about the order or duration [7] of activities are common approaches for segmentation. In this paradigm, ontologies are used to model domain information, whereas rules, widely embraced to compensate for OWL's expressive limitations, aggregate activities, describing the conditions that drive the derivation of complex activities e.g. temporal relations.

Focusing on clinical care through sensing, the work in [8] has deployed infrared motion sensors in clinics to monitor sleep disturbances, limited, though, to a single sensor. Similarly, the work in [9] presents a sensor network deployment in nursing homes in Taiwan to continuously monitor vital signs of patients, using web-based technologies, verifying the system's accuracy, acceptance and usefulness. Nevertheless, it so far lacks the ability to fuse more sensor modalities such as sleep and ambient sensing, with limited interoperability.

Other solutions involve smart home deployments of environmental sensors to observe and assess elder and disabled people activities [10, 11]. The work in [12] monitors the residents' physical activity and vital signs by using wearable sensors, door sensors to measure presence and "fully automated biomedical devices" in the bathroom, while the system presented in [13] provides security monitoring, with actuators to control doors, windows and curtains, but none of the above records sleep. On the other hand, Dem@Home offers a unified view of many life aspects, including sleep and activities, to automatically assess disturbances and their causes, aiding clinical monitoring and interventions.

3   The Dem@Home Framework

Dem@Home proposes a multidisciplinary approach that brings into effect the synergy of the latest advances in sensor technologies addressing a multitude of complementary modalities, large-scale fusion and mining, knowledge representation and intelligent decision-making support. In detail, as depicted in Fig. 1, the framework integrates several heterogeneous sensing modalities, such as physical activity and sleep sensor measurements, combined input from lifestyle sensors and higher-level image analytics, providing their unanimous semantic representation and interpretation.

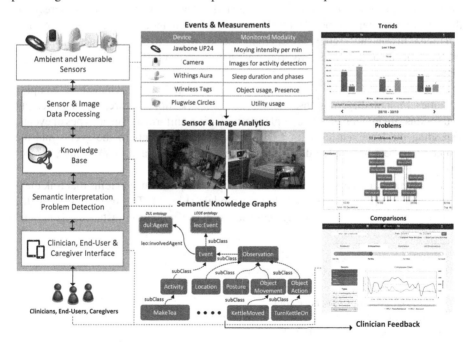

Fig. 1. Dem@Home architecture, sensors and clinical applications

The current selection of sensors is comprised of proprietary, low-cost, ambient or wearable devices, originally intended for lifestyle monitoring, repurposed to a medical context. *Ambient depth cameras*[1] are collecting both image and depth data. The *Plug* sensors[2] are attached to electronic devices, e.g. to cooking appliances, to collect power consumption data. *Tags*[3] are attached to objects of interest, e.g. a drug-box or a watering can, capturing motion events and *Presence* sensors are modified Tags that detect people's presence in a room using IR motion. A wearable *Wristwatch*[4] measures physical activity levels in terms of steps, while a pressure-based *Sleep sensor*[5] is placed underneath the mattress to record sleep duration and interruptions.

Each device is integrated by using dedicated modules that wrap their respective API, retrieve data and process them accordingly to generate atomic events from sensor observations e.g. through aggregation. In the case of image data, computer vision techniques are employed to extract information about humans performing activities, such as opening the fridge, holding a cup or drinking [14]. All atomic events and observations are mapped to a uniform semantic representation for interoperability and stored to the system's Knowledge Base. Dem@Home applies further semantic analysis, activity recognition and detection of problems i.e. anomalies, and then all the derived information can be used by domain-specific applications offering a tailored view to different types of users.

4   Activity Recognition and Problem Detection

To obtain a more comprehensive image of an individual's condition and its progression, driving clinical interventions, Dem@Home employs semantic interpretation to perform intelligent fusion and aggregation of atomic, sensor events to complex ones and identify problematic situations, with a hybrid combination of OWL 2 reasoning and SPARQL queries.

Dem@Home provides a simple pattern for modelling the context of complex activities. First of all, sensor observations, including location, posture, object movement and actions, are integrated with complex activities in a uniform model, as types of events, extending the leo:Event class of LODE[6] (Fig. 1). The agents of the events and the temporal context are captured using constructs from DUL[7] and OWL Time[8], respectively.

Each activity context is described through class equivalence axioms that link them with lower-level observations of domain models (Fig. 1). The instantiation of this pattern is used by the underlying reasoner to classify context instances, generated

---

[1] Xtion Pro - http://www.asus.com/Multimedia/Xtion_PRO/.

[2] Plugwise sensors - https://www.plugwise.nl/.

[3] Wireless Sensor Tag System - http://wirelesstag.net/.

[4] Jawbone UP24 - https://jawbone.com.

[5] Withings Aura - http://www2.withings.com/us/en/products/aura.

[6] LODE - http://linkedevents.org/ontology/.

[7] DUL - http://www.loa.istc.cnr.it/ontologies/DUL.owl.

[8] OWL Time - http://www.w3.org/TR/owl-time/.

during the execution of the protocol, as complex activities. The instantiation involves linking ADLs with context containment relations through class equivalence axioms. For example, given that the activity *PrepareHotTea* involves the observations *TurnKettleOn, CupMoved, KettleMoved, TeaBagMoved* and *TurnKettleOff*, its semantics are defined as:

$$PrepareTea \equiv Context \sqcap \exists contains.TurnKettleOn \sqcap \exists contains.CupMoved$$
$$\sqcap \exists contains.KettleMoved \sqcap \exists contains.TeaBagMoved$$
$$\sqcap \exists contains.TurnKettleOff$$

Other examples in our use case scenario are:

$$Cooking \equiv Context \sqcap \exists contains.TurnCookerOn$$
$$\sqcap \exists contains.KitchenPresence$$

$$PrepareDrugBox \equiv Context \sqcap \exists contains.DrugBoxMoved$$
$$\sqcap \exists contains.DrugCabinetMoved$$
$$\sqcap \exists contains.KitchenPresence$$

$$WatchTV \equiv Context \sqcap \exists contains.TurnTvOn$$
$$\sqcap \exists contains.RemoteControlMoved$$
$$\sqcap \exists contains.LivingRoomPresence$$

$$BathroomVisit \equiv Context \sqcap \exists contains.BathroomPresence$$
$$\sqcap \exists contains.TurnBathroomLightsOn$$

5 Problem Detection

According to clinical experts involved in the development of Dem@Home, highlighting problematic situations next to the entire set of monitored activities and metrics would further facilitate and accelerate clinical assessment. Dem@Home uses a set of predefined rules (expressed in SPARQL) with numerical thresholds that clinicians can adjust and personalize to each of the individuals in their care, through a GUI. This way the thresholds attach to each participant or even to a particular period as well.

Furthermore, analysis is invoked for a period of time allowing different thresholds for different intervals e.g. before and after a clinical intervention. Problematic situations supported so far regard night sleep (short duration, many interruptions, too long to fall asleep), physical activity (low daily activity totals), missed activities (e.g. skipping daily lunch) and reoccurring problems (problems for consecutive days). An instance of a short sleep duration problem in SPARQL is given in Fig. 2.

```
 1: CONSTRUCT {
 2:      ?new a :SleepDurationProblem;
 3:              :duration ?D;
 4:              :date ?date.
 5: }
 6: WHERE {
 7:      ?activity a :Sleep;
 8:              :startTime ?st;
 9:              :endTime ?et.
10:      BIND(:duration(?st, ?et) as ?D)
11:      {
12:        SELECT ?_d ?ActivityType
13:        WHERE {
14:              ?p a :SleepDurationPattern;
15:              :hasDescription
16:              [:definesActivityType
17:                    [:classifiesActivity :Sleep;
18:                          :hasDurationDescription [time:seconds ?_d]
19:                    ]
20:              ].
21:        }
22:      }
23:    FILTER(?D > _d)
24:    BIND (extract_date(?startTime) as ?date)
25: }
```

Fig. 2. SPARQL Problem definition for short sleep duration

6   End-User Assessment Application

At the application level, Dem@Home provides a multitude of user interfaces to assist both clinical staff, summarizing an individual's performance and highlighting abnormal situations, and patients, proposing simplified view of measurements and educational material.

The clinician interface offers four different approaches to monitoring a patient, i.e. *Summary, Comparison,* and *All Observations*, as well as four options of time extent of the data, i.e. *One-Day, Per Day, Per Week* and *Per Month*. In One-Day Summary, sleep measurements are obtained from one single night and are categorized as *Total Time in Bed but Awake, Total Time Shallow Sleep, Total Time Deep Sleep, Total Time Asleep, Number of Interruptions* and *Sleep Latency* (Fig. 3). In Summary Per Day, the clinician is able to select to a time interval between two dates or a single date, to observe sleep stages, physical levels and other activities of daily living, derived from power consumption, moved objects and presence in rooms (Fig. 4).

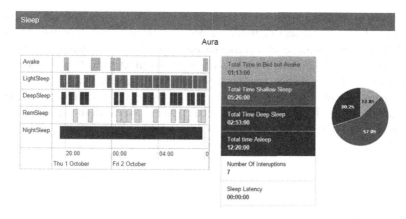

Fig. 3. The clinician interface regarding sleep parameters, in one-day summary session.

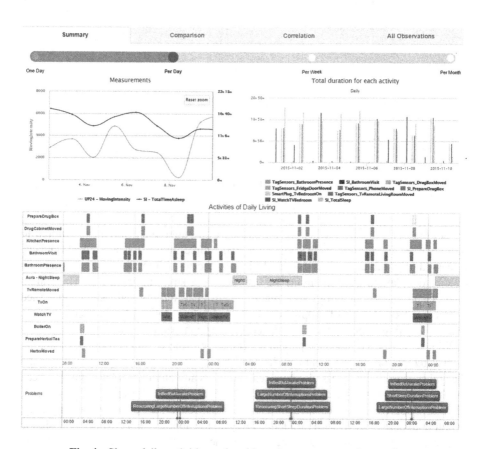

Fig. 4. Sleep, daily activities and problems in summary per day session.

Moreover, the clinician can set specific thresholds about sleeping problems during the night and a problems section will be added (bottom of Fig. 4). In Comparison per Day, different measurements of a particular time period can be combined in the same chart, allowing the clinician to check how observations affect each other, e.g. how physical activity affects sleep or how usage of a device affects a daily activity (Fig. 5). Finally, the Comparison screen shows a scatterplot for two types of measurements, while All Observations shows all collected data in detail. The Per Week/Month options offer the above-mentioned functionalities summarized per week/month.

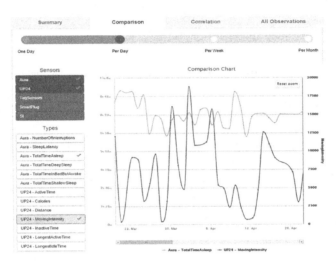

Fig. 5. Comparison Per Day chart between two activities

On the other hand, patients are introduced to an alternative interface, tailored to provide easy monitoring of their daily life and simple interaction with the clinicians. Accessed by a tablet device, a limited view of the most important measurements is displayed, to avoid overwhelming the users or even stressing them out. The patient interface presents 3-day information regarding Physical Activity (daily steps and burned calories), Sleep, Usage of Appliances and Medication. Especially in Sleep section, patient is notified about how many sleep interruptions they had during the night. In addition to sensor readings, the patient interface is enhanced with educational material, such as recipes or instructions to guide them step-by-step to perform routine tasks, and the ability to exchange messages between end-users and clinicians. An example screen of what is shown to end-users can be seen on Fig. 6. Namely, end-users can see a digested view of some metrics, presented in a manner that generates positive thinking only, avoiding to burden them with problems. Overall, the application is explicitly design to help patients feel confident and secure with the system they are using, but also to encourage social interaction between users and clinicians.

Fig. 6. The patient interface, including message inbox (top), chart with sleep metrics for the last 3 days (middle) and alert notification about number of interruptions (bottom).

7   Evaluation

Dem@Home was evaluated in four home installations, in the residences of individuals living alone, clinically diagnosed with mild cognitive impairment or mild dementia, and maintained for four months. Sensors and relevant home areas or devices of the installation (Table 1) were selected after a visit from the clinician to the participants. The majority of deployed sensors covered the areas of kitchen, bathroom and bedroom, since these rooms are strongly linked with most daily activities.

Table 1. Sensors in home installation

Sensor	Home area or device
Camera	Kitchen, Living room, Hall
Plugs	TV, Iron, Vacuum, Cooking device, Boiler, Kettle, Bathroom lights
Tags	TV remote, Iron, Fridge door, Drug cabinet, Drug box, Tea bag, Cup
Presence	Kitchen, Bathroom, Living room
Wristwatch	User's arm
Sleep sensor	Bed

Since the framework embodies an interdisciplinary approach, it was evaluated both from research and clinical perspective. Firstly, we evaluate the effectiveness of activity recognition through fusion of sensor data and existing multimedia analytics. Secondly, clinical results vary and add significant value to monitoring and interventions.

For the evaluation of the ontology-based fusion and activity recognition capabilities of Dem@Home, ground truth has been obtained through annotation (performed once), based on images from ambient cameras. We use the precision and recall measures, to evaluate the performance with respect to ADLs recognized as performed. The clinical expert suggested the monitoring of five activities, namely drug box preparation, cooking, making tea, watching TV and bathroom visit. Table 2 depicts the pertinent context dependency models defined.

Table 2. Context dependency models for the evaluation

Activity Concept	Context dependency set
PrepareDrugBox	DrugBoxMoved, DrugCabinetMoved, KitchenPresence
Cooking	TurnCookerOn, KitchenPresence
PrepareTea	TurnKettleOn, TeaBagMoved, CupMoved, KitchenPresence, TurnKettleOff
WatchTV	TurnTvOn, RemoteControlMoved, LivingRoomPresence
BathroomVisit	BathroomPresence, TurnBathroomLightsOn

Dem@Home's ADL activity recognition performance has been evaluated on a dataset of 31 days, in July 2015. As observed on Table 3, the more atomic and continuous an activity is, the more accurate the detection. *BathroomVisit*, most accurately detected, is never interleaved to do something else. On the contrary, cooking is a long-lasting activity interrupted by instances of other events (e.g. watching TV) and influenced by uncertainty and the openness of the environment. *WatchTV* and *PrepareTea* are fairly short in duration, causing less uncertainty and interleaved events in between, yielding decent precision and recall rates. Therefore, the fact that the worst performance appears in the Cooking activity can be appointed to it being a highly interleaved and long-lasting activity as opposed to the others.

Table 3. Precision and recall for activity recognition

Activity	Recall	Precision
PrepareDrugBox	0.86	0.89
Cooking	0.61	0.68
PrepareTea	0.81	0.86
WatchTV	0.87	0.80
BathroomVisit	0.91	0.94

On the other hand, the clinical evaluation of the framework regards its capabilities and the fulfillment of clinical requirements. With Dem@Home supporting clinical interventions, significant improvement was found in post-pilot clinical assessment in multiple domains, such as increase in physical condition and sleep quality, utterly bringing about positive change in mood and cognitive state, measured objectively by neuropsychological tests. In detail, the first participant has overcome insomnia, the lack of exercise and neglecting daily chores. The second participant has shown improvement in sleep and mood, while the other two users have been benefited with respect to sleep and medication. More elaborate details on the clinical value and outcome of the experiment are ongoing work.

8 Conclusion and Future Work

Dem@Home is an ambient assisted living framework integrating a variety of sensors, analytics and semantic interpretation with a special focus on dementia ambient care. New, affordable sensors have been integrated seamlessly into the framework, along with a set of processing components, ranging from sensor to image analytics. All knowledge is semantically interpreted for further fusion and detection of problematic behaviours, while tailored user interfaces aim to detailed monitoring and adaptive interventions. Evaluation of the framework has yielded valuable and optimistic results with respect to accurate fusion and activity detection and clinical value in care.

Regarding future directions, Dem@Home could be extended for increased portability and installability. Specifically, establishing an open source, IoT-enabled semantic platform, following the latest advances in board computing would allow the platform to be easily deployed in multiple locations. Combined with the infrastructure to push the events on a cloud infrastructure, the framework could constitute a powerful platform for telemedicine and mobile health, combing sensors and sophisticated ambient intelligence techniques such as computer vision.

Acknowledgement. This work has been supported by the H2020-ICT-645012 project KRISTINA: A Knowledge-Based Information Agent with Social Competence and Human Interaction Capabilities.

References

1. Prince, M., Wimo, A., Guerchet, M., Ali, G., Wu, Y.T., Prina, M.: World Alzheimer Report 2015. The global impact of dementia. An analysis of prevalence, incidence, cost and trends. Alzheimers Dis. Int. Lond. (2015)
2. Stavropoulos, T.G., Meditskos, G., Kontopoulos, E., Kompatsiaris, I.: The DemaWare Service-Oriented AAL Platform for People with Dementia. Artif. Intell. Assist. Med. AI-AMNetMed 2014, 11 (2014)
3. Kerssens, C., Kumar, R., Adams, A.E., Knott, C.C., Matalenas, L., Sanford, J.A., Rogers, W.A.: Personalized technology to support older adults with and without cognitive impairment living at home. Am. J. Alzheimers. Dis. Other Demen. **30**(1), 85–97 (2015). doi: 10.1177/1533317514568338
4. Dawadi, P.N., Cook, D.J., Schmitter-Edgecombe, M., Parsey, C.: Automated assessment of cognitive health using smart home technologies. Technol. Health Care Off. J. Eur. Soc. Eng. Med. **21**, 323 (2013)
5. Okeyo, G., Chen, L., Wang, H., Sterritt, R.: Dynamic sensor data segmentation for real-time knowledge-driven activity recognition. Pervasive Mob. Comput. **10**, 155–172 (2014)

6. Riboni, D., Pareschi, L., Radaelli, L., Bettini, C.: Is ontology-based activity recognition really effective? In: Pervasive Computing and Communications Workshops, pp. 427–431. IEEE (2011)
7. Patkos, T., Chrysakis, I., Bikakis, A., Plexousakis, D., Antoniou, G.: A reasoning framework for ambient intelligence. In: Konstantopoulos, S., Perantonis, S., Karkaletsis, V., Spyropoulos, Constantine, D., Vouros, G. (eds.) SETN 2010. LNCS (LNAI), vol. 6040, pp. 213–222. Springer, Heidelberg (2010). doi:10.1007/978-3-642-12842-4_25
8. Suzuki, R., Otake, S., Izutsu, T., Yoshida, M., Iwaya, T.: Monitoring daily living activities of elderly people in a nursing home using an infrared motion-detection system. Telemed. J. E Health. 12, 146–155 (2006)
9. Chang, Y.-J., Chen, C.-H., Lin, L.-F., Han, R.-P., Huang, W.-T., Lee, G.-C.: Wireless sensor networks for vital signs monitoring: Application in a nursing home. Int. J. Distrib. Sens. Netw. 8(11), 685107 (2012)
10. Helal, S., Mann, W., King, J., Kaddoura, Y., Jansen, E.: The gator tech smart house: A programmable pervasive space. Computer 38, 50–60 (2005)
11. Demongeot, J., Virone, G., Duchêne, F., Benchetrit, G., Hervé, T., Noury, N., Rialle, V.: Multi-sensors acquisition, data fusion, knowledge mining and alarm triggering in health smart homes for elderly people. C. R. Biol. 325, 673–682 (2002)
12. Tamura, T., Togawa, T., Ogawa, M., Yoda, M.: Fully automated health monitoring system in the home. Med. Eng. Phys. 20, 573–579 (1998)
13. Bonner, S.G.: Assisted interactive dwelling house. In: Proceedings of the 3rd TIDE Congress: Technology for Inclusive Design and Equality Improving the Quality of Life for the European Citizen, p. 25 (1998)
14. Avgerinakis, K., Briassouli, A., Kompatsiaris, I.: Recognition of activities of daily living for smart home environments. In: 2013 9th International Conference on Intelligent Environments (IE), pp. 173–180. IEEE (2013)

The Resource Action Language: Towards Designing Reactive RDF Stores

Jean-Yves Vion-Dury and Nikolaos Lagos[(⊠)]

Xerox Research Centre Europe, Meylan, France
{jean-yves.vion-dury,nikolaos.Lagos}@xrce.xerox.com

Abstract. In an interconnected world such as the one envisioned by pervasive computing, systems should be able to react to stimuli received from the environment in a streaming fashion. Reactions may include not only performing local updates, but also sending and asking for information from other systems, waiting for responses, and requesting for changes. In this paper we give a short introduction to the main principles of a language we are developing to achieve that, ReAL. Key elements of ReAL in that context include the introduction of explicit operators to deal with concurrency, nested transactions, and streams. Based on these operators we show how interaction with external services could be enabled. In the future we plan to evaluate further the most innovative operators, define the semantics of ReAL, and analyze its relation to SPARQL, the standard Semantic Web query language.

Keywords: Streaming · RDF query · Context · ReAL

1 Introduction

In an interconnected world such as the one envisioned by pervasive computing, systems should be able to react to stimuli received, for instance thanks to sensors, from the environment. Reactions may include not only sending information but also asking for information from other systems, waiting for responses, and requesting for changes in a continuous, streaming fashion. Streaming extensions to the standard Semantic Web query language (SPARQL) have been developed for dealing with continuous data flows [1–5], with the most interesting in our context being EP-SPARQL [3] that uses events as triggers of query execution. However, we observed that interaction with other systems, and the effects on the design of a corresponding query language, have not been explicitly considered up to now.

In this paper we give a short introduction to the main principles of a language we are developing to bridge this gap, **ReAL** (**Re**source **A**ction **L**anguage). The overall objective is to provide a means for describing the dynamic behavior of RDF stores in a streaming fashion, i.e. handle queries within a specific execution context, perform (potentially transformative) actions on the store itself, and allow interaction with external services. The design principles include.

- Offer an explicit mechanism of (nested) transactions, thus allowing the execution context to be clearly defined at query time.

© Springer International Publishing AG 2016
H. Sack et al. (Eds.): ESWC 2016 Satellite Events, LNCS 9989, pp. 369–378, 2016.
DOI: 10.1007/978-3-319-47602-5_50

- Use a concurrency model to allow coordination with other services – we do that based on a "Triple Space" derived from "Tuple Space" as formerly done in coordination languages like Linda [6].
- Follow a streaming execution model to enumerate solutions one by one, thus propagating solutions as soon as possible.
- Allow a synthesis of query and production-rule languages (to define actions and their impacts within the query).
- Aim for modular and highly compositional programming structures (procedures).

In this paper we don't target exhaustivity. In particular, many general purpose primitive actions are missing, as we essentially focus on some of the most interesting features of ReAL in our context. Other higher level actions (e.g. time-oriented and memory-protection-oriented primitives) are work in progress.

We have to note here that ReAL can be seamlessly linked to the LRM upper level ontology [7], being developed in the PERICLES project[1]. An example of such integration will be described in the paper. The LRM OWL ontology has been designed to address dynamicity in the digital preservation field, with a focus on change management through sophisticated model to handle intentional dependencies, versioning mechanisms and reflexive metadata modeling. If ReAL is designed as a "natural infrastructure" to support LRM based services, we do believe that its more fundamental qualities are not bound to any particular data model.

2   Matching, Bindings and Basic Actions

Triples are represented through a syntax similar to the one adopted by the abstract syntax of SWRL [8], using functional notation like *predicate(subject, object)*. where any of the three components can be an IRI using a prefixed form or a variable *?name*. The object component can additionally be a string like "3.1416", a decimal/integral number, or a symbol like true, false. Note that triples extended with language tags or typing IRI are captured by an additional argument (separated by "|"), e.g.:

```
rdf:label(test:c1,"my class" | en)
ex:weight(test:c1,"0.12456" | xsd:decimal)
```

Based on this notation, we introduce next the most basic primitive constructs to perform reading and writing in the RDF store. They constitute what we like to call *basic actions*.

Simple reading. The following reading expression (illustrative)

$$rdf:type(?sub, ?class) \tag{1}$$

will succeed if at least one solution can be read in the triple store. Solution here designates all triples matching the expression. The result is of the form <boolean, Binding>, where boolean (true or false) denotes whether a solution is found, and

---

[1] http://pericles-project.eu/.

Binding denotes the set[2] of pairs (variable, term[3]). Failing queries always return <false, {}>. A new Binding is streamed whenever a matching solution is found, and can be defined as a mapping relating all variables (e.g. ?sub) to subterms such that the filtering terms are made equal to the matching terms. In other words, a Binding represents the substitutive solution that equates the filter to the instance. The substitution operation of an expression e using a binding B is a new expression noted B(e). Note that the expression (1) above, if changed into e.g.

$$\text{rdf:type(ex:nantes-triptych, ex:Artwork)} \qquad (1b)$$

could stream a unique solution (an empty binding {}) in a context where the triple is indeed present in the RDF store.

Destructive reading. To express that you want not only to filter-out the RDF store, but also to withdraw the matching solutions, you may use a "-" operator as a prefix.

$$- \text{rdf:type(?sub, ex:Artwork)} \qquad (2)$$

Note that the store is immediately modified, so that unforeseen "side effects" may occur when such an instruction is combined with others, even if those do not return any solution eventually (this is one reason why nested transactions are relevant in ReAL, as we will see later). Destructive reading may fail if the triple is write-protected (such protection mechanisms will not be presented here).

Explicit inference invocation. When one needs to extract more complex information from the store, he may use inference to stream solutions, thanks to the "!" prefix.

$$! \text{ rdf:type(?sub, ex:Artwork)} \qquad (3)$$

The type of inference is dependent on the context and on the configuration of the corresponding infrastructure, but typically, it could exploit a background taxonomy or ontology. For instance, provided that relations like rdfs:subClassOf (ex:VideoArt, ex: Artwork) and rdfs:subClassOf (ex:SoftwareBasedArt, ex:Artwork) are included in the underlying knowledge base, corresponding inference (based on the rdfs:subClassOf) could be used to infer that the instances of ex:VideoArt and ex:SoftwareBasedArt are solutions of query (3).

Writing triples. In order to insert a new triple inside the store, one may use the "+" prefix:

$$+ \text{ rdf:type(ex:nantes-Triptych,ex:Artwork)} \qquad (4)$$

When one wants to write a triple into the store, the operation might fail if: the triple already exists; the triple is not well formed (remaining unbound variables, bad element organization, such as a string placed at the subject position) - this is an error case; the

---

[2] The set can be empty if the pattern does not involve any variable, or only jokers, noted *?*

[3] Here the term's syntax is defined by the non-terminal BItem of the formal grammar provided in the appendix. .

triple is write-protected (individual triples, or families of triples can be write-protected by a lock - not developed here); the store forbids the writing of such triples (similarly, one can specify access rights; not developed here). If the operation is successful, it will stream a unique solution: the empty binding.

3 Blocking Actions

Reading and writing actions can be suspended until completion when specified with the WAIT primitive. It means that the ReAL process will be suspended until the action can be fulfilled in the current context. This is a powerful way to synchronize and communicate information between concurrent ReAL processes through the intermediary of the RDF triple store (à la LINDA [6]). As an illustration, the three expressions below

```
WAIT rdf:type(ex:nantes-triptych, ex:Artwork)
WAIT -rdf:type(ex:nantes-triptych, ex:Artwork)
WAIT +rdf:type(ex:nantes-triptych, ex:Artwork)
```

will wait for the triple if not initially present at evaluation time. For the last one, a writing action, the waiting process will not start if the problem is linked to a badly formed triple issue (the action will just fail).

4 Stream-Based Logical Connectors

Considering that an expression becomes "true" if at least one solution exists, we propose to consider our set of combinators (as described below) as being dual, each of them being both a logical combinator and a stream-based composition operator as well.

AND. The binary combinator "AND" allows combining solutions from both operands. An (e1 AND e2) expression first looks for solutions of e1; for each corresponding binding B1, it is applied to e2 (applying a binding means doing a substitution: if e2 shares variables with e1, they will be instantiated) and then the operator looks for solutions for B1(e2) and streams them as results. As an illustration,

$$rdf:type(?sub,ex:ArtWork) \text{ AND } ex:creator(?sub,ex:BillViola) \quad (5)$$

will stream the subject IRI for all art works by Bill Viola explicitly known in the RDF store. Now, if we want to do a more powerful operation, for instance replacing the "ex: BillViola" IRI with another one (where the IRI is more abstract and does not mention a name), and specifying the artist's name through the rdfs:label property:

```
$iri AND
+rdfs:label(?iri,"Bill Viola") AND
rdf:type(?sub, ex:ArtWork) AND              Example    (6)
-ex:creator(?sub, ex:BillViola) AND
+ex:creator(?sub, ?iri)
```

The notation $iri is a syntactic sugar for FRESH(?iri), a primitive that streams fresh and unique IRIs bound to the variable ?iri.

OR. This binary combinator propagates only the left substream if any. Otherwise, it propagates the right one, if any.

UNION. This binary combinator propagates first the left substream if any. Afterwards, it propagates the right substream if any (meaning that it fails if both substreams fail). Note that like the OR primitive, no junction is done between left and right terms.

NO. This combinator streams the empty binding if no solution is found for the sub-expression, fails otherwise. Usage example:

```
NO rdf:type(?x,rdfs:Class)
```

FIRST. this unary operator evaluates its subexpression, and just streams the first solution if any. Albeit the RDF store is not ordered, streams can be ordered, especially when yielded by inference based queries.

LAST. Same behavior than FIRST, except that only the last solution is found. Note that (i) it cannot work with infinite streams, and (ii) the substream is delayed until completion since only the last solution is streamed.

REPEAT. Evaluate the subexpression but do not propagate any solutions. Fails if the subexpression fails, returns the empty binding when the stream terminates. Usage example:

```
REPEAT (rdf:type(?x,ex:Book) AND +rdfs:label(?x,"scanned"))
```

Repeat can be parameterized by a counting parameter i.e. number of solutions. If the number cannot be reached, the action will fail. For instance:

```
REPEAT 1 (rdf:type(?x,rdfs:Class) AND + rdfs:label(?x,"scanned"))
```

performs only one action. It is not equivalent to FIRST because REPEAT is opaque. The expression below, will perform as many times as possible the actions of the subexpression, and will return a binding giving the value of ?count, i.e. the number of solutions.

```
REPEAT ?count (rdf:type(?x,rdfs:Class) AND +rdfs:label(?x,"scanned"))
```

TRUE. Streams the empty binding.

FALSE. Always fails (do not stream anything).

CALL some-IRI (i0 ... ik >> o0 ... on). This operator executes the actions defined by some-IRI; parameters i0... ik are passed to the target environment; outputs o0 ... on, when defined, are used to rename the bindings streamed by the action if any.

SPAWN *some-IRI (i0 ... ik)*. This action is similar to the CALL action, except that the action will be executed in a concurrent micro-thread, and cannot stream any solution (one must use synchronized triples to exchange data). This is therefore an asynchronous call, as opposed to CALL which is synchronous.

STOP *some-IRI*. This action stops a process (designated by some-IRI) but fails if the process is not found or is not active anymore.

STOP *some-IRI (msg-IRI)*. Same as the previous combinator, but a message will be associated (msg-IRI, should be an IRI of a lrm:Message instance) to the lrm:ActivityStopped event that will be attached to the RDF activity descriptor (aka some-IRI).

5 Transactions and Sandboxes

Example (6) may raise a problem when the store does not contain any artwork by Bill Viola. In that case, the global action will fail (not returning any solution/binding) when evaluating the third operand rdf:type(?subject, ex:Artwork) but however the store will be eventually modified: a triple rdfs:label(_:b1,"Bill Viola") will be inserted as a side-effect. Indeed, the writing action (as specified by the second operand) is done immediately, as explained in previous sections. One very obvious solution is to reorder the operands:

$$
\begin{array}{ll}
\texttt{rdf:type(?sub, ex:ArtWork) } \textbf{AND} & \\
\texttt{- ex:creator(?sub, ex:BillViola) } \textbf{AND} & \\
\texttt{\$iri } \textbf{AND} & \text{(6b)} \\
\texttt{+ rdfs:label(?iri,"Bill Viola") } \textbf{AND} & \\
\texttt{+ ex:creator(?sub, ?iri)} &
\end{array}
$$

Another more generic solution is to use a transaction: all transformative actions are committed at each streamed solution, if any. If no solution is yielded, the transaction is aborted and the store stays unchanged. The transaction is specified by enclosing square brackets [...], and transactions can be nested. A transaction is transparent (i.e. it always propagates the substream).

$$
\begin{array}{ll}
\texttt{\$iri } \textbf{AND} \texttt{ [} & \\
\texttt{+ rdfs:label(?iri,"Bill Viola") } \textbf{AND} & \\
\texttt{rdf:type(?sub, ex:ArtWork) } \textbf{AND} & \\
\texttt{- ex:creator(?sub, ex:BillViola) } \textbf{AND} & \text{(6c)} \\
\texttt{+ ex:creator(?sub, ?iri)} & \\
\texttt{]} &
\end{array}
$$

A similar mechanism, called the sandbox, allows to confine all transformative actions into a temporary substore which will be forgotten after evaluation, be it a success or not (so it behaves like a transaction that is always aborted). It is denoted by enclosing brackets {...} and like for transactions, it is transparent (it always propagates the substream).

6 Handling Graphs

Graphs can be viewed as a way to modularize RDF stores. We propose two combinators to work with graphs: ON and IN. Their behavior is defined according to a dedicated execution structure, namely, a stack of contexts (an RDF graph for instance can be considered as a context). At the bottom of the stack, there is always the default context (i.e. the context stack is never empty), and transformative actions (addition and deletion of triples) are always performed in the context lying on the top of the stack.

$$\textbf{ON} <\texttt{iri}> \text{ or } <\texttt{var}> \{\texttt{action}\}$$

If the first parameter is an IRI, it must designate an existing graph. If the parameter is a variable, the graph will be created, and in extension to the standard RDF 1.1 semantics, a triple `rdf:type(iri, rdf:Graph)` will be created inside the top context[4]. This (new) graph will be pushed on the stack, and will become the new active context. The action associated with the ON operation will be undertaken and solutions streamed up. Note that transformative actions (insertions and deletions) will only affect the top context, however reading actions will explore the whole context stack in the top-down direction.

$$\textbf{IN} <\texttt{iri}> \text{ or } <\texttt{var}> \{\texttt{action}\}$$

The semantics are pretty much the same, except that IN builds a stack of one unique context, the one given as parameter of the action. Therefore, transformative and reading actions associated with the IN operations are all confined to the same unique graph (in that sense, it is much more restrictive: it locally behaves like if the default store and other graphs do not exist).

7 Summary and Ongoing Work

We have presented the main design principles of a query language for RDF stores based on the notion of actions. We have presented several combinators to handle concurrency, enable interaction with external services, and define the context of execution via the notion of nested transaction.

Currently we are working on:

- Experimenting the most innovative operators, especially the transactional and graph related combinators (we expect the former to simplify greatly concurrent modification and the latter to provide means for simple and efficient safety control mechanisms).
- Decoupling completely ReAL from LRM. The current version of ReAL is still dependent for some operators on the LRM ontology (they are both being developed in the context of the same project, PERICLES). For instance, in the combinator

---

[4] Actually, all named graphs will be associated with such a triple.

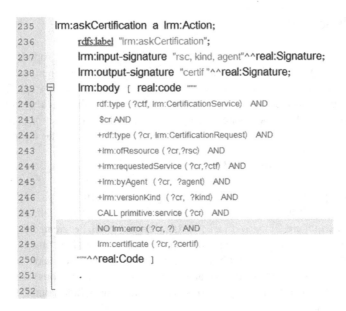

Fig. 1. Example of ReAL and LRM integration

CALL some-IRI (i0 ... ik >> o0 ... on) the reference some-IRI must designate today an instance of a specific LRM class (lrm:Action) which, by design, defines a unique predicate lrm:body where the ReAL code describing the actions is inserted. Figure 1 shows an example of such an instance which is used to invoke a certification service for the versioning of an entity.

Line 237 in Fig. 1 defines the input signature, which must be matched with the input parameters (order and cardinal of the list are both significant and must match; also true for the output) given by the caller; the line 238 defines the output signature: these parameters will be streamed back to the caller if solutions are found.

- Defining the formal semantics for ReAL.
- Analyzing the relation of SPARQL (and streaming variants) to ReAL.

The results of the above actions will be made available in the near future.

Acknowledgments. This work takes place in the framework of the PERICLES project which received funding from the European Union's Seventh Framework Programme for research, technological development and demonstration under grant agreement no. 601138. We thank our colleagues and partners for the fruitful exchanges we shared. We also thank Mehreen Ikram and Stéphane Jean for their valuable collaboration into bridging formally the semantics of the above language with the one of SPARQL.

Appendix: EBNF Grammar

```
ReAL ::=  '[' ReAL ']'    |
          '{' ReAL '}'    |
          'IN' Pattern '{' ReAL '}'    |
          'ON' Pattern '{' ReAL '}'    |
          ReAL 'AND' ReAL  |
          ReAL 'UNION' ReAL  |
          ReAL 'OR' ReAL  |
          'NO' ReAL  |

          'FIRST' ReAL    |
          'LAST' ReAL    |
          'REPEAT' ReAL     |
          'REPEAT' Item ReAL     |
          '(' ReAL ')'     |
          'TRUE' |
          'FALSE' |
          Action

Action  ::=
          'CALL' CallPattern    |
          'SPAWN' CallPattern     |
          'STOP' Pattern    |
          'STOP' CallPattern     |
          'EXPAND' CallPattern    |
          '!' TriplePattern     |
          Iri '!' TriplePattern      |
          '$' <symbol>    |
          'WAIT' BasicAction     |
          BasicAction

BasicAction   ::=
          '+' TriplePattern    |
          '++' TriplePattern     |
          '-' TriplePattern     |
          TriplePattern

TriplePattern  ::=
          Pattern '(' Pattern ',' Item ')'    |
          Pattern '(' Pattern ',' Item, '|' Pattern ')'    |
          Pattern '(' Pattern ',' BTree   ')'

CallPattern  ::=
          Iri '(' ItemList? ')'  |
          Iri '(' ItemList? '>>' ItemList? ')'

Item        ::= Pattern | Atom
Pattern     ::= Iri | Var
Iri         ::=  <symbol> ':' <symbol>
Var         ::=  '?'  |   '?' <symbol>
Atom        ::= <string> |  <number> |  <symbol>
ItemList    ::=  Item ',' ItemList  |  Item
BTree       ::= '[ ' BItem   BTreeList ']'
BTreeList   ::= BItem  BTreeList   |  BItem
BTree       ::= '{ '  BList '}'
BList       ::= BItem  "," BList   |
                BItem  BList   |
                '|' BItem  |  BItem
BItem       ::= BTree | Item
```

References

1. Barbieri, D.F., Braga, D., Ceri, S., Valle, E.D., Grossniklaus, M.: Querying RDF Streams with C-SPARQL. SIGMOD Rec. **39**(1), 20–26 (2010)
2. Calbimonte, J.-P., Jeung, H., Corcho, Ó., Aberer, K.: Enabling Query Technologies for the Semantic Sensor Web. Int. J. Semant. Web Inf. Syst. **8**(1), 43–63 (2012)
3. Anicic, D., Fodor, P., Rudolph, S., Stojanovic, N.: EP-SPARQL: a unified language for event processing and stream reasoning. In: Proceedings of the Twentieth International World Wide Web Conference on WWW 2011, India (2011)
4. Dehghanzadeh, S., Dell'Aglio, D., Gao, S., Della Valle, E., Mileo, A., Bernstein, A.: Approximate continuous query answering over streams and dynamic linked data sets. In: Cimiano, P., Frasincar, F., Houben, G.-J., Schwabe, D. (eds.) ICWE 2015. LNCS, vol. 9114, pp. 307–325. Springer, Heidelberg (2015). doi:10.1007/978-3-319-19890-3_20
5. Dell-Aglio, D., Della Valle, E., Calbimonte, J.P., Corcho, O.: RSP-QL semantics: a unifying query model to explain heterogeneity of RDF stream processing systems. IJSWIS **10**(4), 17–44 (2015)
6. Wells, G.: Coordination languages: back to the future with Linda. In: Proceedings of the Second International Workshop on Coordination and Adaption Techniques for Software Entities (WCAT 2005), pp. 87–98 (2005)
7. Vion-Dury, J.-Y., Lagos, N., Kontopoulos, E., Riga, M., Mitzias, P., Meditskos, G., Waddington, S., Laurenson, P., Kompatsiaris, I.: Designing for inconsistency - the dependency-based PERICLES approach. In: Morzy, T., Valduriez, P., Bellatreche, L. (eds.) New Trends in Databases and Inf. Systems, vol. 539, pp. 458–467. Springer, Heidelberg (2015)
8. Horrocks, I., Patel-Schneider, P.F., Boley, H., Tabet, S., Grosof, B., Dean, M.: SWRL: A semantic web rule language combining OWL and RuleML. W3C Member submission 21, 79 (2004)
9. PERICLES European project. http://www.pericles-project.eu/

International Workshop on Summarizing and Presenting Entities and Ontologies

ABSTAT: Ontology-Driven Linked Data Summaries with Pattern Minimalization

Blerina Spahiu[(✉)], Riccardo Porrini, Matteo Palmonari, Anisa Rula, and Andrea Maurino

University of Milano-Bicocca, Milan, Italy
{Blerina.Spahiu,Riccardo.Porrini,Matteo.Palmonari,
Anisa.Rula,Andrea.Maurino}@disco.unimib.it

Abstract. An increasing number of research and industrial initiatives have focused on publishing Linked Open Data, but little attention has been provided to help consumers to better understand existing data sets. In this paper we discuss how an ontology-driven data abstraction model supports the extraction and the representation of summaries of linked data sets. The proposed summarization model is the backbone of the ABSTAT framework, that aims at helping users understanding big and complex linked data sets. The proposed model produces a summary that is correct and complete with respect to the assertions of the data set and whose size scales well with respect to the ontology and data size. Our framework is evaluated by showing that it is capable of unveiling information that is not explicitly represented in underspecified ontologies and that is valuable to users, e.g., helping them in the formulation of SPARQL queries.

Keywords: Data summarization · Knowledge patterns · Linked data

1 Introduction

As of April 2014 up to 1014 data sets have been published in the Linked Open Data cloud, a number that is constantly increasing[1]. However, a user may find it difficult to understand to what extent a data set covers a domain of interest and structures its content [7,11,15,19,22]. Given a Linked data set, users should be able to answer to questions such as: What types of resources are described in each data set? What properties are used to describe the resources? What types of resources are linked and by means of what properties? How many resources have a certain type and how frequent is the use of a given property? Remarkably, difficulties in answering those questions have several consequences for data consumption, resulting in low adoption of many valuable but unknown data sets [17].

Linked data sets make use of ontologies to describe the semantics of their data. However, answering the above questions by only looking at ontologies is

---

[1] http://linkeddatacatalog.dws.informatik.uni-mannheim.de/state/.

© Springer International Publishing AG 2016
H. Sack et al. (Eds.): ESWC 2016 Satellite Events, LNCS 9989, pp. 381–395, 2016.
DOI: 10.1007/978-3-319-47602-5_51

not easy. Ontologies might be large. For example, at the time of writing DBpedia uses 685 (local) concepts and 2795 properties. Ontologies used to model a large amount of diverse data in a flexible way are often underspecified. The domain is underspecified for 259 properties of the DBpedia Ontology, while the range is underspecified for 187 properties. In relatively expressive ontologies like the Music Ontology, some connections between types may be specified by means of OWL axioms, e.g., qualified range restrictions, which may be difficult to understand for many data practitioners.

Finally, the ontology does not tell how frequently certain modelling patterns occurs in the data set. Answers to the above questions can be collected with explorative queries, but at the price of a significant server overload for data publishers and high response time for data consumers.

ABSTAT is an ontology-driven linked data summarization model proposed to mitigate the data set understanding problem. In our view, a summary is aimed at providing a compact but complete representation of a data set. With complete representation we refer to the fact that every relation between concepts that is not in the summary can be inferred. One distinguishing feature of ABSTAT is to adopt a minimalization mechanism based on *minimal type patterns*. A minimal type pattern is a triple (C, P, D) that represents the occurrences of assertions $<a,P,b>$ in RDF data, such that C is a minimal type of the subject a and D is a minimal type of the object b. Minimalization is based on a *subtype graph* introduced to represent the data ontology. By considering patterns that are based on minimal types we are able to exclude several redundant patterns from the summary and to specify several formal properties of the summaries. As a consequence, summaries based on our model are rich enough to represent adequately the whole data set, and small enough to avoid redundant information. The ABSTAT[2] framework supports users to query (via SPARQL), to search and to navigate the summaries through web interfaces. Other related work on data or ontology summarization have focused on complementary aspects of the summarization, such as the identification of *salient subsets* of knowledge bases using different criteria [7,15,19,22], e.g., connectivity. Other approaches do not represent connections between instance types as our model does [1,8,9].

In this paper we make the following contributions: (i) we describe in detail the summarization model, focusing on the minimalization approach; (ii) we describe the summary extraction workflow; (iii) we provide an experimental evaluation of our approach from two different perspectives, evaluating the compactness and the informativeness of the summaries.

The paper is organized as follows. The summarization model is presented in Sect. 2. The implementation of the model in ABSTAT is given in Sect. 3. Experimental results are presented in Sect. 4. Related work is discussed in Sect. 5 while conclusions end the paper in Sect. 6.

---

[2] http://abstat.disco.unimib.it.

2   Summarization Model

Ontologies (or vocabularies) are often used to specify the semantics of data modelled in RDF. Ontologies, which are usually represented in languages such as RDFS and OWL2, specify the meanings of the elements of the ontology, e.g., concepts, datatypes, properties, individuals, by means of logical axioms [18]. Although we do not focus on a specific ontological language, we borrow the definition of data set from the definition of Knowledge Base in Description Logics (DLs). In a Knowledge Base there are two components: a *terminology* definining the vocabulary of an application domain, and a set of *assertions* describing RDF resources in terms of this vocabulary.

We define a data set as a couple $\Delta = (\mathcal{T}, \mathcal{A})$, where $\mathcal{T}$ is a set of terminological axioms, and $\mathcal{A}$ is a set of assertions. The domain vocabulary of a data set contains a set N^C of *types*, where with type we refer to either a named class or a datatype, a set N^P of named properties, a set of named individuals (resource identifiers) N^I and a set of literals L. In this paper we use symbols like C, C', ..., and D, D', ..., to denote types, symbols P, Q to denote properties, and symbols a,b to denote named individuals or literals. Types and properties are defined in the terminology and occur in assertions.

Observe that different data set adopt different policies with respect to the inclusion of entailed assertions in the published assertions: for example, DBpedia explicitly includes the transitive closure of type inference in the published assertion set, while other data sets do not follow the same policy, e.g., LinkedBrainz. Our summarization model has to handle data sets that may have been published following different inference publication policies. However, we will briefly discuss the impact of different inference publication policies on the summarisation model in the next sections.

Assertions in $\mathcal{A}$ are of two kinds: *typing assertions* of form $C(a)$, and *relational assertions* of form $P(a, b)$, where a is a named individual and b is either a named individual or a literal. We denote the sets of typing and relational assertions by $\mathcal{A}^C$ and $\mathcal{A}^P$ respectively. Assertions can be extracted directly from RDF data (even in absence of an input terminology). *Typing assertions* occur in a data set as RDF triples $< x, \mathtt{rdf:type}, C >$ where x and C are URIs, or can be derived from triples $< x, P, y^{\wedge\wedge}C >$ where y is a literal (in this case y is a typed literal), with C being its datatype. Without loss of generality, we say that x is an instance of a type C, denoted by $C(x)$, either x is a named individual or x is a typed literal. Every resource identifier that has no type is considered to be of type $\mathtt{owl:Thing}$ and every literal that has no type is considered to be of type $\mathtt{rdfs:Literal}$. Observe that a literal occurring in a triple can have at most one type and at most one type assertion can be extracted for each triple. Conversely, an instance can be the subject of several typing assertions. A *relational assertion* $P(x, y)$ is any triple $< x, P, y >$ such that $P \neq Q*$, where $Q*$ is either $\mathtt{rdf:type}$, or one of the properties used to model a terminology (e.g. $\mathtt{rdfs:subClassOf}$).

Abstract Knowledge Patterns (AKPs) are abstract representations of Knowledge Patterns, i.e., constraints over a piece of domain knowledge defined by axioms of a logical language, in the vein of Ontology Design Patterns [18].

For sake of clarity, we will use the term *pattern* to refer to an AKP in the rest of the paper. A pattern is a triple (C, P, D) such that C and D are types and P is a property. Intuitively, an AKP states that there are instances of type C that are linked to instances of a type D by a property P. In ABSTAT we represent a set of AKP occurring in the data set, which profiles the usage of the terminology. However, instead of representing every AKP occurring in the data set, ABSTAT summaries include only a base of minimal type patterns, i.e., a subset of the patterns such that every other pattern can be derived using a subtype graph. In the following we better define these concepts and the ABSTAT principles.

Pattern Occurrence. A pattern (C, P, D) *occurs* in a set of assertions $\mathcal{A}$ iff there exist some instances x and y such that $\{C(x), D(y), P(x, y)\} \subseteq \mathcal{A}$. Patterns will be also denoted by the symbol π.

For data sets that publish the transitive closure of type inference (e.g., DBpedia), the set of all patterns occurring in an assertion set may be very large and include several redundant patterns. To reduce the number of patterns we use the observation that many patterns can be derived from other patterns if we use a *Subtype Graph* that represents types and their subtypes.

Subtype Graph. A *subtype graph* is a graph $G = (\mathsf{N}^C, \preceq)$, where N^C is a set of type names (either concept or datatype names) and $\preceq$ is a relation over N^C.

We always include two type names in N^C, namely `owl:Thing` and `rdfs:Literal`, such that every concept is subtype of `owl:Thing` and every datatype is subtype of `rdfs:Literal`. One type can be subtype of none, one or more than one type.

Minimal Type Pattern. A pattern (C, P, D) is a *minimal type pattern* for a relational assertion $P(a, b) \in \mathcal{A}$ and a terminology graph G iff (C, P, D) occurs in $\mathcal{A}$ and there does not exist a type C' such that $C'(a) \in \mathcal{A}$ and $C' \prec^G C$ or a type D' such that $D'(b) \in \mathcal{A}$ and $D' \prec^G D$.

Minimal Type Pattern Base. A *minimal type pattern base* for a set of assertions $\mathcal{A}$ under a subtype graph G is a set of patterns $\widehat{\Pi}^{\mathcal{A}, G}$ such that $\pi \in \widehat{\Pi}^{\mathcal{A}, G}$ iff π is a minimal type pattern for some relation assertion in $\mathcal{A}$.

Observe that different minimal type patterns (C, P, D) can be defined for an assertion $P(a, b)$ if a and/or b have more than one minimal type. However, the minimal type pattern base excludes many patterns that can be inferred following the subtype relations and that are not minimal type for any assertion. In the graph represented in Fig. 1 considering the assertion set $\mathcal{A} = \{P(a, b), C(a), A(a), F(b), D(b), A(b)\}$, there are six patterns occurring in $\mathcal{A}$, i.e., (C, P, D), (C, P, F), (C, P, A), (A, P, D), (A, P, F), (A, P, A). The minimal type pattern base for the data set includes the patterns (E, Q, D), (E, R, T), (C, Q, D), (C, R, T) and (C, P, D) since E and C are minimal types of the instance c, while excluding patterns like (B, Q, D) or even (A, Q, A) since not B nor A are minimal types of any instance.

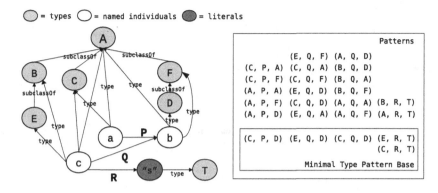

Fig. 1. A small graph representing a data set and the corresponding patterns.

Data Summary. A *summary* of a data set $\Delta = (\mathcal{A}, \mathcal{T})$ is a triple $\Sigma^{\mathcal{A},\mathcal{T}} = (G, \Pi, S)$ such that: G is *Subtype Graph*, $\widehat{\Pi}^{\mathcal{A},G}$ is a *Minimal Type Pattern Base* for $\mathcal{A}$ under G, and S is a set of *statistics* about the elements of G and Π.

Statistics describe the occurrences of types, properties and patterns. They show how many instances have C as minimal type, how many relational assertions use a property P and how many instances that have C as minimal type are linked to instances that have D as minimal type by a property P.

3   Summary Extraction

Our summarization process, depicted in Fig. 2, takes in input an assertion set $\mathcal{A}$ and a terminology $\mathcal{T}$ and produces a summary $\Sigma^{\mathcal{A},\mathcal{T}}$. First, the typing assertion set $\mathcal{A}^C$ is isolated from the relational assertion set $\mathcal{A}^P$, while the subtype graph G is extracted from $\mathcal{T}$. Then, $\mathcal{A}^C$ is processed and the set of minimal types for each named individual is computed. Finally, $\mathcal{A}^P$ is processed in order to compute the minimal type patterns that will form the minimal pattern base $\widehat{\Pi}^{\mathcal{A},G}$. During each phase we keep track of the occurrence of types, properties and patterns, which will be included as statistics in the summary.

Subtype Graph Extraction. The subtype graph G^C is extracted by traversing all the subtype relations in $\mathcal{T}$. The subtype graph will be further enriched with types from external ontologies asserted in $\mathcal{A}^C$ while we compute minimal types of named individuals (i.e., *external* types). The subtype graph does not include equivalence relations.

For what concerning equivalence relations, even if it is possible to convert them in subtype relations, still preserving the partial ordering of the asserted type, we do not follow such approach because this can lead to counterintuitive or undesired inferences. For example, the Village, PopulatedPlace and Place types from the well known DBPedia ontology are equivalent to the type Wikidata:Q532 (i.e., village). By including those equivalence relations in G^C, a correct but undesired inference could be that every Place is also a Village or

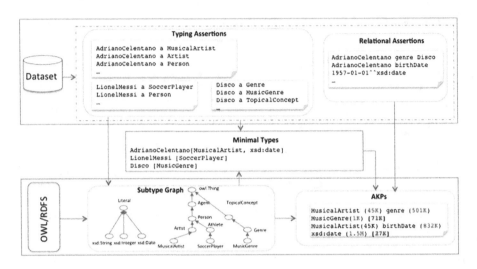

Fig. 2. The summarization workflow.

that `Village`, `PopulatedPlace` and `Place` are equivalent. This real world example highlights an issue related to equivalence relations. Equivalence relations are often used to map named classes from different ontologies (e.g., by leveraging ontology alignment techniques). The existence of mappings leading to counter-intuitive entailments can be explained by precise data management strategies. In the above mentioned equivalence relation, they may have been added to the ontology in order to map `Village` to `Wikidata:Q532` (i.e., villages with villages). For this reason we do not consider equivalence relations in the extraction of G. As a result, the extracted subtype graph G is complete not with respect to the whole terminology $\mathcal{T}$, but with respect to $\mathcal{T}$ without the equivalence relations.

Minimal Types Computation. For each instance x, we compute the set M_x of minimal types with respect to the subtype graph G^C. Given x we select all typing assertions $C(x) \in \mathcal{A}^C$ and form the set $\mathcal{A}_x^C$ of typing assertions about x. Algorithm 1 presents the pseudocode for computing M_x. We first initialize M_x with the type `owl:Thing` (line 1), then we iteratively process all the type assertions. At each iteration we select a type C and remove from M_x all the supertypes of C according to G^C (lines 6–10). Then, if M_x does not contain any subtype of C according to G^C we can add C to M_x (lines 11–14). Notice that one preliminary step of the algorithm is to include C in G^C if it was not included during the subtype graph extraction phase (lines 3–5). Consequently, if a type C is not defined in the input terminology, is automatically considered as a minimal type for all the instances x. This approach allows us to handle instances of types from ontologies not included in the input terminology. At the moment, we do not retrieve such ontologies, but the monotonicity of the minimal type pattern base with respect to the terminology graph G ensures that once they are added to the $\mathcal{T}$, the size of the summary will not increase.

Algorithm 1. Computation of the minimal types set for an instance x.

Input: $\mathcal{A}_x^C$ type assertions about the entity x, G^C subtype graph
Output: the set M_x the minimal types of x

```
1  M_x = {owl:Thing};
2  for C(x) ∈ A_x^C do
3  |   if C ∉ G^C then
4  |   |   G^C = G^C ∪ C
5  |   end
6  |   Sup = findSuperTypes(C, M, G^C);
7  |   if Sup ≠ ∅ then
8  |   |   M_x = M_x \ Sup;
9  |   |   M_x = M_x ∪ {C}
10 |   end
11 |   Sub = findSubTypes(C, M, G^C);
12 |   if Sub = ∅ then
13 |   |   M_x = M_x ∪ {C}
14 |   end
15 end
16 return M_x;
```

Minimal Type Pattern Base Computation. For each relational assertion $P(x, y) \in \mathcal{A}^P$, we get the minimal types sets M_x and M_y. For all $C, D \in M_x, M_y$ we add a pattern (C, P, D) to the minimal types pattern base. If y is a literal value we consider its explicit type if present, `rdfs:Literal` otherwise. In this phase the subproperty graph is enriched with properties that are not defined by the terminology, but still occur in at least one pattern (i.e., *external* properties).

Summary Storing and Presentation. Once extracted, a summary $\Sigma^{\mathcal{A},\mathcal{T}}$ is stored, indexed and made accessible through two user interfaces, i.e., ABSTAT-Browse and ABSTATSearch, and a SPARQL endpoint. SPARQL based access and ABSTATBrowse [3] are described in our previous demo paper [12] and thus is left out of the scope of this paper. ABSTATSearch[4], is a novel interface that implements a full-text search functionality over a set of summaries. Types, properties and patterns are represented by means of their local names (e.g., `Person`, `birthPlace` or `Person birthPlace Country`) and conveniently tokenized, stemmed and indexed to support full text queries. Since the patterns are represented as triples, the study of principled and specialized matching and ranking techniques is an interesting extension that we leave for future work.

4   Experimental Evaluation

We evaluate our summaries from different, orthogonal perspectives. We measure the *compactness* of ABSTAT summaries and compare the number of their patterns to the number of patterns extracted by Loupe [11], an approach similar to ours that does not use minimalization. The *informativeness* of our summaries are evaluated with two experiments. In the first experiment we show that our

---

[3] http://abstat.disco.unimib.it/browse, http://abstat.disco.unimib.it/sparql.
[4] http://abstat.disco.unimib.it/search.

Table 1. Data sets and summaries statistics

	Relational	Typing	Assertions	Types (Ext.)	Properties (Ext.)	Patterns
db2014-core	$\sim$ 40.5M	$\sim$ 29.7M	$\sim$ 70.1M	869 (85)	1439 (15)	**171340**
db3.9-infobx	$\sim$ 96.3M	$\sim$ 19.7M	$\sim$ 116.4M	821 (58)	62572 (14)	**732418**
lb	$\sim$ 180.1M	$\sim$ 39.6M	$\sim$ 221.7M	21 (9)	33 (0)	**161**

summaries provide useful insights about the semantics of properties, based on their usage within a data set. In the second experiment, we conduct a preliminary user study to evaluate if the exploration of the summaries can help users in query formulation tasks. In our evaluation we use the summaries extracted from three linked data sets: DBpedia Core 2014 (**db2014-core**)[5], DBpedia 3.9 (**db3.9-infobox**)[6] and Linked Brainz (**lb**). **db2014-core** and **db3.9-infobox** data sets are based on the DBpedia ontology while the **lb** data set is based on the Music Ontology. DBpedia and LinkedBrainz have complementary features and contain real and large data. For this reason they have been used, for example, in the evaluation of QA systems [10].

4.1 Compactness

Table 1 provides a quantitative overview of data sets and their summaries. To evaluate compactness of a summary we measure the *reduction rate*, defined as the ratio between the number of patterns in a summary and the number of assertions from which the summary has been extracted.

Our model achieves a *reduction rate* of $\sim$0.002 for **db2014-core**, $\sim$0.006 for **db3.9-infobox**, and $\sim$6.72 $\times 10^{-7}$ for **lb**. Comparing the reduction rate obtained by our model with the one obtained by Loupe ($\sim$0.01 for DBpedia and $\sim$7.1 $\times 10^{-7}$ for Linked Brainz) we observe that the summaries computed by our model are more compact, as we only include minimal type patterns. Loupe instead, does not apply any minimalization technique thus its summaries are less compact. The effect of minimalization is more observable on DBpedia data sets, since the DBpedia terminology specifies a richer subtype graph and has more typing assertions. We observe also that 85 external types were added to the **db2014-core** subtype graph and 58 to **db3.9-infobox** subtype graph during the minimal types computation phase as they were not part of the original terminology, and thus are considered by default as minimal types.

4.2 Informativeness

Insights about the semantics of the properties. Our summaries convey valuable information on the semantics of properties for which the terminology does not provide any domain and/or range restrictions. Table 2 provides an overview of the

---

[5] The DBpedia 2014 version with mapping based property only.

[6] The DBpedia Core 3.9 version plus automatically extracted properties.

Table 2. Total number of properties with unspecified domain and range in each data set

	Domain (%)	Range (%)	Domain-Range (%)
db2014-core	259 (~18 %)	187 (~13 %)	48 (~3.3 %)
db3.9-infobox	61368 (~98 %)	61309 (~98 %)	61161 (~97 %)
lb	13 (~39 %)	15 (~45 %)	13 (~39 %)

total number of unspecified properties from the data sets. For example, around 18% of properties from **db2014-core** data set have no domain restrictions while 13% have no range restrictions. Observe that this data set is the most curated subset of DBpedia as it includes only triples generated by user validated mappings to Wikipedia templates. In contrast for **db3.9-infobox** data set which includes also triples generated by information extraction algorithms, most of the properties (i.e., the ones from the `dbpepdia.org/property` namespace) are not specified within the terminology.

In general, underspecification may be the result of precise modelling choices, e.g., the property `dc:date` from the **lb** data set. This property is intentionally not specified in order to favor its reuse, being the Dublin Core Elements (i.e., `dc`) a general purpose vocabulary. Another example is the `dbo:timeInSpace` property from the **db2014-core** data set, whose domain is not specified in the corresponding terminology. However, this property is used in a specific way as demonstrated by patterns (`dbo:Astronaut, dbo:timeInSpace, xsd:double`) and (`dbo:SpaceShuttle dbo:timeInSpace, xsd:double`). Gaining such understanding of the semantics of the `dbo:timeInSpace` property by looking only at the terminology axioms is not possible.

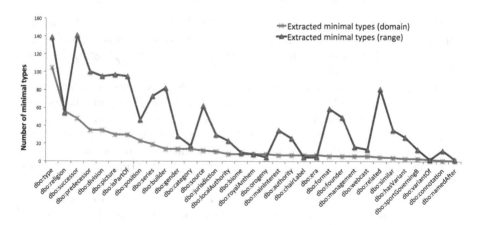

Fig. 3. Distribution of the number of minimal types from the domain and range extracted for not specified properties of the **db2014-core** data set.

We can push our analysis further to a more fine grained level. Figure 3 provides an overview of the number of different minimal types that constitute the domain and range of unspecified properties extracted from the summary of the **db2014-core** data set. The left part of the plot shows those properties whose semantics is less "clear", in the sense that their domain and range cover a higher number of different minimal types e.g., the `dbo:type` property. Surprisingly, the `dbo:religion` property is among them: its semantics is not as clear as one might think, as its range covers 54 disparate minimal types, such as `dbo:Organization`, `dbo:Sport` or `dbo:EthnicGroup`. Conversely, the property `dbo:variantOf`, whose semantics is intuitively harder to guess, is used within the data set with a very specific meaning, as its domain and range covers only 2 minimal types: `dbo:Automobile` and `dbo:Colour`.

Small-scale user study. Formulating SPARQL queries is a task that requires prior knowledge about the data set. ABSTAT could support users that lack such knowledge by providing valuable information about the content of the data set. We designed a user study based on the assignment of cognitive tasks related to query formulation. We selected a set of queries from the *Questions and Answering in Linked Open Data* benchmark[7] [20] to the **db3.9-infobox** data set. The selected queries were taken from logs of the PowerAqua QA system and are believed to be representative of realistic information needs [10], although we cannot guarantee that they cover every possible information need. We provided the participants the query in natural language and a "template" of the corresponding SPARQL query, with spaces intentionally left blank for properties and/or concepts. For example, given the natural language specification *Give me all people that were born in Vienna and died in Berlin*, we asked participants to fill in the blank spaces:

`SELECT DISTINCT ?uri WHERE { ?uri ... <Vienna> . ?uri ... <Berlin> . }`

We selected five queries of increasing length, defined in terms of the number of triple patterns within the `WHERE` clause; one query of length one, two of length two and two of length three. Intuitively, the higher the query length, the more difficult it is to be completed.

We could use a limited number of queries because the tasks are time-consuming and fatigue-bias should be reduced [14]. Overall 20 participants with no prior knowledge about the ABSTAT framework were selected and split into 2 groups: **abstat** and **control**. We profiled all the participants in terms of knowledge about SPARQL, data modelling, DBpedia dataset and ontology, so as to create two homogeneous groups. We trained for about 20 min on how to use ABSTAT only the participants from the first group.

Both groups execute SPARQL queries against the **db3.9-infobox** data set through the same interface and were asked to submit the results they considered correct for each query. We measured the time spent to complete each query and the correctness of the answers. The correctness of the answers is calculated as

---

[7] http://greententacle.techfak.uni-bielefeld.de/~cunger/qald/.

Table 3. Results of the user study

Group	Avg. completion time (s)	Accuracy
query 1 - *How many employees does Google have?* - length 1		
abstat	**358.9**	**0.9**
control	380.6	0.8
query 2 - *Give me all people that were born in Vienna and died in Berlin*- length 2		
abstat	356.3	1
control	**346.9**	0.8
query 3 - *Which professional surfers were born in Australia?*- length 2		
abstat	476.6	0.6
control	**234.24**	**0.7**
query 4 - *In which films directed by Gary Marshall was Julia Roberts starring?* - length 3		
abstat	**333.4**	0.9
control	445.6	0.9
query 5 - *Give me all books by William Goldman with more than 300 pages*- length 3		
abstat	**233.4**	1
control	569.8	0.7

the ratio between the number of correct answers to the given query agains the total number of answers. Table 3 provides the results of the performance of the users on the query completion task[8]. The time needed to perform the 5 queries from all partecipiants in average is 38.6m, while the minimum and the maximum time is 18.4 m and 59.2 m respectively. The independent *t-test*, showed that the time needed to correctly answer Q5, the most difficult query, was statistically significant for two groups. There was a significant effect between two groups, $t(16) = 10.32$, p < .005, with mean time for answering correctly to Q5 being significantly higher (+336s) for the control group than for abstat group. Using 5 queries is coherent with other related work which suggest that the user study would have 20–60 participants, who are given 10–30 minutes of training, followed by all participants doing the same 2–20 tasks, during a 1–3 hour session [14].

Observe that the two used strategies to answer the queries by participants from the **control** group were: to directly access the public web page describing the DBpedia named individuals mentioned in the query and very few of them submitted explorative SPARQL queries to the endpoint. Most of the users searched on Google for some entity in the query, then consulted DBpedia web pages to find the correct answer. DBpedia is arguably the best searchable dataset, which is why this explorative approach was successful for relatively simple queries. However, this explorative approach does not work with other non-indexed datasets (e.g., LinkedBrainz) and for complex queries. Instead, participants of the **abstat** group took advantage of the summary, obtaining huge benefits in terms of average completion time, accuracy, or both. Moreover, they

---

[8] The raw data can be found at http://abstat.disco.unimib.it/downloads/user-study.

achieved increasing accuracy over queries at increasing difficulty, still performing the tasks faster. We interpret the latter trend as a classical cognitive pattern, as the participants became more familiar with ABSTATBrowse and ABSTAT-Search web interfaces.

The noticeable exception is query 3. In particular, participants from the **abstat** group completed the query in about twice the time of participants from **control** group. This is due to the fact that the individual `Surfing` (which is used as object of the property `dbo:occupation`) is classified with no type other than `owl:Thing`. As a consequence, participants from the **abstat** group went trough a more time consuming trial and error process in order to guess the right type and property. Participants from the **abstat** group finally came to the right answer, but after a longer time. This issue might be solved by applying state-of-the-art approaches for type inference on source RDF data [13] and suggest possible improvements of ABSTAT for example including values for concepts that are defined by closed and relatively small instance sets.

5   Related Work

Different approaches have been proposed for schema and data summarization. Most of them identify pieces of knowledge that are more relevant to the user, while the others do not represent the relations among instances but are limited in presenting the co-occurence of the most frequent types and properties. We compare our work to approaches explicitly proposed to summarize Linked Data and ontologies, and to extract statistics about the data set.

A first body of work has focused on summarization models aimed at identifying subsets of data sets or ontologies that are considered to be more relevant. Authors in [22] rank the axioms of an ontology based on their salience to present to the user a view about the ontology. RDF Digest [19] identifies the most salient subset of a knowledge base including the distribution of instances in order to efficiently create summaries. Differently from these approaches, ours aims at providing a complete summary with respect to the data set.

A second body of work has focused on approaches to describe linked data sets by reporting statistics about the usage of the vocabulary in the data. The most similar approach to ABSTAT is Loupe [11], a framework to summarize and inspect Linked Data sets. Loupe extracts types, properties and namespaces, along with a rich set of statistics. Similarly to ABSTAT, Loupe offers a triple inspection functionality, which provides information about *triple patterns* that appear in the data set and their frequency. Triple patterns have the form <*subjectType, property, objectType*> and are equivalent to our patterns. However, Loupe does not apply any mimimalization technique: as shown in Sect. 4.1, summaries computed by our model are significatively more compact.

In [4], authors propose a graph-based approach called ELIS for visualising and exploring induced schema for Linked Open Data. Similarly as in ABSTAT, ELIS also extract patterns, called `schema-level patterns` as a combination between a set of subject types and a set of object types. These subject sets and object sets can be seen as nodes connected by a property. Differently from ABSTAT, ELIS allows users to visualise the induced schema. In ABSTAT, we do not aim at visualising the induced schema, but providing a compact summary which users can browse and search for a pattern, a concept or a property. The summary provided by ABSTAT only includes minimal types, while ELIS does not apply any minimalization technique.

In [2], authors consider vocabulary usage in the summarization process of an RDF graph and use information similar to patterns. A similar approach is also used in MashQL [5], a system proposed to query graph-based data (e.g., RDF) without prior knowledge about the structure of a data set. Our model excludes several redundant patterns from the summary through minimalization, thus producing more compact summaries. Knowledge pattern extraction from RDF data is also discussed in [16], but in the context of domain specific experiments and not with the purpose of defining a general linked data summarization framework. Our summarization model can be applied to any data set that uses a reference ontology and focuses on the representation of the summary.

Other approaches proposed to describe data sets do not extract connections between types but provide several statistics. SchemeEx extracts interesting theoretic measures for large data sets, by considering the co-occurrence of types and properties [8]. A data analysis approach on RDF data based on an warehouse-style analytic is proposed in [3]. This approach focuses on the efficiency of processing analytical queries which poses additional challenges due to their special characteristics such as complexity, evaluated on typically very large data sets, and long runtime. However, this approach differently from ours requires the design of a data warehouse specially for a graph-structured RDF data. Linked Open Vocabularies[9], RDFStats [9] and LODStats [1] provide several statistics about the usage of vocabularies, types and properties but they do not represent the connections between types.

The approach in [21] induces a schema from data and their axioms represent stronger patterns compared to the patterns extracted by our approach. ABSTAT aims to represent every possible connections existing among types while EL axioms aims to mine stronger constraints.

The authors in [6] have a goal even more different than ours. They provide lossless compression of RDF data using inference obtaining thus a reduction rate of 0.5 in best cases. Our approach loses information about instances because aims at representing schema-level patterns, but achieves a reduction rate of 0.002.

6   Conclusion and Future Work

Getting an understanding of the shape and nature of the data from large Linked Data sets is a complex and a challenging task. In this paper, we proposed a minimalization-based summarization model to support data set understanding. Based on the experimentation we show that our summarization framework is able

---

[9] http://lov.okfn.org/.

to provide both *compact* and *informative* summaries for a given data set. We showed that using ABSTAT framework the summaries are more compact than the ones generated from other models and they also help the user to gain insights about the semantics of underspecified properties in the ontology. The results of our preliminary experiment showed that ABSTAT help users formulating SPARQL queries both in terms of time and accuracy.

We plan to run the experiment in large scale, thus including more users with different background characteristics in order to analyse in details which is the target group of users for which ABSTAT is more useful. Several are the future research directions. We plan to complement our coverage-oriented approach with relevance-oriented summarization methods based on connectivity analysis. Another interesting direction was highlighted by our user study, that is the inference of specific types for untyped instances found in the data set. We are also planning to consider the inheritance of properties to produce even more compact summaries. Finally, we envision a complete analysis of the most important data set available in the LOD cloud.

References

1. Auer, S., Demter, J., Martin, M., Lehmann, J.: LODStats – an extensible framework for high-performance dataset analytics. In: ten Teije, A., Völker, J., Handschuh, S., Stuckenschmidt, H., d'Acquin, M., Nikolov, A., Aussenac-Gilles, N., Hernandez, N. (eds.) EKAW 2012. LNCS, vol. 7603, pp. 353–362. Springer, Heidelberg (2012)
2. Campinas, S., Perry, T.E., Ceccarelli, D., Delbru, R., Tummarello, G.: Introducing RDF graph summary with application to assisted SPARQL formulation. In: DEXA (2012)
3. Colazzo, D., Goasdoué, F., Manolescu, I., Roatiş, A.: RDF analytics: lenses over semantic graphs. In: WWW (2014)
4. Gottron, T., Knauf, M., Scherp, A., Schaible, J.: ELLIS: interactive exploration of linked data on the level of induced schema patterns. In: Proceedings of the 2nd International Workshop on Summarizing and Presenting Entities and Ontologies (SumPre 2016) Co-located with the 13th Extended Semantic Web Conference (ESWC 2016), Anissaras, Greece, 30 May 2016 (2016)
5. Jarrar, M., Dikaiakos, M.: A query formulation language for the data web. IEEE Trans. Knowl. Data Eng. **24**(5), 783–798 (2012)
6. Joshi, A.K., Hitzler, P., Dong, G.: Logical linked data compression. In: Cimiano, P., Corcho, O., Presutti, V., Hollink, L., Rudolph, S. (eds.) ESWC 2013. LNCS, vol. 7882, pp. 170–184. Springer, Heidelberg (2013). doi:10.1007/978-3-642-38288-8_12
7. Khatchadourian, S., Consens, M.P.: ExpLOD: summary-based exploration of interlinking and RDF usage in the linked open data cloud. In: Aroyo, L., Antoniou, G., Hyvönen, E., ten Teije, A., Stuckenschmidt, H., Cabral, L., Tudorache, T. (eds.) ESWC 2010, Part II. LNCS, vol. 6089, pp. 272–287. Springer, Heidelberg (2010)
8. Konrath, M., Gottron, T., Staab, S., Scherp, A.: SchemEX - efficient construction of a data catalogue by stream-based indexing of linked data. J. Web Sem. **16**, 52–58 (2012)
9. Langegger, A., Wöß, W.: RDFStats - an extensible RDF statistics generator and library. In: DEXA (2009)

10. Lopez, V., Unger, C., Cimiano, P., Motta, E.: Evaluating question answering over linked data. Web Seman. Sci. Serv. Agents WWW **21**, 3–13 (2013)
11. Mihindukulasooriya, N., Poveda Villalon, M., Garcia-Castro, R., Gomez-Perez, A.: Loupe - an online tool for inspecting datasets in the linked data cloud. In: ISWC Posters & Demonstrations (2015)
12. Palmonari, M., Rula, A., Porrini, R., Maurino, A., Spahiu, B., Ferme, V.: ABSTAT: linked data summaries with ABstraction and STATistics. In: Gandon, F., Guéret, C., Villata, S., Breslin, J., Faron-Zucker, C., Zimmermann, A. (eds.) ESWC 2015. LNCS, vol. 9341, pp. 128–132. Springer, Heidelberg (2015). doi:10.1007/978-3-319-25639-9_25
13. Paulheim, H., Bizer, C.: Type inference on noisy RDF data. In: Alani, H., Kagal, L., Fokoue, A., Groth, P., Biemann, C., Parreira, J.X., Aroyo, L., Noy, N., Welty, C., Janowicz, K. (eds.) ISWC 2013, Part I. LNCS, vol. 8218, pp. 510–525. Springer, Heidelberg (2013)
14. Perer, A., Shneiderman, B.: Integrating statistics and visualization: case studies of gaining clarity during exploratory data analysis. In: Proceedings of the SIGCHI Conference of Human Factors in Computing Systems, pp. 265–274. ACM (2008)
15. Peroni, S., Motta, E., d'Aquin, M.: Identifying key concepts in an ontology, through the integration of cognitive principles with statistical and topological measures. In: Domingue, J., Anutariya, C. (eds.) ASWC 2008. LNCS, vol. 5367, pp. 242–256. Springer, Heidelberg (2008)
16. Presutti, V., Aroyo, L., Adamou, A., Schopman, B.A.C., Gangemi, A., Schreiber, G.: Extracting core knowledge from linked data. In: COLD 2011 (2011)
17. Schmachtenberg, M., Bizer, C., Paulheim, H.: Adoption of the linked data best practices in different topical domains. In: Mika, P., Tudorache, T., Bernstein, A., Welty, C., Knoblock, C., Vrandečić, D., Groth, P., Noy, N., Janowicz, K., Goble, C. (eds.) ISWC 2014, Part I. LNCS, vol. 8796, pp. 245–260. Springer, Heidelberg (2014)
18. Staab, S., Studer, R.: Handbook on Ontologies. Springer, Heidelberg (2010)
19. Troullinou, G., Kondylakis, H., Daskalaki, E., Plexousakis, D.: RDF digest: efficient summarization of RDF/S KBs. In: Gandon, F., Sabou, M., Sack, H., d'Amato, C., Cudré-Mauroux, P., Zimmermann, A. (eds.) ESWC 2015. LNCS, vol. 9088, pp. 119–134. Springer, Heidelberg (2015)
20. Unger, C., Forascu, C., Lopez, V., Ngomo, A.N., Cabrio, E., Cimiano, P., Walter, S.: Question answering over linked data (QALD-4). In: CLEF (2014)
21. Völker, J., Niepert, M.: Statistical schema induction. In: Antoniou, G., Grobelnik, M., Simperl, E., Parsia, B., Plexousakis, D., De Leenheer, P., Pan, J. (eds.) ESWC 2011, Part I. LNCS, vol. 6643, pp. 124–138. Springer, Heidelberg (2011)
22. Zhang, X., Cheng, G., Qu, Y.: Ontology summarization based on rdf sentence graph. In: WWW (2007)

2nd Int. Workshop on Semantic Web for Scientific Heritage (SW4SH)

Text Encoder and Annotator: An All-in-one Editor for Transcribing and Annotating Manuscripts with RDF

Fabio Valsecchi[1]([✉]), Matteo Abrate[1], Clara Bacciu[1], Silvia Piccini[2], and Andrea Marchetti[1]

[1] Institute of Informatics and Telematics (IIT),
National Research Council (CNR), Via G. Moruzzi 1, Pisa, Italy
{fabio.valsecchi,matteo.abrate,clara.bacciu,andrea.marchetti}@iit.cnr.it
[2] Institute for Computational Linguistics (ILC),
National Research Council (CNR), Via G. Moruzzi 1, Pisa, Italy
silvia.piccini@ilc.cnr.it

Abstract. In the context of the digitization of manuscripts, transcription and annotation are often distinct, sequential steps. This could lead to difficulties in improving the transcribed text when annotations have already been defined. In order to avoid this, we devised an approach which merges the two steps into the same process. Text Encoder and Annotator (TEA) is a prototype application embracing this concept. TEA is based on a lightweight language syntax which annotates text using Semantic Web technologies. Our approach is currently being developed within the *Clavius on the Web* project, devoted to studying the manuscripts of Christophorus Clavius, an influential 16th century mathematician and astronomer.

Keywords: Manuscript transcription · Annotation · RDF · Semantic Web

1 Introduction

Within the field of Digital Humanities, several projects are devoted to preserving, analyzing and studying the large amounts of manuscripts, books, newspapers, maps, photos and paintings stored in archives, museums and libraries around the world.

Transcription and annotation are widely used methods to make these sources accessible to a wider audience, by elucidating the context and the content of this significant cultural heritage.

In this area, we think that Semantic Web technologies, such as the Resource Description Framework (RDF), could make possible to approach text annotation in an innovative way. RDF-based annotations provide a method for enriching texts using structured data already described in details and maintained by the

© Springer International Publishing AG 2016
H. Sack et al. (Eds.): ESWC 2016 Satellite Events, LNCS 9989, pp. 399–407, 2016.
DOI: 10.1007/978-3-319-47602-5_52

Semantic Web community (i.e., Linked Data sets). In addition, due to the interlinked structure of Linked Data, RDF-based annotations produce valuable annotated documents, characterized by a strong connection with external resources.

In this work we present the Text Encoder and Annotator (TEA), a web based tool for transcribing and annotating digital facsimiles of manuscripts or printed texts. It was designed as a fundamental tool for the *Clavius On the Web*[1] project [1], which aims to restore and enrich the manuscripts written by Christophorus Clavius (1538-1612), one of the most respected and influential mathematicians and astronomers of his time. Among the manuscripts analysed is his correspondence with other famous scientists of his era as well as his own scientific works. His *Commentaries on Sacrobosco's De Sphaera mundi* or on *Euclid's Elements* were a particular focus, since major historical figures such as René Descartes, Marin Mersenne and Johannes Kepler are known to have built their knowledge on them.

1.1 Background

In Hemminger and TerMat [4] annotations are defined as "markings (e.g., highlights) or comments made by a human agent that exist within or are attached to the text, whether in paper or digital format". In other words, annotation consists in attaching additional information such as comments, tags or links, to specific portions of a text. The process of annotating is a practice that dates back to ancient times: while studying and transcribing manuscripts, men of science tended to add explanatory notes which could take the form of interlinear or marginal glosses, brief scholia for personal purposes or postils. In the 5th century BC scholars were already adding glosses to Homer's works, and this was common practice for medieval scribes when copying biblical and legal manuscripts. The advent of the digital age has not changed this, on the contrary attention is nowadays directed to the development of tools to assist scholars in this task. Annotations can be mainly performed using two methods: *inline* and *standoff*, which reflects the ancient dichotomy between interlinear and marginal glosses, mutatis mutandis. The inline method includes annotations directly within the text, while standoff stores them separately from the text. Inline markup keeps the annotations and the annotated text close together, but it has the drawback of weighing the document down. Moreover, depending on the complexity of the markup language, the text could become hard to read. This aspect is crucial and must be taken into account when developing manual annotation tools, as users need to be able to read the annotated text with ease. Last but not least, a complex and heavyweight markup language could make the manual annotation process even more difficult for two reasons. Users have to firstly know all the syntax rules and secondly write a considerable amount of additional markup. In contrast, the standoff approach does not have markup overloading problems due to its total independence from the resource text. Annotations are in fact separately defined in a different location where the relative text offsets are stored and

---

[1] http://claviusontheweb.it/.

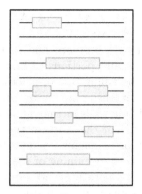

(a) Inline markup allows annotations to be included directly within the text keeping close the text fragments that have to be annotated with their annotations. A heavy markup language could make the text both heavy and hard to read. Instead a complex annotation language could make the annotation process difficult.

(b) Standoff annotations have no markup overloading problem since annotations are not stored within the text but in a different location. If transcription and annotation are treated as different steps, standoff could raise problems during the latter step if a transcription error is found.

Fig. 1. .

kept up to date. In addition, standoff markup has the advantage of allowing overlapping annotations. Nevertheless, this approach has some drawbacks related to the sequential process of transcribing and annotating. Typically, the available tools of this type separate the transcription and the annotation phases. However, if a transcription error is found during the subsequent annotation phase, it is necessary to recompute the offsets in order to reflect the changes in the transcribed text. This implies an automatic recomputation of the offsets, a process that could be complex and costly. The logical conclusion is therefore to make transcription and annotation a joint process (Fig. 1).

1.2 Our Approach

In literature, several projects and initiatives have been already conducted in order to develop tools for annotating texts and also images. For instance, Pundit [3] and Refer.cx [14] provide RDF-based annotation features but limited only to web pages and without the possibility of transcribing text. Brat [13] allows text annotation supported by Natural Language Processing technology, however it does not provide a transcription feature. RDFaCE [8] is a text editor for annotating text of web pages using a graphical UI and displaying results with different views such as WYSIWYG and WYSIWYM (i.e., What You See Is What You Get/Mean). Amaya[2], developed in the W3C Annotea project [7], is

---

[2] https://dev.w3.org/Amaya/doc/WX/Annotations.html.

a Web browser and editor for annotating text within web pages. Among the online resources provided by the collaborative initiative of Pelagios [6,12], the Recogito tool [11] allows places to be annotated both on maps and texts. The tool provides features for marking text fragments and portions of images but the current version has been devised mainly for the annotation of places. Other examples of annotation tools can be found in the extensive survey by Uren [15].

To the best of our knowledge, none of the existing tools include an approach such the one we are proposing. From the analysis of the state of the art we conducted, it seems that there is no specific tool for managing documents (e.g., texts, images) which can assist scholars from the very beginning of the process. In particular, most of the tools lack a transcription feature and support the user only in annotation. The correct tool should support users in studying their documents by providing the digitized version of their manuscripts thus allowing for transcription and annotation.

Therefore, the core idea of this work is to combine transcription and annotation of text, thus streamlining the workflow process. Hence, we devised a lightweight language that enables this continuous and mixed process. Another keypoint is that we propose to treat every textual phenomenon as an annotation independently of its specific type (e.g., semantic, syntactic, lexical). Every portion of text is treated in the same way, and RDF-based annotations are used to describe their content. RDF supports our purpose by allowing any possible annotation to be specified using the enormous amount of ontologies, vocabularies and Linked Data sets available on the Web.

2   Text Encoder and Annotator

In the light of the above, this article presents the Text Encoder and Annotator, a Web application, which provides an editor to transcribe texts as well as a lightweight language for annotating them with RDF (all within the same environment). It was envisaged for linguists, historians and more generally for scholars and students. We devised a layout composed of three main views, horizontally placed along the interface of the application (Fig. 2):

1. *Image box*: it displays the digitized image of a specific manuscript to be transcribed and annotated. Zoom and pan mechanisms are available to correctly select the portion of manuscript a user is interested in.
2. *Editor box*: it is the main component of the tool. It is used to write text and enrich it with RDF-based annotations, which follow the specific language syntax described below. Text highlighting identifies the markup and improves its legibility. Moreover, a top bar contains shortcut buttons for inserting some basic annotations, characterized by common RDF predicates, such as *rdfs:seeAlso*, which can be used to link to external resources, *rdfs:comment* to specify text comments and *foaf:page* to include hyperlinks related to the topic of the annotation.
3. *Diagram box*: it is an optional view that can be activated on-demand if the user needs a summary of their annotations. The node-link diagram contains a

white node representing the whole text of the document, blue nodes identifying the portion of text referred to in an annotation, orange nodes displaying the identifiers of the annotation and gray nodes describing the objects of the RDF triples specified. The edges between orange and gray nodes represent the predicates of the triples defined in the annotations.

The tool has been implemented using different client-side libraries based on Javascript. Backbone.js[3] has been used for structuring the web application with different modules according to the Model-View-star (MV*) paradigm, the Jison library[4] for generating a parser able to analyze the lightweight language, CodeMirror[5] for including the text editor and D3.js[6] combined with Cola.js[7] for creating the summarizing visualization.

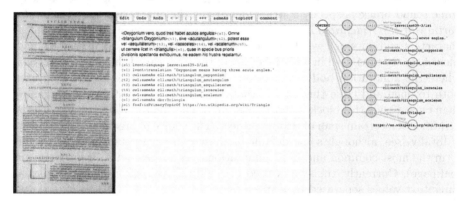

Fig. 2. The interface of the application is mainly composed of three views. From left to right there is a box containing the image of a document, an editor allowing the transcription and annotation and a diagram displaying a visual summary of the annotations. The prototype tool is available on github at http://github.com/nitaku/TEA

2.1 Lightweight Language

Considering the pros and cons of inline and standoff annotations discussed above, we think that a hybrid lightweight syntax is a suitable solution to fulfill the requirement for a simultaneous workflow of transcription and annotation.

We propose a Lightweight Markup Language[8] (LML) employed only within the interface of TEA, in order to provide an easy and quick way of transcribing and annotating text using RDF. The main reason behind the adoption of an LML is that common markup languages based on XML (e.g., TEI) are not easy to write and read in their raw form, due to their complex syntax. Moreover, the

---

[3] http://backbonejs.org/.
[4] http://zaa.ch/jison/.
[5] http://codemirror.net.
[6] http://d3js.org.
[7] http://marvl.infotech.monash.edu/webcola/.
[8] http://en.wikipedia.org/wiki/Lightweight_markup_language.

use of LMLs has already proven to be beneficial in other systems such as the Leiden-plus[9] language employed in the papyri.info editor. It is worth clarifying that we do not propose our approach as a format for the representation and interchange of texts, consequently it cannot be compared to standards such as the Text Encoding Initiative (TEI) [5]. Furthermore our language markup should be considered as distinct from semantic markup languages like Microformat [9], RDFa [2] and Microdata [10] since it has not been not devised for annotating HTML and XML documents.

Our language is used for marking portions of text and assigning them identifiers that will be used in a different, reserved and *"standoff-like"* section of the text where the details of the annotations are specified (Fig. 3). More precisely, a portion of text can be annotated by enclosing it in a *span* using angle brackets, while round brackets specify a string identifying the annotation (i.e., ⟨*annotated portion of text*⟩ *(identifier)*). This inline syntax uses a very limited amount of characters and does not weigh the text down too much, keeping it easy-to-read. Identifiers are then used in a distinct part of the text, called the *directive section*, where the annotation body is specified. Three plus signs (i.e., $+++$) are used both to open and close this section that can be repeated within the text more than once. Inside this block, annotations can be specified as RDF triples with the identifier of a certain span of text as subject. The choice of predicates and objects is totally free, although some default predicates are suggested in order to perform the most common and basic annotations (e.g., rdfs:seeAlso, rdfs:comment, foaf:page). Currently, the syntax used in the directive section defines triples as three text values separated by a space.

2.2   Example

We here provide an example of annotation performed on a portion of text extracted from Clavius *Euclidis Elementorum Libri XV. Accessit XVI de solidorum regularium comparatione. Omnes perspicuis demonstrationubus, accuartisque scholiis illustrati.* [1] (paragraph 30). It is an annotated translation from Greek into Latin of Euclid's Elements. This text was considered one of the most comprehensive and authoritative of the XVI century. A free English translation of the Latin text fragment, shown in Fig. 3, follows: *"It is called Oxigonium as it has three acute angles. Every Oxigonium triangle, or Acutangle triangle, could be either Equilateral, or Isosceles or Scalene as you can see from the classification provided above and not reported here"*. Figure 3 shows the code resulting from the text encoding and annotation process. Portions of text are marked using spans while the body of the annotations is specified within the directive section. The annotation identifiers (e.g., s1, t1, t2) are used as the subjects of the triples while predicates and objects are freely chosen by annotators. The Latin language (i.e., lexvo:iso639-3/lat) and the translation have been specified in the first annotation $s1$ using the Lexvo ontology[10]. Lexical entries of the

---

[9] http://papyri.info/editor/documentation?docotype=text.
[10] http://lexvo.org/ontology.

mathematical lexicon of Clavius[11] (e.g., cll:math/triangulum_oxygonium) and the DBpedia Triangle resource (i.e., dbr:Triangle) have been linked through the *seeAlso* predicate of the RDF Schema[12]. The triangle entry of Wikipedia has been specified as an interesting web page (i.e., foaf:page) for the annotation *e1* using the FOAF vocabulary[13].

<Oxygonium vero, quod tres habet acutos angulos>(s1). Omne <triangulum Oxygonium>(t1), sive <acutangulum>(t2), potest esse vel <aequilaterum>(t3), vel <isosceles>(t4), vel <scalenum>(t5), ut cernere licet in <triangulis>(e1), quae in speciebus prioris divisionis spectanda exhibuimus, ne eadem hic frustra repetantur.

```
+++
(s1)  lvont:language lexvo:iso639-3/lat
(s1)  lvont:translation "..."
(t1)  rdfs:seeAlso cll:math/triangulum_oxygonium
(t2)  rdfs:seeAlso cll:math/triangulum_acutangulum
(t3)  rdfs:seeAlso cll:math/triangulum_aequilaterum
(t4)  rdfs:seeAlso cll:math/triangulum_isosceles
(t5)  rdfs:seeAlso cll:math/triangulum_scalenum
(e1)  rdfs:seeAlso dbr:Triangle
(e1)  foaf:page https://en.wikipedia.org/wiki/Triangle
+++
```

Fig. 3. An annotated fragment of the *Euclidis Elementorum Libri XV. Accessit XVI.* Text is marked with spans highlighted in blue. Identifiers in orange are used in the directive section where they correspond to the subjects of RDF triples (Color figure online)

3   Conclusion and Future Works

This article describes an approach for merging the distinct steps of transcription and annotation as a single process. We implemented a tool based on a lightweight syntax language that allows RDF annotations to be performed. We conducted some preliminary tests, which involved 50 students, who were asked to use the prototype, and provide feedback. Future works will consist in developing an improved language with a syntax capable of handling nested as well as overlapping (i.e., not hierarchically nested) annotations. A preliminary analysis showed us that the syntax presented above must be slightly changed in order to treat nested and overlapping cases. These particular phenomena will be handled by writing the identifier of an annotation on both sides of the span marking the text. In this way, it will not be possible to misinterpret the start and closure of a certain annotation span. New syntax operators will be introduced: *milestone*

---

[11] http://claviusontheweb.it/lexicon/math/.
[12] http://www.w3.org/TR/rdf-schema/.
[13] http://xmlns.com/foaf/spec/.

elements, for annotating a single location in the text (e.g., a gap), *partition* elements, for the identification of phenomena such as line, page or sentence breaks. Additional syntax will be introduced to provide shortcuts ("syntactic sugar") to the most common annotations. The Turtle syntax[14] will also be taken into account for the RDF triples specification. Finally, different formats (e.g., turtle, json, csv, xml) will be chosen for exporting annotations according to various data models (e.g., Open Annotation, NLP Interchange Format (NIF)).

References

1. Abrate, M., Del Grosso, A.M., Giovannetti, E., Duca, A.L., Luzzi, D., Mancini, L., Marchetti, A., Pedretti, I., Piccini, S.: Sharing cultural heritage: the clavius on the web project. In: Language Resources and Evaluation Conference (2014)
2. Adida, B., Birbeck, M., McCarron, S., Pemberton, S.: Rdfa in xhtml: syntax and processing. Recommendation W3C (2008)
3. Grassi, M., Morbidoni, C., Nucci, M., Fonda, S., Piazza, F.: Pundit: augmenting web contents with semantics. Literary Linguist. Comput. **28**, 640 (2013)
4. Hemminger, B.M., TerMaat, J.: Annotating for the world: attitudes toward sharing scholarly annotations. J. Assoc. Inform. Sci. Technol. **65**(11), 2278–2292 (2014)
5. Ide, N., Véronis, J.: Text Encoding Initiative: Background and Contexts. Springer Science & Business Media, New York (1995)
6. Isaksen, L., Simon, R., Barker, E.T.E., de Soto Cañamares, P.: Pelagios and the emerging graph of ancient world data. In: Proceedings of the 2014 ACM conference on Web science, pp. 197–201. ACM (2014)
7. Kahan, J., Koivunen, M.-R., Prud'Hommeaux, E., Swick, R.R.: Annotea: an open rdf infrastructure for shared web annotations. Comput. Netw. **39**(5), 589–608 (2002)
8. Khalili, A., Auer, S., Hladky, D.: The rdfa content editor-from wysiwyg towysiwym. In: Computer Software and Applications Conference. IEEE (2012)
9. Khare, R., Çelik, T.: Microformats: a pragmatic path to the semantic web. In: Proceedings of the 15th International Conference on World Wide Web (2006)
10. Ruggles, S., Sobek, M., Fitch, C.A., Hall, P.K., Ronnander, C.: Integrated public use microdata series: Version 2.0. Historical Census Projects, Department of History, University of Minnesota (1997)
11. Simon, R., Barker, E., Isaksen, L., de Soto Cañamares, P.: Linking early geospatial documents, one place at a time: annotation of geographic documents with recogito. e-Perimetron **10**(2), 49–59 (2015)
12. Simon, R., Isaksen, L., Barker, E., de Soto Cañamares, P.: The pleiades gazetteer and the pelagios project (2015)
13. Stenetorp, P., Pyysalo, S., Topić, G., Ohta, T., Ananiadou, S., Tsujii, J.: Brat: a web-based tool for nlp-assisted text annotation. In: Conference of the European Chapter of the Association for Computational Linguistics. Association for Computational Linguistics (2012)
14. Tietz, T., Waitelonis, J., Jäger, J., Sack, H.: Smart media navigator: visualizing recommendations based on linked data. In: 13th International Semantic Web Conference, Industry Track (2014)

---

[14] https://www.w3.org/TR/turtle/.

15. Uren, V., Cimiano, P., Iria, J., Handschuh, S., Vargas-Vera, M., Motta, E., Ciravegna, F.: Semantic annotation for knowledge management: requirements and a survey of the state of the art. Web Semant. Sci. Serv. Agents World Wide Web **4**(1), 14–28 (2006)

Extending the Lemon Model for a Dictionary of Old Occitan Medico-Botanical Terminology

Anja Weingart[1(✉)] and Emiliano Giovannetti[2]

[1] Seminar für Romanische Philologie,
Georg-August-Universität Göttingen, Göttingen, Germany
aweinga@gwdg.de
[2] Istituto Di Linguistica Computazionale,
Consiglio Nazionale delle Ricerche, Pisa, Italy
emiliano.giovannetti@ilc.cnr.it

Abstract. The article presents the adaptation of the lemon model (a model for lexica as RDF data) for a multilingual and multi-alphabetical lexicon of Old Occitan medico-botanical terminology. The lexicon is the core component of an ontology-based information system that will be constructed and implemented within the DFG-funded project "Dictionnaire des Termes Médico-botaniques de l'Ancien Occitan" (DiTMAO). The difficulties for the lemmatization raised by the particularities of the corpus (terms in Latin, Hebrew and Arabic script and corresponding terms in other ancient languages, mostly Hebrew and Arabic) can be perfectly solved by extending the basic properties of lemon and introducing domain specific vocabulary.

Keywords: Lemon model · RDF · Multilingual · Multi-alphabetical · Historical lexicon · Medico-Botanical terminology · Old occitan · Hebrew · Arabic

1 Introduction

The project "Dictionnaire de Termes Médico-botaniques de l'Ancien Occitan" (DiTMAO)[1] aims at constructing an ontology-based information system for Old Occitan medico-botanical terminology. The article shows the application of the lemon model[2] to the lexicon component and focuses on the modelling of the historical, multilingual terminology.

1.1 Aims, Background and Structure of the Article

Old Occitan is the medieval stage of Occitan, the autochthonous Romance language spoken in Southern France, today regional minority language with several dialects.

---

[1] DiTMAO is a joint project of the PIs Gerrit Bos (Universität zu Köln), Andrea Bozzi (Istituto di Linguistica Computazionale "Antonio Zampolli" of the CNR), Maria Sofia Corradini (Università di Pisa) and Guido Mensching (Georg-August-Universität Göttingen). The project is funded by the Deutsche Forschungsgemeinschaft (DFG).

[2] http://lemon-model.net/ (last access: 30/06/2016).

© Springer International Publishing AG 2016
H. Sack et al. (Eds.): ESWC 2016 Satellite Events, LNCS 9989, pp. 408–421, 2016.
DOI: 10.1007/978-3-319-47602-5_53

During the Middle Ages, the region and its language played a significant role in medical science due to the medical schools of Toulouse and Montpellier and the strong presence of Jewish physicians and scholars. For this reason, Old Occitan medico-botanical terminology is documented both in Latin and in Hebrew characters (cf. [3]). The DiTMAO project aims at making this terminology accessible to several scientific communities, such as those of Romance and Semitic studies, as well as that of the history of medicine.

The textual basis[3] of the lexicon, as described in [2, 9, 10], consists of medico-botanical texts in Latin and in Hebrew script. Among the sources in Hebrew script, the most prominent text type are so-called synonym lists, which contain a large amount of Old Occitan medical and botanical terms in Hebrew characters with equivalents or explanations in other languages (also spelled in Hebrew characters), mostly in (Judaeo-)Arabic, but also in Hebrew, Latin, or other Romance languages and sometimes in Greek, Aramaic or Persian. These lists can be described as ancient multilingual dictionaries, which are of particular importance for Old Occitan lexicography for two main reasons: (i) the synonym lists of the Jewish tradition include vernacular (Old Occitan) terms already from the 13th century on, hence these lists contain very early testimonies of Old Occitan technical terms. (ii) The corresponding terms in other ancient languages help to determine the meaning of otherwise opaque Old Occitan terms (cf. [3, 18, 19, 21]). A special difficulty of medieval texts in vernacular languages is that most terms are documented in a large number of variants (reflecting different spellings, dialects, or historical stages of the languages at issue). Thus the dictionary will include all variants of Old Occitan terms, together with the corresponding terms in at least six other ancient languages. Whenever possible, also a translation to modern French and English will be provided. The dictionary aims to be useful not only for users interested in Old Occitan but also in reading the numerous Medieval Hebrew medico-botanical texts written or translated in Southern France, since these texts are full of Occitan terminology and thus partially inaccessible even for readers with a good knowledge of Hebrew (cf. [22]).

After introducing the lemon model and our extensions, the article primarily deals with the lemmatization of simple and multiword terms and their representation in lemon. Furthermore, we will show how the corresponding terms in other ancient languages can be integrated and we will propose a way to resolve polysemy[4].

---

[3] The corpus consists of 11 texts in Latin script, which are mostly books of prescriptions, herbals and books about medical practices, and nine texts in Hebrew or Arabic script, which are mostly synonym lists, anonymous or contained in medico-botanical books. Each text is represented by up to four manuscripts. The corpus of DiTMAO combines already edited manuscripts ([7, 8] for texts in Latin script and [3, 4] for texts in Hebrew script. In addition, terms from several unedited manuscripts will be included.

[4] In lemon a lexicon is restricted, by definition, to exactly one language. Besides a lexicon for terms in Old Occitan, labeled ditmao, we define a lexicon for each of the other languages: ditmao_hebrew, ditmao_arabic, ditmao_latin, ditmao_greek, ditmao_aramaic and ditmao_persian.

1.2   The Ontological Conception and the Lemon Model

Current trends in linguistic and lexical resources show a growing interest towards the publishing in the context of the Semantic Web [14–16]. The sharing of lexica in accordance with linked data principles is, nowadays, mandatory: a resource (not only of linguistic nature) that cannot be accessed, shared and reused as a dataset is basically considered unreachable, and, thus, pretty much useless from a semantic web perspective. The lemon model has been developed as a standard for publishing lexica as RDF data. More precisely, lemon should be considered as an Ontology-Lexicon model for the Multilingual Semantic Web [11] and its nature and purpose perfectly satisfy our needs of representing the DiTMAO lexicon and the relative ontologies. DiTMAO consists of three main domains: (i) the lexicographic domain, including the lemmatized forms (lemma, variants and corresponding terms in other ancient languages) and their linguistic and lexicographic description. (ii) The conceptual domain, describing the meaning of each term by means of subontologies for the fields of botany, zoology, mineralogy, human anatomy, diseases and therapy (medication, medical instruments). We aim to complement the onomasiological description, if possible, with a modern scientific classification, for at least most of the plant names, and a medieval classification[5] of plants and other simple drugs. (iii) The documentation domain, giving the source for each form of a term and its meaning. The documentation is indispensable for a historical (diachronic) dictionary.

The lemon model will be extended with a documentation domain and new vocabulary that is necessary for the lemmatization of a historical multilingual and multi-alphabetical dictionary[6].

2   The Lexicographic Component

In the following sections, we describe the lemmatization of simple and multiword terms in Latin and Hebrew script and their representation in lemon. The representation will be illustrated by some representative examples from our corpus. The fact that we use just a few terms should not obscure the fact that our corpus contains about 5800 Old Occitan forms in Latin script and 3200 forms in Hebrew script. Furthermore, the corresponding terms in the other ancient languages amount to 3050 terms.

---

[5] The medieval classification follows the Galenic system of four basic body humors (blood, yellow bile, black bile and phlegm). The humors are associated with the two primary qualities by cross-combining the pairs HOT–COLD and DRY–WET (cf. [6]) The simple drugs are classified by these quality pairs together with a certain degree of intensity, which varies from one to four (cf. [13]). In order to ensure that the categorization is in conformity with the classification used in medieval Southern France, we will only introduce the classification provided in the texts of our corpus.

[6] The full extension of the lemon model, together with all data (without copyright restrictions) will be published on the project web site: https://www.uni-goettingen.de/en/487498.html (last access: 30/06/2016).

2.1 Lemmatization and Determination of Variants

As a general criterion of lemmatization, it has been decided for DiTMAO that a *lemma* is a term in Latin characters. All forms that differ from the lemma are classified as *variants*. Among the forms in Latin script the lemma is determined following a set of criteria[7] and the form of an Old Occitan lemma is the oblique[8] singular form for nouns, the oblique singular masculine for adjectives, and the infinitive for verbs. For example, the corpus contains the following variants for the word meaning 'hemp seed': *canabo*, *canebe*, *canabos*, and variants in Hebrew characters (represented here together with the transliterated forms[9]): קנבוש/QNBWŠ, קנבוש/QiNaBWuŠ, קנבונש/QNBWNŠ. The form *canabo* is taken as lemma or leading variant. The form *canabos* is the plural form of the lemma *canabo*. It is classified as morphological variant. The form *canebe* differs with respect to spelling and pronunciation. The form is thus classified as grapho-phonetic variant. As a general definition, the variants in Hebrew characters are all alphabetical variants. The forms קנבוש/QNBWŠ and קנבוש/QiNaBWuŠ are alphabetical variants of the plural form *canabos*. In this sense they are variants of a variant. The form קנבוש/QiNaBWuŠ additionally differs with respect to phonology. As indicated by the vowel signs, the initial syllable has to be interpreted as [ki] instead of [ka]. The form קנבונש/QNBWNŠ (read: "canabons") has no corresponding form in Latin script in our corpus. It is thus classified as alphabetical variant of the lemma, and additionally as grapho-phonetic[10] and morphological variant. Furthermore, concerning variants in Latin characters, there are pure graphic variants, where the spelling does not reflect a difference in pronunciation e.g. *alcanna* and *alquana*.

A certain difficulty for lemmatization lies in the fact that about 40 % of the terms are only documented in Hebrew characters. Nevertheless, the general criterion for lemmatization (a lemma is a term in Latin script) has been established for two main reasons. First of all, it is not possible to uniquely link a Hebrew character to a Latin character. For example the letter Alef (א - ᾿) may represent different vowels e.g. it stands for /e/in אשפרמא/᾿ŠPRM᾿ (read: "esperma", 'sperm'), for /a/in ארמולש/᾿RMWLŠ (read "armols", 'orache'). The combinations of initial Alef with Yod or Waw can be interpreted as /i/or /e/like in אינגיליש/᾿YNGYLŠ (read: "enguilas", 'eels') or as /o/o / u/like in אורטיגש/᾿WRṬYGŠ (read "ortigas", 'stinging nettles'). Thus, having lemmata in two alphabets would additionally complicate the string search and the display of the

---

[7] The criteria are hierarchically: (i) the simple term is chosen over the compound term, e.g. *oli*, not *oli rossat*; (ii) the form that corresponds to the lemma in most of the standard dictionaries is chosen. e.g., *bleda* is chosen over *bleta* (the form *bleta* is considered a cultism) (iii) the form that is closer to the etymon is chosen, e.g., *oli* not *holi* (< Lat. oleum); (iv) the most frequent form is chosen.

[8] Old Occitan preserved the Vulgar Latin two-case system (nominative vs. oblique case) which was lost by the fourteenth century and the nominative forms have been abandoned in favor of the oblique forms (cf. [3]).

[9] For the transliteration of Hebrew characters, we use the system described in [3, 20]: Alef (א - ᾿), Bet (ב - B), Gimel (ג - G), Gimel (ג׳- Ǧ), Dalet (ד - D), He (ה - H), Waw (ו - W), Zayin (ז - Z), Ḥet (ח - Ḥ), Ṭet (ט - Ṭ), Yod (י - Y), Kaf (כ ך - K), Lamed (ל - L), Mem (מ ם - M), Nun (נ ן - N), Samekh (ס - S), Ayin (ע - ᾿), Pe (פ ף - P), Ṣade (צ ץ - Ṣ), Qof (ק - Q), Resh (ר - R), Shin (ש - Š), Tav (ת - T).

[10] The form קנבונש/QNBWNŠ contains a so-called *n-mobile*, a particular phonological characteristic of Old Occitan (cf. [3]).

results in alphabetical order. In case a term is only documented in Hebrew characters, a corpus-external lemma, a form documented in other dictionaries, will be included. But in some cases, there is no such corpus-external lemma (so the variant in Hebrew spelling is the only documented form), and we have to introduce a hypothetical or reconstructed form. For example for the term אנאקירד - 'N'QYRD (read "anacard"), we introduce the form *anacard* as hypothetical Old Occitan form with the meaning 'marking nut', fruit of Semecarpus anacardium L. . The meaning is documented for the Arabic term בלאדר/BL'DR that features as its synonym in the lists edited in [4]. Thus, we need to indicate for a lexical entry whether the lemma is corpus-external, a reconstructed or a hypothetical from.

2.2   Modelling the Lemma and Its Variants

A lexicon entry in lemon consists of a `Form` and a `LexicalSense`. For the lemmatization, the class `Form` and its relations with `LexicalEntry` (lexi-calForm and its subproperties `canononicalForm` and `otherForm`) are relevant. In lemon the lemma *canabo* will have the following shape:

```
:canabo a lemon:LexicalEntry;
lemon:canonicalForm [lemon:writtenRep "canabo" @aoc-Latn;
    lexinfo:partOfSpeech lexinfo:noun ;
    lexinfo:gender lexinfo:masculine ;
    lexinfo:number lexinfo:singular ] .
```

The lemma is represented by the `canonicalForm` of the entry and its realization is the written representation (`writtenRep`). The language, although inferable from the lexicon, will be represented together with the ISO 15924 script code: `Latn` for Latin, `Arab` for Arabic, and `Hebr` for Hebrew. This is an elegant way to avoid the definition of a property specifying the script type. The linguistic information like part of speech, gender and number will be integrated as attribute-value pairs from the Lexinfo ontology[11], an extension of lemon that provides data categories for linguistic annotations. These will be defined as subproperties of the property `lemon:property`. In a similar vein, the labels for corpus-external lemmata and hypothetical and reconstructed forms can be added to the `canonicalForm`.

```
ditmao:lemmaInfo rdfs:subPropertyOf lemon:property .
```

The subproperty `ditmao:lemmaInfo` will have the following values: `dit-mao:corpusExternalLemma`, `ditmao:hypotheticalForm` and `ditmao:reconstructedForm`. For the representation of variants, the lemon model only provides the relation `otherForm`. The variant *canabos* has the following entry:

```
lemon:otherForm [lemon:writtenRep "canabos" @aoc-Latn ;
lexinfo:number lexinfo:plural] .
```

---

[11] http://www.lexinfo.net/ontology/2.0/lexinfo.owl (last access: 30/06/2016).

The fact that *canabos* is a morphological variant can be inferred from the value of `lexinfo:number`. An alphabetical variant can be formalized by adding a script tag to the language tag e.g. aoc[12]-Hebr or aoc-Arab. In order to give the transliteration, we adopted `lexinfo:transliteration` which is defined as a subproperty of `lemon:representation` (the superproperty of `lemon:writtenRep`), in accordance to the Lemon Cookbook [17]. The specific transliteration alphabets are defined as subproperties of `lexinfo:transliteration`. For the DiTMAO, a transliteration of Hebrew and Arabic is needed. The former is labelled `HebrTransliteration` and the latter `ArabTransliteration` with the respective abbreviations `HebrTrsl` and `ArabTrsl`.[13] The entry for קנבונש/ QNBWNŠ (read "canabons") would have the following shape.

```
lemon:otherForm [lemon:writtenRep "קנבונש" @aoc-Hebr ;
ditmao:HebrTransliteration "QNBWNŠ" @aoc-HebrTrsl ;
lexinfo:number lexinfo:plural] .

lexinfo:transliteration rdfs:subPropertyOf
lemon:representation .

ditmao:HebrTransliteration rdfs:subPropertyOf
lexinfo:transliteration .
```

A problem is the formalization of the graphic and grapho-phonetic variants. Only users who are familiar with Old Occitan phonology and dialectology may distinguish graphic from grapho-phonetic variants. But as the dictionary also wants to reach researchers from other domains, an indication of these types of variants is desired. We propose to specify all types of variants (morphological, alphabetical, grapho-phonetic and graphic variants) as values of `ditmao:variant`, defined as a subproperty of `lemon:property`. This subproperty will take the following values: `ditmao:alphabeticalVariant`, `ditmao:graphicVariant`, `ditmao:morphologicalVariant`, and `ditmao:graphophoneticVariant`. The form *canebe* bears only the value `ditmao:graphophoneticVariant`. Additionally to the marking of the script and grammatical number, the entry קנבונש/QNBWNŠ has the following shape:

---

[12] For Old Occitan, ISO proposes the language tag *pro*, which is derived from the term Provençal. But Provençal, like Gascon, Limousin, Languedocian, and Auvergnat, has to be considered a dialect of Old Occitan (cf. [1]). Thus, we take the name Old Occitan (French Ancien Occitan) to be the correct hyperonym and define a new language tag *aoc* for DiTMAO.

[13] The alternative option is to label the transliteration alphabet as Latin script, but this would not be correct, because the transliteration alphabets contain special phonetic symbols e.g. the symbols ' and ' (replacing Alef and Ayin, respectively).

```
lemon:otherForm [lemon:writtenRep "קנבוש" @aoc-Hebr ;
  ditmao:HebrTransliteration "QNBWNŠ" @aoc-HebrTrsl ;
  lexinfo:number lexinfo:plural;
  ditmao:variant ditmao:alphabeticalVariant;
  ditmao:variant ditmao:morphologicalVariant;
  ditmao:variant ditmao:graphophoneticVariant ]
```

The other variants in Hebrew characters have been classified as variants of a variant. The terms קנבוש/QNBWŠ and קָנָבּוּש/QiNaBWuŠ are alphabetical variants of the morphological variant *canabos*. In order to represent a relation between two forms of one lexical entry, lemon provides the property formVariant. A symmetric subproperty of formVariant, ditmao:varOfVar, will be defined:

```
ditmao:varOfVar rdfs:subPropertyOf lemon:formVariant .
```

The subproperty ditmao:varOfVar will be added to the variant in Hebrew characters. An exemplary entry is shown below for the form קָנָבּוּש/QiNaBWuŠ.

```
lemon:otherForm :canabos ;
lemon:otherForm : קָנָבּוּש ;

:canabos [lemon:writtenRep "canabos" @aoc-Latn;
  lexinfo:number lexinfo:plural;
  ditmao:variant ditmao:graphophoneticVariant ] .

: קָנָבּוּש [lemon:writtenRep "קנבוש" @aoc-Hebr;
  ditmao:HebrTransliteration "QiNaBWuŠ" @aoc-HebrTrsl ;
  lexinfo:number lexinfo:plural ;
  ditmao:varOfVar :canabos ;
  ditmao:variant ditmao:graphophoneticVariant ;
  ditmao:variant ditmao:alphabeticalVariant ] .
```

2.3 Modeling Multiword Expressions

The multiword expressions contained in our corpus are mostly noun-adjective expressions, like *goma arabica*, 'arabic gum' or syntagmatic noun-preposition-noun expressions, like *goma de gingibre*, 'ginger gum'. Multiword terms are classified as sublemma in the sense of a strict alphabetical macrostructure of a dictionary. Both nouns, *goma arabica* and *gomma de ginibre*, are sublemmata of the lemma *goma*. Sublemmata are modeled as a relation between two lexical entries by means of the property LexicalVariant. For DiTMAO, a sub-property of LexicalVariant, sublemmaOf, will be defined. The entry of the term *goma arabica* will have the following entry:

```
:goma_arabica a lemon:LexicalEntry;
  ditmao:sublemmaOf :goma .
```

For a description of the internal structure of multiword expressions, lemon provides a phrase structure module. Multiword terms can be decomposed into their components by means of an ordered list, the `lemon:componentList`. A list consists of components, which are linked by means of the property `lemon:element` to the lexical entries. Each component can be associated to a leaf of a tree structure, representing the internal structure of the phrases *goma arabica* and *goma de gingibre*. The determinatum *goma* is the head of the noun phrase and the determinans is the adjective phrase or the prepositional phrase, which are themselves decomposed into an adjective and a preposition + noun phrase. Each component is linked to its lemma, which is unproblematic for the noun *goma de gingibre*, because the components correspond to the canonical form of the lemmata at issue. However, the term *arabica* is inflected for feminine and the canonical form of an adjective is, per definition, the masculine singular form. For relating such components, we cannot use the `lemon:element` property since it is defined to have the class `LexicalEntry` as range. For this reason, we chose to define a specific property, whose range is the `lemon:Form`:

```
ditmao:formElement rdfs:subPropertyOf lemon:property .
```

The decomposition of *goma arabica* is shown in Fig. 1.

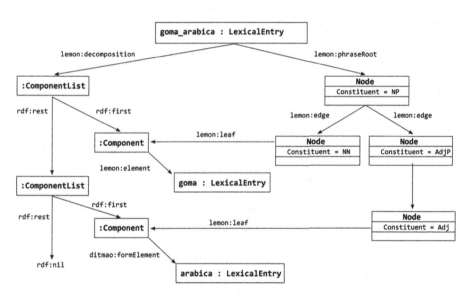

Fig. 1. Decompositon of *goma arabica*

A particularity of our corpus is multiword expressions, consisting of an Old Occitan and a Hebrew word e.g. חתום בול/ BWL ḤTWM meaning 'sealed clay/earth' and אגוז מושקאדא/'GWZ MWŠQ'D', meaning 'nutmeg'. The former consists of an Old Occitan head noun, בול/BWL, an alphabetical variant of the term *bol*, followed by a Hebrew participle passive *ḥatum*. The latter has a Hebrew head noun אגוז/'GWZ, meaning 'nut', followed by an alphabetical variant of the Old Occitan adjective *muscada*. These mixed terms mostly occur in Hebrew prose texts or in Hebrew translations and should be considered as foreign technical terms of Jewish physicians living in the Southern France. As for the lemmatization, the terms are taken to be lexical entries of the ditmao_hebrew lexicon, irrespective of the language of the head noun. Due to the decomposition function of lemon, we can preserve the information that the components בול/BWL and מושקאדא/MWŠQ'D' are variants of the Old Occitan terms *bol* and *muscat*[14], respectively. How these terms can be represented in lemon will be discussed in the following subsection.

The term אגוז מושקאדא/'GWZ MWŠQ'D' is a sublemma of the Hebrew entry אגוז/'GWZ, meaning 'nut'. The adjective מושקאדא/MWŠQ'D' is the alphabetical variant of the feminine form *muscada*, hence a variant of a variant.

```
:ditmao_hebrew lemon:entry :אגוז_מושקאדא
ditmao:sublemmaOf :אגוז
: אגוז_מושקאדא lemon:canonicalForm [lemon:writtenRep "אגוז
מושקאדא" @heb-Hebr ;
     ditmao:HebrTransliteration "'GWZ MWŠQ'D'" @hebr-
HebrTrsl ].
```

In order to decompose the term, the relations lemon:element and dit-mao:formElement are needed, because the head noun corresponds to a lemma of the ditmao_hebrew lexicon and the adjective is a variant.

```
lemon:decomposition ( [lemon:element :אגוז ]
               [ditmao:formElement :מושקאדא ] ) .
```

The representations for the terms אגוז/'GWZ and מושקאדא/MWŠQ'D' have the following shape.

```
:ditmao_hebrew lemon:entry :אגוז
: אגוז lemon:canonicalForm [lemon:writtenRep "אגוז" @heb-Hebr
;
ditmao:HebrTransliteration "'GWZ" @heb-HebrTrsl ].

:ditmao lemon:otherForm : מושקאדא
: מושקאדא lemon:otherForm [lemon:writtenRep          @aoc-Heb
;
ditmao:HebrTransliteration "מושקאדא" @aoc-HebrTrsl ;
  ditmao:varOfVar :muscada ;
  ditmao:variant ditmao:alphabeticalVariant ] .
```

As for the term בול חתום/BWL ḤTWM, which consists of an Old Occitan head noun and a Hebrew participle passive, lemmatization is more problematic. In order to define a sublemma relation we would need to assume, contrary to fact, that the simple term בול/BWL was a Hebrew medical term. An equally undesired solution would be to allow

---

[14] מושקאדא / MWŠQ'D' is an alphabetical variant of the feminine singular form *muscada*. The lemma of adjectives is per definition the the masculin singular form, here *muscat*.

the sublemma relation to be valid across the lexica. Thus, multiword terms with an Old Occitan head noun will not be lemmatized with respect to the sublemma relation, but the information that the word בול/BWL is an alphabetical variant of the Old Occitan term *bol* may be preserved, due to the decomposition, as shown below.

```
:ditmao_hebrew lemon:entry בול_חתום:
בול_חתום lemon:canonicalForm [lemon:writtenRep " בול_חתום "
@heb-Hebr ;
ditmao:HebrTransliteration "BWL ḤTWM" @hebr-HebrTrsl ].


lemon:decomposition ( [lemon:element בול: ]
[ditmao:formElement חתום: ]  ) .


:ditmao lemon:otherForm בול:
בול: lemon:otherForm [lemon:writtenRep "בול" @aoc-Hebr
ditmao:HebrTransliteration "BWL" @aoc-HebrTrsl;
ditmao:varOfVar :bol ; ditmao:variant
ditmao:alphabeticalVariant  ] .
```

Further we preserve the information that the word חתום/ḤTWM is a morphological variant of the lemma חתם/ḤTM

```
:ditmao_hebrew lemon:otherForm חתום:
חתום: lemon:otherForm [lemon:writtenRep "חתום" @hebr-Hebr
;
ditmao:HebrTransliteration "ḤTWM"@hebr-HebrTrsl;
ditmao:variant ditmao:morpholgicalVariant ].
```

In some cases a mixed term is documented in a synonym list together with a term in Old Occitan. The mixed term will be classified as corresponding term, in the same way as simple terms or other monolingual multiword expressions. E.g. the term אגוז מושקאדא /'GWZ MWŠQ'D' appears together with the Old Occitan term נוץ מושקאדא/NWṢ MWŠQ'D' (read: "noz muscada") and the Arabic term גוז בוי/GWZ BWY (read: "ğawz bawwā"). The mixed term and the Arabic term will be linked as correspondence to the sublemma in Latin script: *noze moscada*. How these corresponding terms are modeled in lemon will be discussed in the next section.

2.4 Corresponding Terms and Other Sense Relations

As mentioned in the introduction, our corpus contains corresponding terms in other ancient languages, which have been considered as synonyms by the authors of the manuscripts. For example the term ליטוגא/LYṬWG' (a variant of *laytugua*) figures as synonym of the Aramaic term חסא/ḤS' and the Arabic term כס/KS in the synonym lists

edited in [3]. The meaning of all three terms is documented[15] as 'lettuce' (in particular
Lactuca sativa L.). But even if the terms have exactly the same meaning, they should
not be considered as synonyms in the modern understanding of the term, because they
do not belong to the same language (cf. [5]). In order to model this relation in lemon,
we propose the property ditmao:correspondence, as a subproperty of
senseRelation. It links the senses of two lexical entries that belong to distinct
lexica of ancient languages. In order to give a corresponding term in modern French
and modern English, the subproperty lemon:translationOf will be used. The
relations have to be kept apart for mainly two reasons: corresponding terms and
translations belong to different historical stages and to different registers. The former
are medieval technical terms and the latter are modern common names. Furthermore,
the corpus contains Old Occitan terms that are synonyms in the modern understanding
of the term, e.g. the terms *litargia* and *mal de dormir* have the meaning: 'fatigue'. The
corresponding LexicalSense of both terms is linked via the subproperty lemon:
equivalent. The relations are represented in Fig. 2.

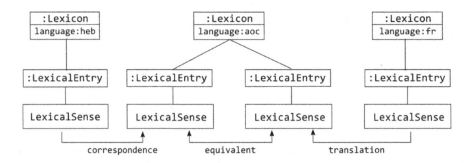

Fig. 2. Relating lexical senses in DiTMAO

But about 20 % of the lemmata in our corpus have more than one meaning. For
example, we often find polysemic plant names which designate several species of a
genus, e.g. the term *laureola* is documented with the names for the species Daphne
oleoides Schreb., Daphne gnidium L., and Daphne sericea Vahl. In lemon, polysemy
will be formalized as follows: a LexicalEntry has several instances of LexicalSense. The Arabic and Hebrew corresponding terms that feature in the synonym
lists, give an additional meaning: Daphne mezereum L. The entry of *laureola* has four
instances of LexicalSense. Each LexicalSense has a translation into modern
French and English and the LexicalSense referring to Daphne mezereum L. will be
linked via ditmao:correspondence to the respective Arabic and Hebrew entries.
Furthermore, each LexicalSense of *laureola* has a referent in the botanical branch
of the ontology, giving a general description of the plant e.g. that it is a kind of shrub.
These entities are linked to the modern classification, here the binominal plant names,

---

[15] For a complete documentation see pp. 225/226 of [3].

and to a medieval classification. The term *laureola* is described as HOT and DRY in the third degree (see [12] and fn. 5). The general division of the conceptual subontology into an onomasiological subontology, a medieval and a modern classification system allows us to provide a description of the term's concepts independently from a modern or a medieval classification. This division is necessary for terms that designate e.g. medical instruments or substances whose composition is uncertain.

3 Conclusion and Outlook

We have shown how the lemon model can be adapted to the needs of historical lexicography, by defining subproperties of the basic lemon properties: `lemon:senseRelation`, `lemon:formVariant`, `lemon:element` and `lemon:property`. Furthermore, we introduced our own, domain specific, vocabulary for the description of form variants. In the spirit of lemon, and, in general, of the Semantic Web, we plan to link the dictionaries to other resources. However, at the moment the most important resource related to Old Occitan (i.e. DOM[16]) is a database and it's not exposed as a linked data. Among the resources we are planning to use to provide the conceptual references of lexical senses we cite DBpedia[17], Wikidata[18] and more domain-specific datasets, such as TDWG[19] or the Biological Taxonomy Vocabulary[20].

To ease the process of modelling of the various lexica in lemon and the construction of the ontologies of reference, we are also working on a web editor. As a matter of fact, none of the currently available tools for the editing of lexica and ontologies appears suited to our purpose. Protégé[21], probably the most used tool for the construction of ontological resources, is general enough to allow the building of lemon resources. However, the process can be quite tedious, requiring the manual construction of instances of entries, senses, forms and relations among them. In addition, it is a stand-alone tool which cannot be used collaboratively by a team of users (its Web version[22] has several limitations, as the lack of support for reasoning mechanisms and plug-in extensions). We also plan to develop a controlled natural language querying interface to ease the access to the resources.

---

[16] Dictionnaire de l'occitan médiéval - http://www.dom.badw.de/indexde.htm (last access: 30/06/2016).

[17] http://wiki.dbpedia.org/ (last access: 30/06/2016).

[18] http://www.wikidata.org (last access: 30/06/2016).

[19] http://rs.tdwg.org/dwc/rdf/dwctermshistory.rdf (last access: 30/06/2016).

[20] http://lov.okfn.org/dataset/lov/vocabs/biol (last access: 30/06/2016).

[21] http://protege.stanford.edu/ (last access: 30/06/2016).

[22] http://webprotege.stanford.edu (last access: 30/06/2016).

References

1. Bolduc, M.: Occitan Studies. In: Classen, A. (ed.) Handbook of medieval studies: terms, methods trends, vol. 2, pp. 1023–1038. De Gruyter, Berlin (2010)
2. Bos, G., Corradini, M.S., Mensching, G.: Le DiTMAO (Dictionnaire des Termes Médico-botaniques de l'Ancien Occitan): caractères et organisation des données lexicales. In: Proceedings of the XIen Congrès de l'Asociacion Internacionala d'Estudis Occitans (AIEO) (in press)
3. Bos, G., Hussein, M., Mensching, G., Savelsberg, F.: Medical synonym lists from Medieval provence: Shem Tov ben Isaak of Tortosa: Sefer ha-Shimmush. Book 29, Part1: Edition and Commentary of List 1 (Hebrew-Arabic-Romance/Latin). Brill, Leiden (2011)
4. Bos, G., Kley, J., Mensching, G., Savelsberg, F.: Medical synonym lists from medieval provence: Shem Tov ben Isaak of Tortosa: Sefer ha-Shimmush. Book 29, Part2: Edition and Commentary of List 2 (Romance/Latin-Arabic-Hebrew). Brill, Leiden (in prep)
5. Bos, G., Mensching, G.: Arabic-romance medico-botanical glossaries in hebrew manuscripts from the Iberian Peninsula and Italy. In: Aleph, vol. 15.1, pp. 9–61. Indiana University Press (2015)
6. Bruchhausen, W., Schott, H.: Geschichte, Theorie und Ethik der Medizin. Vandenhoeck/Ruprecht, Göttingen (2008)
7. Corradini, M.S.: Ricettari medico-farmaceutici medievali nella Francia meridionale. Olschki, Florence (1997)
8. Corradini, M.S.: Per l'edizione del corpus delle opere mediche in occitanico e in catalano: nuovo bilancia della tradizione manoscritta e analisi linguistica dei testi. In: Rivista di Studi Testuali, vol III, pp. 127–195. Università di Torino (2001)
9. Corradini, M.S., Mensching, G.: Les méthodologies et les outils pour la rédaction d'un Lexique de la terminologie médico-botanique de l'occitan du Moyen Âge. In: Iliescu, M., Siller-Runggaldier, H., Danler, P. (eds.) Actes du XXVe Congrès International de Linguistique et de Philologie Romanes, vol. 6, pp. 87–96. Max Niemeyer, Tübingen (2010)
10. Corradini, M.S., Mensching, G.: Nuovi aspetti relativi al Dictionnaire de Termes Médico-botaniques de l'Ancien Occitan (DiTMAO): creazione di una base di dati integrata con organizzazione onomasiologica. In: Herrero, E.C., Rigual, C.C. (eds.) Actas del XXVI Congreso Internacional de Lingüística y de Filología Románicas, vol. VIII, pp. 113–124. Max Niemeyer, Tübingen (2013)
11. Declerck, T., Buitelaar, P., Wunner, T., McCrae, J.P., Montiel-Ponsoda, E., de Cea, G.A.: Lemon: an ontology-lexicon model for the multilingual semantic web. In: W3C Workshop: The Multilingual Web - Where Are We? Madrid, 26/10/2010–27/10/2010
12. González, A.G.: Alphita. Edición crítica y comentario, SISMEL-Edizioni del Galluzzo, Florence (2007)
13. Gerabek, W.: Enzyklopädie Medizingeschichte. Walter de Gruyter, Berlin (2005)
14. Hayashi, Y.: Direct and indirect linking of lexical objects for evolving lexical linked data. In: Proceedings of the 2nd International Workshop on the Multilingual Semantic Web (MSW), vol. 775, pp. 62–67. Bonn (2011)
15. Hellmann, S., McCrae, J.P., Del Gratta, R., Frontini, F., Khan, F., Monachini, M.: Converting the parole simple clips lexicon into RDF with lemon. Semant. Web 6(4), 387–392 (2015)
16. Lezcano, L., Sánchez-Alonso, S., Roa-Valverde, A.J.: A survey on the exchange of linguistic resources: publishing linguistic linked open data on the web. Program 47(3), 263–281 (2013)

17. McCrae, J.P., Aguado-de-Cea, G., Buitelaar, P., Cimiano, P., Declerck, T., Pérez, A.G., Gracia, J., Hollink, L., Montiel-Ponsoda, E., Spohr, D., Wunner, T.: Lemon Cookbook. http://lemon-model.net/lemon-cookbook/index.html
18. Mensching, G.: Per la terminologia medico-botanica occitana nei testi ebraici: le liste di sinonimi di Shem Tov Ben Isaac di Tortosa. In: Corradini, M.S., Periñán, B. (eds.) Atti del convegno internazionale: Giornate di studio di lessicografia romanza, pp. 93–109. ETS, Pisa (2006)
19. Mensching, G.: Listes de synonymes hébraïques-occitanes du domaine médico-botanique au Moyen Âge. In: Latry, G. (ed.) La voix occitane. Actes du VIIIe Congrès Internationale d'Études Occitanes, vol. I, pp. 509–526. Presses universitaires de Bordeaux, Bordeaux (2009)
20. Mensching, G.: Éléments lexicaux et textes occitans en caractères hébreux ». In: Trotter, D. (ed.) Manuel de la philologie de l'édition, pp. 237–264. De Gruyter, Berlin (2015)
21. Mensching, G., Savelsberg, F.: Reconstrucció de la terminologia mèdica occitanocatalana dels segles XIII i XIV a través de llistats de sinònims en lletres hebrees. In: Actes del congrés per a l'estudi dels Jueus en territori de llengua catalana, pp. 69–81. Universidad de Barcelona (2004). http://www.institutmonjuic.googlepages.com/2.ACTESPDF.pdf
22. Mensching, G., Zwink, J.: L'ancien occitan en tant que langage scientifique de la médecine. Termes vernaculaires dans la traduction hébraique du Zad al-musafir wa-qut al-hadir (XIIIe). In: Garabato, C.A., Torreilles, C., Verny, M.-J. (eds.) Los que fan viure e tresluire l'occitan (AIEO 2011), pp. 226–236. Lambert-Lucas, Limoges (2014)

1st Workshop on Humanities in the Semantic Web (WHiSe 2016)

An Ecosystem for Linked Humanities Data

Rinke Hoekstra[1,5(✉)], Albert Meroño-Peñuela[1,4], Kathrin Dentler[1],
Auke Rijpma[2,7], Richard Zijdeman[2,6], and Ivo Zandhuis[3]

[1] Department of Computer Science,
Vrije Universiteit Amsterdam, Amsterdam, Netherlands
{rinke.hoekstra,albert.merono,k.dentler}@vu.nl
[2] International Institute of Social History, KNAW, Amsterdam, Netherlands
{auke.rijpma,richard.zijdeman}@iisg.nl
[3] Ivo Zandhuis Research and Consultancy, Haarlem, Netherlands
ivo@zandhuis.nl
[4] Data Archiving and Networked Services, KNAW, The Hague, Netherlands
[5] Faculty of Law, University of Amsterdam, Amsterdam, Netherlands
[6] University of Stirling, Stirling, UK
[7] Utrecht University, Utrecht, Netherlands

Abstract. The main promise of the digital humanities is the ability to perform scholarly studies at a much broader scale, and in a much more reusable fashion. The key enabler for such studies is the availability of sufficiently well described data. For the field of socio-economic history, data usually comes in a tabular form. Existing efforts to curate and publish datasets take a top-down approach and are focused on large collections. This paper presents QBer and the underlying structured data hub, which address the *long tail* of research data by catering for the needs of individual scholars. QBer allows researchers to publish their (small) datasets, link them to existing vocabularies and other datasets, and thereby contribute to a growing collection of interlinked datasets. We present QBer, and evaluate our first results by showing how our system facilitates three use cases in socio-economic history.

Keywords: Digital humanities · Structured data · Linked Data · QBer

1 Introduction

In a 2014 article in CACM, [10] describes digital humanities as a "movement and a push to apply the tools and methods of computing to the subject matter of the humanities." As the fuel of the computational method, the key enabler for digital humanities research is the availability of data in digital form. At the inauguration of the Center for Humanities and Technology (CHAT), José van Dijck, the president of the Dutch Royal Academy of Sciences, characterizes progress in this field as the growing ability to tremendously increase the scale at which humanities research takes place, thereby allowing for much *broader* views on the subject matter [29]. Tackling this important challenge for the digital

© Springer International Publishing AG 2016
H. Sack et al. (Eds.): ESWC 2016 Satellite Events, LNCS 9989, pp. 425–440, 2016.
DOI: 10.1007/978-3-319-47602-5_54

humanities requires straightforward *transposition* of research queries from one humanities dataset to another, or even allow for direct *cross-dataset querying*. It is widely recognized that Linked Data technology is the most likely candidate to fill this gap. We argue that current efforts to increase the availability and accessibility of this data do not suffice. They do not cater for the "long tail of research data" [8], the large volumes of small datasets produced by *individual* researchers; and existing Linked Data tooling is too technology-oriented to be suitable for humanities researchers at large.

This paper presents QBer and the underlying CLARIAH Structured Data Hub (CSDH),[1] whose aim is to address the limitations of current data-publishing practice in the digital humanities, and socio-economic history in particular. The CSDH integrates a selection of large datasets from this domain, while QBer is a user-facing web application that allows *individual* researchers to upload, convert and link 'clean' data to existing datasets and vocabularies in the hub without compromising the detail and heterogeneity of the original data (see Sect. 2). Under the hood, we convert all data to RDF, but QBer does not bother scholars with technical aspects. An inspector-view displays the result of the mappings – a growing network of interconnected datasets – in a visually appealing manner (See Fig. 2). The visualization is just one of the incentives we are developing. The most important incentive will be the ability to allow for transposing research queries across datasets, and the ability to perform cross-dataset querying. According to Wikipedia, an ecosystem is "a community of living organisms in conjunction with the nonliving components of their environment (things like air, water and mineral soil), interacting as a system".[2] Translating that to our situation, it means that QBer and the CSDH should interact with users that have different roles, that mutually benefit from each other, and that collectively – as a system – exhibit behavior that would not otherwise emerge.

Section 4 describes two use-cases that evaluate the ability of QBer and the CSDH to fulfill that promise. We first discuss related work in Sect. 2 and describe the QBer and CSDH systems in Sect. 3.

2   Related Work

Historical data comprises text, audiovisual content or – in our case – data in the more traditional sense: structured data in tabular form. Preparing historical data for computational analysis takes considerable expertise and effort. As a result, digital data curation efforts are organized (and funded) in a top-down fashion, and focus on the enrichment of individual datasets and collections of sufficient importance and size. Examples are the North Atlantic Population Project (NAPP) [31], the Clio-Infra repository [5], and the Mosaic project.[3] Such projects face three important issues:

---

[1] A screencast of the system is available at https://vimeo.com/158153564.
[2] See http://en.wikipedia.org/wiki/Ecosystem.
[3] See https://www.clio-infra.eu and http://www.censusmosaic.org/.

1. They often culminate in a website where *subsets* of the data can be *down-loaded*, but cannot be programmatically accessed, isolating the data from efforts to cross-query over multiple datasets.
2. They enforce commitment to a shared standard: standardization leads to loss of detail, and thus information. The bigger a project is, the higher the cost of reconciling heterogeneity – in time, region, coding etc. – between the large number of sources involved.
3. Their scale is unsuited for the large volumes of important – but sometimes idiosyncratic – smaller datasets created by individual researchers: the long tail of research data [8].

For this last reason, it is difficult for individual researchers to make their data available in a sustainable way [33]. Despite evidence that sharing research data results in higher citation rates [28], researchers perceive little incentive to publish their data with sufficiently rich, machine interpretable metadata. Data publishing and archiving platforms such as EASY[4] (in the Netherlands), Dataverse[5] or commercial platforms such as Figshare[6] and Dryad[7] aim to lower the threshold for data publishing, and cater for increasing institutional pressure to archive research data. However, as argued in [14], the functionality of these platforms – data upload, data landing page, citable references, default licensing, long term preservation – is limited with respect to the types of *provenance* and *content* metadata that can be associated with publications, and they do not offer the flexibility of the Linked Data paradigm. This has a detrimental effect on both findability and reusability of research data.

The FAIR guiding principles for data management and stewardship [35], designed and endorsed by "a diverse set of stakeholders – representing academia, industry, funding agencies, and scholarly publishers", corroborate this view. Data should be *findable, accessible, interoperable* and *reusable*. These goals can only be achieved when, among others, data is (globally) uniquely identifiable using URIs, accessible at these identifiers, described using rich metadata in a formal language (RDF), using vocabularies and attributes that use FAIR principles, and published with a clear data usage license and provenance [35, Box 2.].

In socio-economic history, a central challenge is to query data combined from multiple tabular sources: spreadsheets, databases and CSV files. The multiple benefits of Linked Data as a data integration method [11] encourage the representation of tabular sources as Linked Data.[8] CSV and HTML tables can be represented in RDF using CSV2RDF and DRETa [18,26]. For other tabular formats, like Microsoft Excel, Google Sheets, and tables encoded in JSON or XML, larger frameworks are needed, like Opencube [15], Grafter [30], and the

---

[4] See http://easy.dans.knaw.nl.
[5] See http://dataverse.harvard.edu and http://dataverse.nl.
[6] See http://figshare.com.
[7] See http://datadryad.org.
[8] For a comprehensive list, see e.g. https://github.com/timrdf/csv2rdf4lod-autom ation/wiki and http://www.w3.org/wiki/ConverterToRdf.

combination of OpenRefine and DERI's RDF plugin [7,25]. The RMLEditor[9] [13] is an editor for the R2RML mapping language,[10] and allows users to map legacy data to an RDF-based schema by bringing elements from a tabular data representation to the familiar graph-based visualization of RDF.

Perhaps closest to our work is the Karma [16,32, a.o.] tool for bringing structured data in JSON, CSV, XML and other formats to the Semantic Web.[11] Karma's user interface combines a tabular representation of the source data with a graph-based schema view for mapping the schema of legacy data files to an ontology. This mapping can then be used to transform data to RDF or to produce a R2RML mapping file. To assist users, it can automatically provide suggestions for mappings and rewrites, both at a schema level and at the data level.

The above tools cater for tabular data that represents one observation (record) per row. Datasets in social history, however, are often presented as multidimensional views that use other tabular layout features, such as hierarchical headers, column and row spanning cells, and arbitrary data locations [1,21] TabLinker [22] uses a semi-automatic approach to represent multidimensional tables as RDF Data Cube [6]. As in TopBraid Composer,[12] TabLinker uses external mapping files rather than an interactive interface.

An important question is: how can mappings be created? Work in ontology and vocabulary alignment, as in the OAEI,[13] or identity reconciliation, aim to perform *automatic* alignments. Given the often very specific (historic) meaning of terms in our datasets, these techniques are likely to be error-prone, hard to optimize (given the heterogeneity of our data) and unacceptable to scholars. Interactive alignment tools, such as Amalgame [27], developed for the cultural heritage and digital humanities domains, are more promising, but treat the alignment task in isolation rather than as part of the data publishing process. Anzo for Excel[14] is an extension for Microsoft Excel for mapping spreadsheet data to ontologies. Similarly, [2] and RightField[15] allow for selecting terms from an ontology from within Excel spreadsheets, but these require the data to conform to a pre-defined template. TabLinker processes mappings expressed in Excel worksheets.

To summarize, existing tools are not suitable for two reasons:

1. they are targeted to relatively tech-savvy users; data engineers that understand the RDF data model, the use of RDF Schema, and for whom the conversion to RDF is a goal in itself. In our case, prospective users will benefit from interlinked data, but are unlikely to have any interest in the underlying technology.

---

[9] See http://rml.io.

[10] See http://www.w3.org/TR/r2rml/.

[11] See https://github.com/usc-isi-i2/Web-Karma.

[12] https://www.w3.org/2001/sw/wiki/TopBraid.

[13] The Ontology Alignment Evaluation Initiative, see oaei.ontologymatching.org/.

[14] https://www.w3.org/2001/sw/wiki/Anzo.

[15] https://www.sysmo-db.org/rightfield.

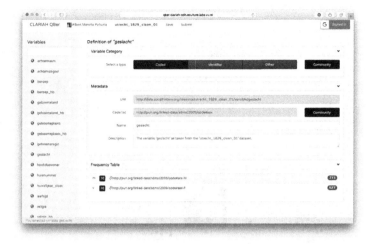

Fig. 1. Variable mapping screen of QBer with the variable 'geslacht' (sex) selected.

2. they focus on *generating* mappings, isolate the mapping task from the data, or do not provide users with sufficient support for selecting the right URI to map against.

3   QBer and the Structured Data Hub

To create a viable research ecosystem for Linked Humanities Data of all sizes, we need to combine expert knowledge with automated Linked Data generation. It should be *easy* and *profitable* for individual researchers to enrich and publish their data via our platform.

To achieve the first goal, we developed QBer;[16] an interactive tool that allows non-technical scholars to convert their data to RDF, to map the 'variables' (column names) and values in tabular files to Linked Data concept schemes, and to publish their data on the structured data hub. What sets QBer apart is that all Linked Data remains under the hood.

To achieve the second goal, we build in direct feedback (reuse of existing content, visualizations, etc.) on top of the CSDH and demonstrate the research benefits of contributing data to it (see Sect. 4). We illustrate QBer by means of a walk-through of the typical usage of the tool, and then summarize its connection with the CSDH.

Using QBer consists of interacting with three main views: the *welcome screen*, the *mapping screen*, and the *inspector*. In the welcome screen, users first authenticate with OAuth compatible services (e.g. Google accounts), and then select a raw dataset to work with. Datasets can be selected directly from the CSDH

---

[16] See https://github.com/CLARIAH/qber.

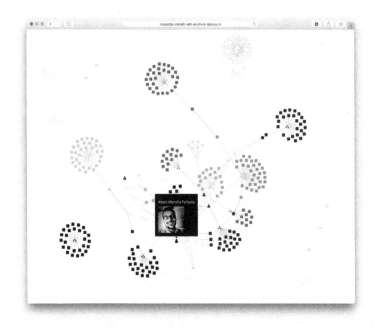

Fig. 2. The inspector view over the datasets currently in the CSDH

versioned file store, uploaded from Dropbox, or imported from a Dataverse collection by providing a DOI.

Once a dataset is loaded, QBer displays the mapping screen (Fig. 1). This screen is divided into the *variables sidebar* (left) and the *variable panel* (right). The sidebar allows the user to search and select a variable (i.e. column) from the dataset. Once the user clicks on one variable, the variable panel will show that variable's details: the *variable category*, the *variable metadata*, and the value *frequency table*.

We distinguish between three *variable categories*: *coded, identifier* and *other*. Values for coded variables are mapped to corresponding concepts (skos:Concept) within a skos:ConceptScheme, which establishes all possible values the variable can take. If the variable is of type *identifier*, its values are mapped to dataset specific minted URIs. Finally, the values of variables of type *other* are mapped to literals instead of URIs. The 'Community' button gives access to all known predefined datacube dimensions. These come from LSD Dimensions, an index of dimensions used in Data Structure Definitions of RDF Data Cubes on the Web [20] and from datasets previously processed by QBer that now reside on the CSDH.

The *variable metadata* panel can be used to change the label and the description of the variable. If the variable has been specified to be "coded" in the previous pane, it can be linked to existing code lists curated by the Linked Open Data community. QBer provides access to all concept schemes from the Linked

Data cache[17] and the CSDH itself. If the variable is of type "other", this panel lets users define their own transformation function for literals.

The *frequency table* panel has three purposes. First, it allows for quick inspection of the distribution of all values of the selected variable, by displaying their frequency. Second, if the variable type is "coded", it lets the user map the default minted URI for the chosen value to any `skos:Concept` within the selected `skos:ConceptScheme` in the variable metadata panel. QBer also has a batch mapping mode that prompts the user to map all values of the variable interactively. Third, if the variable type is "other", users can specify their own literal-to-literal transformations by providing their own transformation functions; this is useful e.g. if the original data needs to be expressed in different units of measure, or if strings need a systematic treatment. Finally, the panel shows the current mappings for values of the selected variable.

Mappings can be materialized in two ways. Users can click on *Save* in the navigation bar, which stores the current mapping status of all variables in their local cache. Clicking on *Submit* sends the mappings to the CSDH API, which integrates them with other datasets in the hub. Under the hood, the data is converted to a Nanopublication [9] with provenance metadata in PROV, where the assertion-graph is an RDF Data Cube representation of the data [6]. The RDF representation is a verbatim conversion of the data; mappings between the original values and pre-existing vocabularies are explicitly represented using SKOS mapping relations. This scheme allows for the co-existence of alternative interpretations (by different scholars) of the data, thus overcoming the standardization-limitation alluded to in Sect. 2.

The *Inspector*, shown in Fig. 2, allows users to explore the contents of the CSDH. The visualization shows a graph of nodes and edges, with different icons representing different node types. *User* nodes represent users that have submitted data to the hub, according to their provided OAuth identities in the welcome screen. *Dataset nodes* are shown as stacked boxes, and represent Data Structure Definitions[18] (DSD) submitted by these users. *Dimension nodes*, shown as rounded squares, represent dimensions (i.e. variables, columns in the raw data) within those DSD. Dimensions that are externally defined (e.g. by SDMX or some other external party) and are thus not directly used in datasets, are represented as cloud-icons. The color of nodes is based on the source dataset from which the information was gleaned. Users can interact with the inspector in several ways:

- hovering over a node displays a summary of its metadata, e.g. for person nodes this includes a depiction of the author;
- dragging nodes around redraws the force layout

---

[17] See http://lod.openlinksw.com, we aim to extend this with schemes from the LOD-Laundromat, http://www.lodlaundromat.org.

[18] According to [6], a Data Structure Definition "defines the structure of one or more datasets. In particular, it defines the dimensions, attributes and measures used in the dataset along with qualifying information such as ordering of dimensions and whether attributes are required or optional".

– scrolling zooms in and out
– clicking a node opens up a new browser tab with a tabular representation of
  the resource (using brwsr[19]).

QBer and the Inspector work on top of the CSDH API,[20] which carries out
the backend functionality. This includes converting, storing, and managing POST
and GET requests over datasets. The CSDH API functionality will be extended
to cover standard data search, browsing and querying tasks. All data is currently
stored in an OpenLink Virtuoso CE triplestore,[21] but since CSDH communicates
through the standard SPARQL protocols it is not tied to one triple store vendor.

4   Evaluation

In this section, we evaluate the results of our approach by means of three use
cases in socio-economic history research. The first use case investigates the question as to whether the CSDH indeed allows research to be carried out on a
broader scale. In this case, we transpose a query that was built to answer a
research question aimed at a dataset of one country to a dataset that describes
another country. The second use case investigates the question as to whether
our system facilitates the workflow of a typical individual researcher. A third
use case shows how the CSDH can speed up the research cycle.

4.1   Use Case 1: Born Under a Bad Sign

Economic and social history takes questions and methods from the social sciences
to the historical record. An important line of research focuses on the determi-
nants of historical inequality. One hypothesis here is that prenatal [3] and early-
life conditions [12] have a strong impact on socioeconomic and health outcomes
later in life. For example, a recent study on the United States found that people
born in the years and states hit hardest during the Great Depression of the 1930s
had lower incomes and higher work disability rates in 1970 and 1980 [34]. This
study inspired this use case.

Most studies on the impact of early life conditions are case studies of single
countries. Therefore, the extent to which results can be generalized – their external validity – is difficult to establish (e.g., differing impact of early life conditions
in rich and poor countries). Moreover, historical data is often idiosyncratic. This
means that dataset-specific characteristics such as sampling and variable coding
schemes might influence the results (see Sect. 2).

In this use case, we explore the relation between economic conditions in individuals' birth year and occupational status in the historical census records of
Canada and Sweden in 1891. In many cases it would be necessary to link the
two census datasets so that they can be queried in the same way. Here, however,

---

[19] See http://github.com/Data2Semantics/brwsr.
[20] https://github.com/CLARIAH/wp4-csdh-api.
[21] See http://www.openlinksw.com.

we use two harmonized datasets from the North Atlantic Population Project (NAPP) [31]. Economic conditions are measured using historical GDP per capita figures from the Clio-Infra repository [5]. Because our outcome is occupational status, we have to enrich the occupations in the census with occupational codes and a status scheme. Because the NAPP-project uses an occupational classification that cannot provide internationally comparable occupational status scores, we have to map their occupational codes to the HISCO system,[22] so that we can use the HISCAM cross-nationally comparable occupational status scheme [17,19].[23]

In general terms, the data requirements are typical of recent trends in large database usage in economic and social history:

1. the primary unit of analysis is the individual (microdata);
2. a large number of observations is analyzed;
3. multiple micro-datasets are analyzed;
4. microlevel observations are linked to macro-level data through the dimensions time and geographical area;
5. qualitative data is encoded to extract more information from it.

Current Workflow. The traditional workflow to do this would include the following steps. First, the researcher has to find and download the datasets from multiple repositories. The datasets, which come in various formats, then have to be opened, and, if necessary, the variables have to be renamed, cleaned, and re-encoded to be able to join them with other datasets. We can rely on previous cleaning and harmonization efforts of the NAPP project, but in many other situations the researcher would have to do this manually. Finally, the joined data has to be saved in a format that can be used by a statistical program.

New Workflow. Using QBer and the CSDH, the workflow is as follows. Linked-data tools are used to discover data on the hub. In our case, we used the CSDH Inspector, a linked data browser[24] and exploratory SPARQL queries. The Inspector provides a simple overview of all datasets in the CSHD.[25] Note that to discover datasets and especially linked datasets on the CSDH, it is necessary that someone uploaded the datasets and created the links in the first place, for example by linking datasets to a common vocabulary. While it is unavoidable that someone has to do this at some point, the idea behind the hub is that if it is done once, the results can be re-used by other researchers.

Next, queries are built and stored on GitHub. The result sets that these queries produce against the data hub are then used to create the dataset that is to be analyzed. `grlc`, a tool we developed for creating Linked Data APIs using

---

[22] HISCO: Historical International Standard Classification of Occupations.

[23] https://github.com/rlzijdeman/o-clack and http://www.camsis.stir.ac.uk/hiscam/.

[24] https://github.com/Data2Semantics/brwsr.

[25] Currently at http://inspector.clariah-sdh.eculture.labs.vu.nl/overview, but this may change.

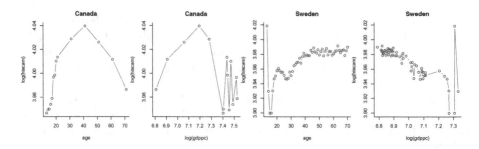

Fig. 3. HISCAM scores versus log (GDP per capita) in Canada (1891) and Sweden (1891)

SPARQL queries in GitHub repositories, was helpful in exploring the data on the hub and executing the eventual query [24].[26] Among others, it populates the API specification with allowed values for request parameters based on data available in the underlying repository. This tool can also be used to download the data directly into a statistical environment like R via HTTP requests, for example using curl. Alternatively, the CSDH can be queried directly from a statistical environment using SPARQL libraries.

Observations. While more sophisticated models are required to disentangle cohort, period and age effect [4], the results suggest that in Canada in 1891 the expected effects of early life-conditions are found: higher GDP per capita in a person's birth year was associated with higher occupational status at the time of the census. However, in Sweden, the opposite was the case (see Fig. 3). This shows the relative ease at which the CSDH facilitates reusable research questions by means of query transposition.

4.2   Use Case 2: Railway Strike

The second use case takes the form of a user study. It is about the "Dwarsliggers"[27] dataset by Ivo Zandhuis that collects data pertaining to a solidarity strike at the maintenance workshop of the Holland Railway Company (*Hollandsche IJzeren Spoorweg-Maatschappij*), in the Dutch city of Haarlem in 1903. From a sociological perspective, strikes are of interest for research on social cohesion as it deals both with the question of when and why people live peaceful together (even when in disagreement) and the question of how collective action is successfully organized, a prerequisite for a successful strike. The Dwarsliggers dataset is one of the few historical cases where data on strike behavior is available at the *individual* level.

The creation and use of this dataset is exemplary of the workflow of small to medium quantitative historical research projects in the sense that it relies on

---

[26] See http://grlc.io/ and https://github.com/CLARIAH/grlc.

[27] In Dutch, a "dwarsligger" can mean both a railroad tie, and an obstructive person.

multiple data sources that need to be connected in order to answer the research questions. We briefly discuss this workflow, and then show the impact that QBer and the CSDH have.

Current Workflow. Zandhuis' current workflow is very similar to the one reported in the first use case. He first digitized the main dataset on the strike behaviour of employees at the maintenance workshop of the railway company ($N = 1163$). Next, he gathered data from multiple sources in which these employees also appear, adding individual characteristics that explain strike behaviour. For example, he derived family situations from the Dutch civil registers, and the economic position from tax registers, resulting in a separate dataset per source. Next, he inserted these datasets into a SQL database. In order to derive a concise subset to analyze his research questions, using e.g. QGIS, Gephi or R, he wrote SQL queries to extract the relevant information. These queries are usually added as an appendix to his research papers.

New Workflow. In collaboration with Zandhuis, we revisited this workflow using QBer. Zandhuis, as many historians do, uses spreadsheets to enter data, and uses a specific layout to enhance the speed and quality of data entry. The first step was to convert the data to a collection of .csv files. This is just a temporary limitation, as the CSDH is not necessarily restricted to CSV files. It uses the Python Pandas library[28] for loading tabular files into a data frame.

The second step involves visiting each data file in turn, and linking the data to vocabularies and through them to other datasources. Data about the past often comes with a wide variety of potential values for a single variable. Religion, for example, can have dozens of different labels as new religions came about and old religions disappeared. As described in Sect. 3, QBer provides access to a large range of such classifications, basically all those available in the Linked Data cloud and the CSDH. For example, QBer provides all occupation concepts from the HISCO classification used in the first use case [19]. Researchers can use occupational labels to get the correct codes from the latest version of this classification and, eventually, concepts linked to it. QBer however also shows the results of earlier coding efforts, so that historians can benefit from these (e.g. another dataset may have the same literal value already mapped to as HISCO code). This step is new compared to Zandhuis' original workflow. The linking of occupational labels now enables him to combine an employee with his social status (HISCAM). This allows him to directly include a new, relevant, aspect in his study. Moreover, since QBer makes coding decisions explicit, they can be made subject to the same peer review procedure used to assess the quality of a research paper. In the CSDH, original values of the dataset and the mapped codings (potentially by different researchers) live side-by-side. Thus QBer adds to the ease of use in coding variables, increases flexibility by allowing for multiple interpretations, and allows for more rigorous evaluation of coding efforts. The inspector graph of Fig. 2 depicts the result of the new workflow.

---

[28] See http://pandas.pydata.org.

The third step was then to query the datasets in order to retrieve the subset of data needed for analysis. As in the first use case, we design SPARQL queries that, when stored on GitHub, can be directly executed through the `grlc` API. This makes replication of research much easier: rather than including the query as an appendix of a research paper, the query is now a first order citizen and can even be applied to other datasets that use the same mappings. Again, through the API, these queries can easily be accessed from within R, in order to perform statistical analysis. Indeed, the `grlc` API is convenient, but it is a lot to ask non-computer science researchers to design SPARQL queries. However, as we progress, we expect to be able to identify a collection of standard SPARQL query templates that we can expose in this manner (see also [24]).

To illustrate this, consider that since the Dwarsliggers collection contains multiple datasets on the same individuals at the same point in time, there are multiple observations of the same characteristics (e.g. age, gender, occupation, religion). However, the sources differ in accuracy. For example, measuring marital status is one of the key aims of the civil registry, while personnel files may contain information on marital status, but it is not of a key concern for a company to get this measurement right. By having all datasets mapped to vocabularies through QBer and having the queries stored in GitHub and executed by `grlc`, each query can readily be repeated using different sources on the same variables. This is useful as a robustness check of the analysis or even be used in what historians refer to as a 'source criticism' (a reflection of the quality and usefulness of a source). This, again, is similar to the first use case, but it emphasizes an additional role for the queries as so-called 'edit rules' [23].

Observations. To conclude, this use case shows that the QBer tool and related infrastructure provides detailed insight in how the data is organized, linked and analyzed. Furthermore, the data can be queried live. This ensures reusable research *activities*; not just reusable *data*.

4.3   Use Case 3: An Ecosystem?

The researchers in the two use cases correspond to two roles. The railway strike shows a data owner who wants to publish and analyze his data and benefits from pre-existing data in the CSDH. The inequality use case illustrates a user who is primarily interested in CSDH data for the purpose of comparative and cross-dataset research. Although both use cases show benefit for both roles, they reflect fairly traditional data driven processes.

Our third use case only emerged after one of the authors of this paper decided to have a closer look at the results of Use Case 2. In just under 15 min, he was able to reproduce the analysis for Canada and Sweden, and show that by adding an additional correction for age, the respective positive and negative correlation apparent in respectively Canada and Sweden are not only both negative, but also not significant. Essential in this reproduction of research is that the queries used in Use Case 2 were available on GitHub, and exposed as a RESTful API through `grlc`. This highlights the third role: a user who wants to build on earlier work (not just data), and thus has the most to gain from our approach. The CSDH plays an essential part in making sure that these different users meet, and collectively increase both the speed and quality of research.

5   Conclusion

The preceding sections presented QBer and the CSDH to address the limitations of existing digital humanities data curation projects in facilitating (1) the *long tail* of research data and (2) research at a broader scale, enabling cross-dataset querying and reuse of queries. We argued that existing Linked Data publishing and mapping tools do not meet the needs of scholars that are not technologically versed (or interested).

QBer and the CSDH enable individual scholars to publish and use their data in a flexible manner. QBer allows researchers to publish their (small) datasets, link them to existing vocabularies and other datasets, and thereby contribute to a growing collection of interlinked datasets hosted by the CSDH. The CSDH offers services for inspecting data, and (in combination with `grlc`) reusable querying across multiple datasets. We illustrated these features by means of two use cases. The first shows the ability of the Linked Data paradigm used in the CSDH to significantly lower the effort needed to do comparative research (even when the data was published as part of the same larger standardization effort). The second use case shows how publishing data through QBer allows individual researchers to have more grip on their data, to be more explicit regarding data interpretation (coding) and, via the CSDH, to be able to answer more questions for free (e.g. the mapping through HISCO to HISCAM). The third use case shows how the hard work of other scholars, both in data curation and in the formulation of queries over the data, can be readily reproduced and used to further the field.

Of course, there still is room for expansion. To ensure uniqueness of identifiers, historical 'codes' need to be mapped to URIs. This is technically trivial, but historians are not used to these lengthy identifiers in their statistical analyses. Secondly, formulating research questions as queries requires an understanding of the structure of the data. Given the large numbers of triples involved, this can be difficult. As said above, standard APIs based on SPARQL query templates should solve some of this problem, but offering a user-friendly data inspection tool is high on our list. SPARQL templates allow us to solve another issue: allowing for free-form querying can have a detrimental effect on the performance of the CSDH. The use of templates enables more efficient use of caching strategies. A building block for this approach is the `grlc` service, which serves SPARQL queries as Linked Data APIs using Swagger, an API specification format and user interface also for non API experts. But even without such improvements, we believe that the use cases show that QBer and the CSDH already broaden the scope of supported workflows and data in our ecosystem, and bring the benefits of Linked Data and the Semantic Web at the fingertips of humanities scholars; an important step in towards FAIR data management.

Acknowledgments. This work was funded by the CLARIAH project of the Dutch Science Foundation (NWO) and the Dutch national programme COMMIT.

References

1. Ashkpour, A., Meroño-Peñuela, A., Mandemakers, K.: The Dutch historical censuses: harmonization and RDF. Hist. Methods J. Quant. Interdisc. Hist. **48** (2015)
2. van Assem, M., Rijgersberg, H., Wigham, M., Top, J.: Converting and annotating quantitative data tables. In: Patel-Schneider, P.F., Pan, Y., Hitzler, P., Mika, P., Zhang, L., Pan, J.Z., Horrocks, I., Glimm, B. (eds.) ISWC 2010, Part I. LNCS, vol. 6496, pp. 16–31. Springer, Heidelberg (2010). doi:10.1007/978-3-642-17746-0_2
3. Barker, D.J.: The fetal and infant origins of adult disease. BMJ Br. Med. J. **301**(6761), 1111 (1990)
4. Bartels, L.M., Jackman, S.: A generational model of political learning. Electoral. Stud. **33**, 7–18 (2014)
5. Bolt, J., Timmer, M., van Zanden, J.L.: GDP per capita since 1820. In: How Was Life? Global well-being since 1820, pp. 57–72. Organisation for Economic Co-operation and Development, October 2014
6. Cyganiak, R., Reynolds, D., Tennison, J.: The RDF data cube vocabulary. Technical report, W3C (2013). http://www.w3.org/TR/vocab-data-cube/
7. DERI: RDF Refine - a Google Refine extension for exporting RDF. Technical report, Digital Enterprise Research Institute (2015). http://refine.deri.ie/
8. Ferguson, A.R., Nielson, J.L., Cragin, M.H., Bandrowski, A.E., Martone, M.E.: Big data from small data: data-sharing in the 'long tail' of neuroscience. Nat. Neurosc. **17**(11), 1442–1447 (2014)
9. Groth, P., Gibson, A., Velterop, J.: The anatomy of a nanopublication. Inf. Serv. Use **30**(1–2), 51–56 (2010)
10. Haigh, T.: We have never been digital. Commun. ACM **57**(9), 24–28 (2014)
11. Heath, T., Bizer, C.: Linked Data: Evolving the Web into a Global Data, 1st edn. Morgan and Claypool, Palo Alto (2011)
12. Heckman, J.J.: Skill formation and the economics of investing in disadvantaged children. Science **312**(5782), 1900–1902 (2006). http://www.sciencemag.org/content/312/5782/1900
13. Heyvaert, P., Dimou, A., Herregodts, A.-L., Verborgh, R., Schuurman, D., Mannens, E., Van de Walle, R.: RMLEditor: a graph-based mapping editor for linked data mappings. In: Sack, H., Blomqvist, E., d'Aquin, M., Ghidini, C., Ponzetto, S.P., Lange, C. (eds.) ESWC 2016. LNCS, vol. 9678, pp. 709–723. Springer, Heidelberg (2016). doi:10.1007/978-3-319-34129-3_43
14. Hoekstra, R., Groth, P.: Linkitup: link discovery for research data. In: AAAI Fall Symposium Series Technical Reports (FS-13-01), pp. 28–35 (2013)
15. Kalampokis, E., Nikolov, A., et al.: Exploiting linked data cubes with opencube toolkit. In: Posters and Demos Track, 13th International Semantic Web Conference (ISWC2014), vol. 1272. CEUR-WS, Riva del Garda, Italy (2014). http://ceur-ws.org/Vol-1272/paper_109.pdf
16. Knoblock, C.A., et al.: Semi-automatically mapping structured sources into the semantic web. In: Simperl, E., Cimiano, P., Polleres, A., Corcho, O., Presutti, V. (eds.) ESWC 2012. LNCS, vol. 7295, pp. 375–390. Springer, Heidelberg (2012). doi:10.1007/978-3-642-30284-8_32

17. Lambert, P.S., Zijdeman, R.L., Van Leeuwen, M.H., Maas, I., Prandy, K.: The construction of HISCAM: a stratification scale based on social interactions for historical comparative research. Hist. Methods J. Quant. Interdisc. Hist. **46**(2), 77–89 (2013)
18. Lebo, T., McCusker, J.: csv2rdf4lod. Technical report, Tetherless World, RPI (2012). https://github.com/timrdf/csv2rdf4lod-automation/wiki
19. van Leeuwen, M., Maas, I., Miles, A.: HISCO: Historical International Standard Classification of Occupations. Leuven University Press, Leuven (2002)
20. Meroño-Peñuela, A.: LSD dimensions: use and reuse of linked statistical data. In: Lambrix, P., Hyvönen, E., Blomqvist, E., Presutti, V., Qi, G., Sattler, U., Ding, Y., Ghidini, C. (eds.) EKWA 2014 Satellite Events. LNCS, vol. 8982, pp. 159–163. Springer, Heidelberg (2015). doi:10.1007/978-3-319-17966-7_22
21. Meroño-Peñuela, A., Ashkpour, A., van Erp, M., Mandemakers, K., Breure, L., Scharnhorst, A., Schlobach, S., van Harmelen, F.: Semantic technologies for historical research: a survey. Seman. Web Interoperability Usability Applicability **6**(6), 539–564 (2015)
22. Meroño-Peñuela, A., Ashkpour, A., Rietveld, L., Hoekstra, R., Schlobach, S.: Linked humanities data: the next frontier? In: 2nd International Workshop on Linked Science (LISC2012), ISWC, vol. 951. CEUR-WS (2012). http://ceur-ws.org/Vol-951/
23. Meroño-Peñuela, A., Guéret, C., Schlobach, S.: Linked edit rules: a web friendly way of checking quality of RDF data cubes. In: 3rd International Workshop on Semantic Statistics (SemStats 2015), ISWC. CEUR (2015)
24. Meroño-Peñuela, A., Hoekstra, R.: grlc makes GitHub taste like linked data APIs. In: Sack, H., Rizzo, G., Steinmetz, N., Mladenić, D., Auer, S., Lange, C. (eds.) ESWC 2016 Satellite Events. LNCS, vol. 9989, pp. 342–353. Springer, Heidelberg (2016)
25. Morris, T., Guidry, T., Magdinie, M.: OpenRefine: a free, open source, powerful tool for working with messy data. Technical report, The OpenRefine Development Team (2015). http://openrefine.org/
26. Muñoz, E., Hogan, A., Mileo, A.: DRETa: extracting RDF from Wikitables. In: International Semantic Web Conference, Posters and Demos, pp. 92–98. CEUR-WS (2013)
27. van Ossenbruggen, J., Hildebrand, M., de Boer, V.: Interactive vocabulary alignment. In: Gradmann, S., Borri, F., Meghini, C., Schuldt, H. (eds.) TPDL 2011. LNCS, vol. 6966, pp. 296–307. Springer, Heidelberg (2011). doi:10.1007/978-3-642-24469-8_31
28. Piwowar, H.A., Day, R.S., Fridsma, D.B.: Sharing detailed research data is associated with increased citation rate. PloS one **2**(3), e308 (2007). http://dx.plos.org/10.1371/journal.pone.0000308
29. Renckens, E.: Digital humanities verfrissen onze blik op bestaande data. E-Data Res. **10** (2016)
30. Roman, D., Nikolov, N., et al.: DataGraft: one-stop-shop for open data management. Sem. Web Interoperability Usability Applicability (2016, under review). http://www.semantic-web-journal.net/content/datagraft-one-stop-shop-open-data-management
31. Ruggles, S., Roberts, E., Sarkar, S., Sobek, M.: The North Atlantic population project: progress and prospects. Hist. Methods J. Quant. Interdisc. Hist. **44**(1), 1–6 (2011)

32. Szekely, P., Knoblock, C.A., Yang, F., Zhu, X., Fink, E.E., Allen, R., Goodlander, G.: Connecting the Smithsonian American art museum to the linked data cloud. In: Cimiano, P., Corcho, O., Presutti, V., Hollink, L., Rudolph, S. (eds.) ESWC 2013. LNCS, vol. 7882, pp. 593–607. Springer, Heidelberg (2013). doi:10.1007/978-3-642-38288-8_40
33. Tenopir, C., Allard, S., Douglass, K., Aydinoglu, A.U., Wu, L., Read, E., Manoff, M., Frame, M.: Data sharing by scientists: practices and perceptions. PLoS ONE 6(6), e21101 (2011). http://dx.doi.org/10.1371/journal.pone.0021101
34. Thomasson, M.A., Fishback, P.V.: Hard times in the land of plenty: the effect on income and disability later in life for people born during the great depression. Explor. Econ. Hist. 54, 64–78 (2014)
35. Wilkinson, M., et al.: The fair guiding principles for scientific data management and stewardship. Sci. Data (160018) (2016). http://www.nature.com/articles/sdata201618

Erratum to: Edinburgh Associative Thesaurus as RDF and DBpedia Mapping

Jörn Hees[1,2(✉)], Rouven Bauer[1,2], Joachim Folz[1,2], Damian Borth[1,2], and Andreas Dengel[1,2]

[1] Computer Science Department, University of Kaiserslautern, Kaiserslautern, Germany
[2] Knowledge Management Department, DFKI GmbH, Kaiserslautern, Germany
{joern.hees,rouven.bauer,joachim.folz,damian.borth, andreas.dengel}@dfki.de

Erratum to:
Chapter "Edinburgh Associative Thesaurus as RDF and DBpedia Mapping" in: H. Sack et al. (Eds.):
The Semantic Web, LNCS 9989,
https://doi.org/10.1007/978-3-319-47602-5_4

Two percentages in the listing on page 19 were incorrect in the original publication. The paper should read as follows:

- Composite phrases (22%...
- Plural words (20%...

The updated online version of this chapter can be found at
https://doi.org/10.1007/978-3-319-47602-5_4

H. Sack et al. (Eds.): ESWC 2016 Satellite Events, LNCS 9989, p. E1, 2016.
https://doi.org/10.1007/978-3-319-47602-5_55

Author Index

Printed in the United States
By Bookmasters